URANUS
The Planet, Rings and Satellites

THE ELLIS HORWOOD LIBRARY OF SPACE
SCIENCE AND SPACE TECHNOLOGY

SERIES IN ASTRONOMY

Series Editor: JOHN MASON
Consultant Editor: PATRICK MOORE

This series aims to coordinate a team of international authors of the highest reputation, integrity and expertise in all aspects of astronomy. It will make a valuable contribution to the existing literature, encompassing all areas of astronomical research. The titles will be illustrated with both black and white and colour photographs, and will include many line drawings and diagrams, with tabular data and extensive bibliographies. Aimed at a wide readership, the books will appeal to the professional astronomer, undergraduate students, the high-flying 'A' level student, and the non-scientist with a keen interest in astronomy.

PLANETARY VOLCANISM: A Study of Volcanic Activity in the Solar System
PETER CATTERMOLE, formerly Lecturer in Geology, Department of Geology, Sheffield University, UK, now Freelance Writer and Consultant and Principal Investigator with NASA's Planetary Geology and Geophysics Programme

DIVIDING THE CIRCLE: The development of critical angular measurement in astronomy 1500–1850
ALLAN CHAPMAN, Centre for Medieval and Renaissance Studies, Oxford, UK

SATELLITE ASTRONOMY: The Principles and Practice of Astronomy from Space
JOHN K. DAVIES, Royal Observatory, Edinburgh, UK

THE ORIGIN OF THE SOLAR SYSTEM: The Capture Theory
JOHN R. DORMAND, Department of Mathematics and Statistics, Teesside Polytechnic, Middlesbrough, UK, and MICHAEL M. WOOLFSON, Department of Physics, University of York, UK

THE DUSTY UNIVERSE
ANEURIN EVANS, Department of Physics, University of Keele, UK

SPACE-TIME AND THEORETICAL COSMOLOGY
MICHEL HELLER, Department of Philosophy, University of Cracow, Poland

ASTEROIDS: Their Nature and Utilization
CHARLES T. KOWAL, Space Telescope Institute, Baltimore, Maryland, USA

COMET HALLEY — Investigations, Results, Interpretations
Volume 1: Organisation, Plasma, Gas
Volume 2: Dust, Nucleus, Evolution
Editor: J. W. MASON, Scientific and Technical Consultant

ELECTRONIC AND COMPUTER-AIDED ASTRONOMY: From Eyes to Electronic Sensors
IAN S. McLEAN, Joint Astronomy Centre, Hilo, Hawaii, USA

URANUS: The Planet, Rings and Satellites
ELLIS D. MINER, Jet Propulsion Laboratory, Pasadena, California, USA

THE PLANET NEPTUNE
PATRICK MOORE, CBE

ACTIVE GALACTIC NUCLEI
IAN ROBSON, Director of Observatories, Lancashire Polytechnic, Preston, UK

ASTRONOMICAL OBSERVATIONS FROM THE ANCIENT ORIENT
RICHARD F. STEPHENSON, Department of Physics, Durham University, Durham, UK

EXPLORATION OF TERRESTRIAL PLANETS FROM SPACECRAFT: Instrumentation,
Investigation, Interpretation
YURI A. SURKOV, Chief of the Laboratory of Geochemistry of Planets, Vernandsky Institute of Geochemistry, USSR Academy of Sciences, Moscow, USSR

THE HIDDEN UNIVERSE
ROGER J. TAYLER, Astronomy Centre, University of Sussex, Brighton, UK

AT THE EDGE OF THE UNIVERSE
ALAN WRIGHT, Australian National Radio Astronomy Observatory, Parkes, New South Wales, Australia, and HILARY WRIGHT

URANUS
The Planet, Rings and Satellites

ELLIS D. MINER B.Sc., Ph.D.
Assistant Project Scientist (Voyager Project)
Jet Propulsion Laboratory
California Institute of Technology, Pasadena, USA

ELLIS HORWOOD
NEW YORK   LONDON   TORONTO   SYDNEY   TOKYO   SINGAPORE

First published in 1990 by
ELLIS HORWOOD LIMITED
Market Cross House, Cooper Street,
Chichester, West Sussex, PO19 1EB, England

A division of
Simon & Schuster International Group
A Paramount Communications Company

Typeset in Times by Ellis Horwood Limited
Printed and bound in Great Britain
by Hartnolls, Bodmin, Cornwall

British Library Cataloguing in Publication Data

Miner, Ellis D.
Uranus.
1. Uranus
I. Title
523.4'7
ISBN 0–13–946880–3

Library of Congress Cataloging-in-Publication Data

Miner, Ellis D.
Uranus: the planets, rings, and satellites / Ellis D. Miner.
p. cm. — (The Ellis Horwood library of space science and
space technology. Series in astronomy)
ISBN 0–13–946880–3
1. Uranus (planet). 2. Project Voyager. I. Title. II. Series.
QB681.M56   1990
523.4'7–dc20
90–26779
CIP

Table of contents

*To my dear wife, Beverly, for the many hours she has
sacrificed during the course of the Voyager encounters
and the writing of this book*

Preface

More than 12 years of my life have been closely tied to the Voyager missions to the outer planets of our Solar System. It has been my pleasure to serve as Assistant Project Scientist since shortly after the launches of Voyagers 1 and 2 in the late summer of 1977. Prior to that time, I had been associated with missions to Mercury (Mariner 10), Venus (Mariner 10), and Mars (Mariners 6, 7, and 9, and Vikings 1 and 2), but none of these earlier missions can compare to the continued thrill of discovery associated with the Voyager encounters.

Much has been written about the Voyager missions, generally by authors far more eloquent and experienced in writing than I. However, no comprehensive book on any part of the missions had yet been written by an author who was a member of the Voyager Project during the long years of planning, analyzing, revising, and polishing of the complex encounter events, and publishing of the scientific results. It was my desire to present an 'insider's' point of view that prompted me to accept the challenge to write this book.

It is worth mentioning that the task was undertaken at a time when I was heavily involved in preparations for the Voyager 2 encounter with Neptune. The press of those responsibilities, combined with a one-year delay (from June 1987 to June 1988) in an international colloquium on Uranus, resulted in more than a year's delay in my completion of the manuscript. In August of 1986, only a few months after I agreed to write about the Voyager findings at Uranus, I was called to serve as a Bishop in the Church of Jesus Christ of Latter-day Saints, and I have continued to enjoy service in that arena ever since. Extension of normal working hours into evenings or weekends on a regular basis was therefore impractical. The continued encouragement of my family and co-workers and the patience of the publishers, Messrs Ellis Horwood, are gratefully acknowledged.

I have attempted to present the facts and resulting theories about Uranus as clearly and understandably as possible. My goal was to make the Voyager findings understandable to interested individuals not readily conversant in the terminology used by planetary scientists. My good friend and co-worker Margaret Reisdorf very kindly consented to read the manuscript; her many excellent editorial suggestions have undoubtedly made the text more understandable for non-scientists. Ann Harch, Mona Delitsky, and Randii Wessen have also read portions of the manuscript and offered helpful suggestions. Evora Simien and Anita Sohus have assisted in the acquisition of many of the figures. Aloma Law, Laurie Lincoln, Lynn Eckerle, and Victoria Leavitt have also assisted in correspondence matters.

Detailed study of the Voyager data is continuing, and some modifications of the interpretations presented herein are likely. I have relied heavily on the manuscripts submitted for publication in the more technical book, *Uranus* (edited by Jay Bergstralh and myself), which is soon to be published by the University of Arizona Press. Permission to use of the visual material generated by the chapter authors is gratefully acknowledged.

Pasedena, California   Ellis D. Miner
September 1989

1

The discovery of Uranus

1.1 SIR WILLIAM HERSCHEL

> XXXII. Account of a Comet. By Mr Herschel, F.R.S.; communicated by Dr Watson, Jun. of Bath, F.R.S. Read April 26, 1781.
> On Tuesday the 13th of March, between ten and eleven in the evening, while I was examining the small stars in the neighbourhood of H Geminorum, I perceived one that appeared visibly larger than the rest: being struck with its uncommon magnitude, I compared it to H Geminorum and the small star in the quartile between Auriga and Gemini, and finding it so much larger than either of them, suspected it to be a comet. [1]

So began Sir William Herschel's account of his discovery of the Solar System's seventh planet. It was to be several years before 'Herschel's Comet' was officially recognized as a major planet and given the name Uranus. Herschel's discovery was nevertheless the first planet unknown to ancient observers of the heavens. This and other discoveries earned him recognition as the foremost astronomer of his time.

Friedrich Wilhelm Herschel (Fig. 1.1) was born November 15th, 1738, in Hanover, now in West Germany. During this period, the city of Hanover was a possession of King George II of England. Herschel's father was a musician in the Hanoverian Guards, and Wilhelm entered the same profession in May of 1753. In 1756 Herschel was sent to England for several months with his regiment of the Guard. Herschel learned to speak English during this short period. Because of changes caused by the Seven Years' War musicians were no longer needed or wanted in the Hanoverian Guards, so on his father's advice, he asked for and was granted a discharge. By the late autumn of 1757, Herschel had returned to England, where he lived for the rest of his life. Shortly after his arrival he anglicized his name to William Herschel. His made his living by giving recitals and copying music; by 1766 his musical talents were recognized in the resort city of Bath, where he was appointed organist at the Octagon Chapel on Milsom Street. He provided additional income by teaching music to as many as 35 pupils a week.

Fig. 1.1 — Portrait of Sir William Herschel in 1794 (age 56) by J. Russell. (P-29487A)

By 1772 his brother Alexander and his sister Caroline (Fig. 1.2) had joined him in Bath to pursue musical careers of their own. Within a few years William had become Director of Public Concerts in Bath, and was seemingly launched into a successful musical career. But William's destiny lay in a different direction. His study of music theory led him into mathematics. He studied the works of Newton, and continued to have a great desire to see the heavens for himself. This desire also led him to study optics. His meager income would not permit him to purchase a good telescope, so he decided to grind his own mirrors and make his own telescope. At first, Caroline was not happy with these distractions from music, but she eventually became as enthusiastic and talented in telescope making and astronomical observations as her brother. She worked side by side with William for the remainder of his life. With the

Fig. 1.2 — Portrait of Caroline Herschel in 1829 (age 79) by Tielmann (figure from Moore, 1983).

help of his good friend Sir William Watson, Herschel was granted an annual pension from George III in 1782. The pension was large enough that he could give up his music career and devote his full attention to astronomy. He and Caroline left Bath, eventually taking up residence outside Windsor at Observatory House in Slough. His marriage in 1788 to a wealthy widow (Mary Pitt of Upton) was a happy one and released him from the necessity of building telescopes to provide additional income. William and Mary Herschel's only child, John Frederick William Herschel, became a renowned astronomer in his own right by extending to the southern hemisphere his father's studies of the heavens.

Bolstered by a very large additional grant from George III, Herschel constructed the largest telescope of his time. This enormous reflecting telescope was 40-ft in length and of 48-in diameter (Fig. 1.3). His discoveries in the Saturn system were made using this instrument.

Fig. 1.3 — Herschel's 40-ft long, 48-in aperture reflecting telescope near Windsor, England. It was with this telescope that he discovered two of Saturn's satellites, Mimas and Enceladus.

Herschel's discovery of Uranus was only one of many accomplishments in the field of astronomy. He and Caroline made an exhaustive study of the heavens. They compiled a descriptive list of 2500 nebulous objects in the heavens, more than ten times the number in any prior catalogs. His interest in 'the construction of the heavens' led also to a comprehensive study of double stars. He believed double stars offered a means to determine the distances of stars from the Earth. The method had first been suggested by Galileo, who reasoned that if one were to look at stars appearing in nearly the same direction in the sky, but which were in reality widely separated in distance from the Earth, that the motion of the Earth around the Sun would cause the closer of the stars to appear to move back and forth relative to the more distant star. Herschel was unable to verify such motions, but he did correctly conclude that most doubles consisted of stars which were actually the same distance from Earth and circled about one another due to their mutual gravitational attraction. During his lifetime he discovered about 800 of these 'binary' stars. His observations also confirmed that Sir Isaac Newton's laws of motion were applicable at stellar distances as well as within the smaller scale of our Solar System.

Herschel's studies of the Solar System did not end with his discovery of the planet Uranus. Six years later, with an improved telescope, also of his own making, he discovered the two largest moons of Uranus. These were later to be known as Titania and Oberon, king and queen of the fairies in William Shakespeare's 'A Midsummer Night's Dream'. Convinced that Uranus must have more satellites, Herschel continued for more than a decade to search for these elusive bodies. In 1798 he reported four other satellites, but those reports proved spurious. However, his 1789 observations with the 40-ft telescope enabled him to add Saturn's satellites Mimas and Enceladus to his list of discoveries. He measured the rotation period of Saturn and showed that Saturn's rings also rotated. Herschel's studies of star motions also led him to another startling discovery about the Solar System: that the Sun is not stationary in the heavens, but is moving toward a point in the constellation Hercules. In a manner similar to Nicolaus Copernicus's conclusion that the Earth was not the center of the universe, Herschel concluded that the Sun also could not claim that distinction.

Another major discovery of Herschel came as a result of his experiments on the temperature of the different colors contained in sunlight. He was amazed to find the highest temperatures occurred beyond the red end of the visible spectrum (Fig. 1.4), where no light could be seen. He rightly concluded that a form of sunlight which is invisible to the eye was responsible for the effect. Such light is now known as infrared light.

Herschel was elected to membership of the Royal Society in 1781, was involved in founding the Royal Astronomical Society (originally known as the Astronomical Society of London), and became its first president in 1820. Because of poor health he was never able to take an active role in the Royal Astronomical Society. In 1816 he was knighted by the Prince Regent for his scientific accomplishments. He died August 25, 1822, at Observatory House in Slough, Buckinghamshire, England, where he had lived for more than 36 years. It is interesting to note that his lifetime (just under 84 years) spanned almost precisely the time it takes the planet he discovered to circle the Sun.

1.2   THE CITY OF BATH, ENGLAND

When William Herschel arrived in Bath late in the year 1766, it was a small resort town with a population of about 2000. The town is named for the hot springs around which the Romans built baths and a temple to the goddess Minerva in the first century AD. The Celts attributed healing powers to the gushing waters and maintained that the springs were sacred to their deity Sul. In recognition of this, the Romans named the town Aquae Sulis. The town flourished for about 400 years, but quickly fell into disrepair after the decline of the Roman Empire.

Bath did not long remain dormant. During the reign of the Saxons, Christian monks rediscovered the springs and built new baths around them. Late in the seventh century the first Abbey was constructed nearby. In that Saxon Abbey in the year 973

Fig. 1.4 — Herschel's study of the temperatures of different colors of sunlight led to the discovery of infrared light, which could not be seen by the human eye, but which registered the highest temperatures in his experiment. (P-20751)

Edgar was crowned first king over all of England. The Abbey in existence in 1781 was constructed during the sixteenth century.

By the time William Herschel arrived, Bath had been transformed into a prosperous and beautiful Georgian city, primarily by three men: Beau Nash, Ralph Allen, and John Wood. Nash set the social tone by establishing a program of nightly concerts, held mainly at the famous Pump House overlooking the restored Roman baths. William's service as Bath's Director of Public Concerts involved him in that aspect of the town's activities. Allen provided stone from nearby quarries. Wood's architectural talents were responsible for the classic style that set the standard for the growing city.

Queen Square, named after Queen Caroline, wife of George II, was one of John Wood's first major building projects. With imposing structures on all four sides of the park-like square, it was designed to resemble a palace with a courtyard. A large obelisk in the center of Queen Square commemorates the visit of Frederick, Prince of Wales. In April of 1770, William moved to 7 New King Street, a two-minute stroll from Queen Square. There he began his career as astronomer and telescope maker. Four years later William moved with his brother and sister to a house near Walcot turnpike. It was a site better suited for telescope making than the New King Street house, but it was inconveniently far from the concert halls in the center of town.

The three Herschels moved back to New King Street in 1777, to No. 19. It was there that they lived for most of the remainder of their tenure in Bath. The Herschels moved for a year and a half to 5 Rivers Street, possibly to be closer to the newly-formed Philosophical Society of Bath. However, the Rivers Street home had no garden for his observations, so William had to set up his telescope in front of the house. It was on one such night that he met his life-long friend, Dr (later Sir) William Watson.

By early 1781, William and Caroline had established a routine of working together on their observations of the heavens. For Caroline, it was perhaps an unfortunate coincidence that on the night of March 13th, 1781, their move back to New King Street was only partially complete. Caroline was at the house on Rivers Street, while William was at 19 New King Street making perhaps the most important observation of his life, the discovery of a new planet.

1.3   THE DISCOVERY OF HERSCHEL'S COMET

Herschel's discovery of Uranus could scarcely be considered an accident. He had already completed a survey of all stars of 4th magnitude [2] or brighter, and had begun a more ambitious program of studying all stars brighter than 8th magnitude. Uranus is an object of 6th magnitude, and was therefore one of the brighter objects observed by Herschel. Normally, Caroline would assist him by making notes and organizing data to more easily facilitate later comparisons. Not even the rigors of changing residences dampened William's ardor for the work. On the night of March 13th, 1781, Caroline was attending to some last-minute business at their prior residence. William had already set up his new 6.2-in-diameter, 7-ft-long homemade reflecting telescope (Fig. 1.5) in the back yard of the home at 19 New King Street (Fig. 1.6). It is not known whether the precise site was in the garden or on the flat 'porch' area that formed the roof over the workshop. His journal contains the following entry:

> Tuesday, March 13. In the quartile near zeta Tauri the lowest of two is a curious either Nebulous Star or perhaps a Comet. A small star follows the Comet at 2/3 of the field's distance. [3]

William's next journal entry on the 'comet' was four nights later:

> I looked for the Comet or Nebulous Star and found that it is a Comet, for it has changed its place. [4]

Fig. 1.5 — Replica of Herschel's 7-ft long, 6.2-in aperture reflecting telescope in Herschel House Museum, Bath, England. (P-29489A)

It seems very likely that Herschel was convinced the object was not a star even on the night of the 13th of March. In his first paper to the Royal Society he explained:

I was then engaged in a series of observations on the parallax of the fixed stars, which I hope soon to have the honour of laying before Royal Society; and those observations requiring very high powers, I had ready at hand the several magnifiers of 227, 460, 932, 1536, 2010, &c. all which I have successfully used upon that occasion. The power I had on when I first saw the comet was 227. From experience I knew that the diameters of the fixed stars are not proportionally magnified with higher powers, as the planets are; therefore I now put on the powers of 460 and 932, and found the diameter of the comet increased in proportion to the power, as it ought to be, on a supposition of its not being a fixed star, while the diameters of the stars to which I compared it were not increased in the same ratio. Moreover, the comet being magnified much beyond what its light would admit of, appeared hazy and ill-defined with these great powers, while the stars preserved that lustre and distinctness which from many thousand observations I knew they

Fig. 1.6(a) — Front view of Herschel House at 19 New King Street, Bath, England. (P-29488A)

would retain. The sequel has shown that my surmises were well founded, this proving to be the Comet we have lately observed. [5]

After 'verifying' on March 17th that the 'comet' indeed moved, Herschel hastened to notify the Astronomer Royal, Nevil Maskelyne, at Greenwich, who replied on April 23rd as follows:

I am to acknowledge my obligations to you for the communication of your discovery of the present Comet, or planet, I don't know which to call it. It is as likely to be a regular planet moving in an orbit nearly circular round the sun as a Comet moving in a very eccentric ellipse. I have not yet seen any Coma or tail to it. [6]

Fig. 1.6(b) — Rear view of Herschel House. It was here that Herschel discovered Uranus.
(P-29488B)

It is obvious from the foregoing statement that Maskelyne suspected the object might not be a comet at all, but rather a 'new' planet, a far more significant discovery. Nineteen days earlier, he had written to Sir William Watson, referring to Herschel's discovery as a 'comet or new planet' [7]. Herschel's friend Watson was a member of both the Royal Society and Bath's Philosophical Society.

In spite of the growing consensus that he had discovered a planet, Herschel himself continued to refer to the object as a comet. His measurements of its diameter had convinced him that it was growing in apparent size and was likely approaching the inner Solar System. He wrote to Watson, '... my apparatus being but ill-adapted to such observations as are necessary to settle the orbit of a Comet, which may be much better done in a regular Observatory, I resign it to abler hands' [8]. Herschel presented a paper entitled, 'Account of a Comet,' to the Philosophical Society of Bath Philosophical Society Bath late in the month of March, 1781. His friend

Watson, as a member of the Royal Society, communicated Herschel's paper to that august group on April 26, 1781.

1.4   FROM COMET TO NEW PLANET

Official recognition of Herschel's 'comet' as a previously unknown planet came slowly. Maskelyne seemed more convinced that it was a planet than many others. The Reverend Thomas Hornsby at the Oxford Observatory wrote Herschel: 'I do not in the least question but this is the comet of 1770 [9], but whether it has passed its Perihelion [10] or has not yet come to it, is more than I can say at present. I will very soon try to construct its orbit.' [11] Johann Elert Bode of Berlin, on page 210 of his *Berliner Astronomisches Jahrbuch (Astronomical Yearbook)* for 1784 [12], described the outstanding astronomical event of 1781 as 'the discovery of a moving star that can be deemed a hitherto unknown planet-like object circulating beyond the orbit of Saturn.' He did not know the name of the discoverer, but described him as a 'very attentive lover of astronomy, probably a German, living at Bath in England.' He gave information on eight observations of the object between March 17 and May 28, after which it was too near the direction of the Sun to observe it. Bode began telescopic observations in early July 1781, but was unable to see Herschel's object until August 1st. He made six additional observations in August and early September, plotting the positions on a map of the heavens (Fig. 1.7). Bode's plot

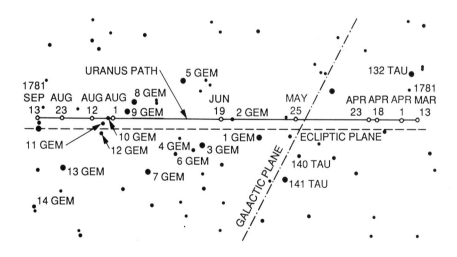

Fig. 1.7 — The path of Uranus between March 13 and September 13, 1781. Adapted from Bode's plate II in the *Berliner Astronomisches Jahrbuch* for 1784.

confirmed that the path of Herschel's object was parallel to and slightly north of the ecliptic [13], and concluded that its orbit was consistent with its being a previously unknown planet of the Solar System, orbiting the Sun at a distance approximately twice that of Saturn's orbit. Spurred on by these reports, Herschel carefully repeated his earlier observations of the diameter of Uranus several times over a 12-month period. He concluded that Uranus was not growing and had a mean value of 4.18″ (seconds of arc [14]). This was only slightly higher than the presently accepted value of 3.88″ at mean opposition [15].

By November, 1781, Herschel's discovery had been judged sufficiently meritorious for him to be awarded the Royal Society's Copley Medal. Although no official astronomical position was available, George III was persuaded to grant Herschel an annual stipend. Conditions of the stipend required that he move to a location close to Windsor so members of the royal family might occasionally look through his telescopes. The move from Bath occurred in early August 1782. On the 8th of August, Maskelyne wrote a letter inviting Herschel to name the new planet:

> Astronomy and Mechanics are equally indebted to you for what you have done; the first for your shewing to artists to what degree of perfection telescopes may be wrought; and the latter for your discovering to Astronomers a number of hitherto hidden wonders of the heavens, which could not be explored before for want of telescopes equal to yours; and they are both likely to receive equal improvement from it in the construction of better telescopes, and in the application that may be made of them to the heavens for repeating and extending your observations. I hope you will do the astronomical world the faver to give a name to your planet, which is entirely your own, & which we are so much obliged to you for the discovery of. [16]

Bode apparently was the first to suggest the name Uranus for the new planet [17] in late 1781. On page 191 of his *Berliner Astronomisches Jahrbuch* for 1785 (published near the end of 1782), Bode said he had suggested the name because Uranus was the mythological father of Saturn, just as Saturn was the mythological father of Jupiter. Other astronomers had proposed the name Herschel's Planet, after the new planet's discoverer. In response to Maskelyne's request in the summer of 1782, Herschel proposed the name 'Novum sidus Georginum' (George's new star), after George III. His friend Watson assured Herschel that the correct Latin form would be Georgium Sidus, and it was that name, or its anglicized 'Georgian Planet', that Herschel used throughout his life.

By the end of 1782 most astronomers were convinced that Georgium Sidus was indeed a major planet. In a 1783 letter to Sir Joseph Banks, President of the Royal Society, Herschel first acknowledged that 'By the observation of the most eminent Astronomers in Europe it appears that the new star, which I had the honour of pointing out to them in March, 1781, is a Primary Planet of our Solar System.' [18] He also put forth his arguments for the name 'Georgium Sidus' as follows:

> In the fabulous ages of ancient times the appellations of Mercury, Venus, Mars, Jupiter and Saturn, were given to the Planets, as being the names of their principal heroes and divinities. In the present more philosophical era it would hardly be allowable to have recourse to the same method, and call on Juno, Pallas, Apollo, or Minerva, for a name to our new heavenly body. The

first consideration in any particular event, or remarkable incident, seems to be its chronology: if in any future age it should be asked, when this last-found Planet was discovered? It would be a very satisfactory answer to say, 'in the Reign of King George the Third.' As a philosopher then, the name of Georgium Sidus presents itself to me, as an appellation which will conveniently convey the information of the time and country where and when it was brought to view. But as a subject of the best of Kings, who is the liberal protector of every art and science; — as a native of the country from whence this Illustrious Family was called to the British throne; — as a member of that Society, which flourishes by the distinguished liberality of its Royal Patron; — and, last of all, as a person now more immediately under the protection of this excellent Monarch, and owing every thing to His unlimited bounty: — I cannot but wish to take this opportunity of expressing my sense of gratitude, by giving the name Georgium Sidus to a star, which (with respect to us) first began to shine under His auspicious reign. [19]

The name Uranus proposed by Bode was not universally accepted until after Herschel's death in 1822. The Nautical Almanac, in fact, continued to call it 'Georgian' until about 1850. In a letter to Bode on 13 August 1783, Herschel philosophically acknowledged that '... When I named it Geo-Sidus I hardly expected that this name would become generally accepted, because we already know by experience that the first names of the satellites of Jupiter and Saturn were soon changed...' [20]. It seems that Galileo Galilei and Jean Dominique Cassini had also proposed naming their discoveries after their respective sovereigns, only to have the names of the satellites later changed to their presently accepted mythological ones.

1.5 CONTROVERSY LEADS TO THE DISCOVERY OF NEPTUNE

In the year of Herschel's arrival in Bath (1766), Johann Daniel Titius, Professor of Mathematics in Wittenberg, was struck by the orderly progression of distances of the planets from the Sun. He noted that their distances in astronomical units [21] could be approximated by adding Mercury's distance (0.4 AU) to the expression 0.3×2^n, where $n=0$ for Venus, $n=1$ for Earth, $n=2$ for Mars, $n=4$ for Jupiter, and $n=5$ for Saturn (see Table 1.1). The fact that $n=3$ was missing was interpreted by Titius as an indication that there was an undiscovered celestial body between the orbits of Mars and Jupiter, most likely a missing moon of one of the two planets. In 1772, Bode championed the suggestion of Titius, but argued that the gap must be filled by a planet, not by anything as insignificant as a moon. This numerical progression has come to be known as Bode's Law (or, more historically accurate, the Titius–Bode Law).

Most astronomers of the day dismissed Bode's Law as pure coincidence. They argued that there was no physical basis for the law, that there was no planet for $n=3$, and that there were significant departures from the predicted distances for the cases of Mars and Saturn. Bode's realization that Herschel's new planet fit the progression ($n=6$ for Uranus) provided added credence for Bode's Law. Bode and five other astronomers met in September 1800 to work out a plan that would lead to the

Table 1.1 — Bode's Law of planetary distances (distance in AU $0.4+0.32\times2^n$)

| Planet | n | Distance in AU | | |
		Predicted	Observed	Pred.−Obs.
Mercury	—	0.400	0.387	+0.013
Venus	0	0.700	0.723	−0.023
Earth	1	1.000	1.000	0.000
Mars	2	1.600	1.524	+0.076
Ceres	3	2.800	2.766	+0.034
Jupiter	4	5.200	5.203	−0.003
Saturn	5	10.000	9.555	+0.445
Uranus	6	19.600	19.218	+0.382
Neptune	7	38.800	30.110	+8.690
Pluto	8	77.200	39.44	+37.76

discovery of the missing planet between Mars and Jupiter. But on January 1st, 1801, before Bode and his colleagues could implement their plan, Giuseppe Piazzi, Director of the Palermo Observatory, discovered what would later become known as the first of the minor planets, Ceres. An orbit was calculated, and Ceres' mean distance was found to be in excellent agreement with Bode's Law!

The euphoria felt by Bode and his colleagues was destined to be short-lived. Only 14 months after the discovery of Ceres, German astronomer Heinrich Wilhelm Matthaeus Olbers, a colleague of Bode, found what seemed to be another planet at the same distance as Ceres. The new planet was named Pallas. At first, supporters of Bode's Law were dumbfounded, but Olbers himself suggested the solution. In June 1802 he proposed to Herschel, 'What if Ceres and Pallas were just a pair of fragments, or portions of a once greater planet which at one time occupied its proper place between Mars and Jupiter and was in size analogous to the outer planets, and perhaps millions of years ago, had, either through the impact of a comet, or from an internal explosion, burst into pieces?' [22]

Herschel was to provide further evidence in support of Olbers' proposal. His micrometer measurements of Ceres and Pallas [23] yielded diameters of 261 km (162 miles) and 237 km (147 miles), respectively. More recent observations [24] show that the respective diameters of Ceres and Pallas are 1025 km and 583 km, but Herschel's measurements nevertheless served to confirm that they were much smaller than the major planets. Herschel argued that such small bodies should not reasonably have the same status as the primary planets, and he named them 'asteroids', a title still used today. Olbers' hypothesis and Herschel's supporting measurements spurred further searches for asteroids, and two more (Juno and Vesta) were discovered by 1807. There are now several thousand known asteroids.

Perhaps even more important than the continued search for asteroids was the suggestion that a major trans-Uranian (beyond Uranus) planet might exist. Astronomers were no longer bound by the assumption that the planets formed a closed set known to ancient observers. In fact, some were so convinced by the successes of Bode's Law that they chose a tentative name, Ophion, for the undiscovered planet that was expected to be about 38.8 AU from the Sun.

An additional factor, more firmly based in scientific theory than Bode's Law, led to the search for a trans-Uranian planet. Attempts to fit an elliptical orbit to the path of Uranus around the Sun met with only limited success. Even when scientists included the perturbing effects of the known planets, the discrepancies between predicted and observed positions of Uranus were far too large. Searches through the astronomical literature yielded more than 20 pre-discovery positions of Uranus [25] as far back as 1690. The predicted positions for the planet for these earlier times also disagreed with the actual observations.

The details of Neptune's discovery are told by Grosser [26]. In 1843 John Couch Adams of England began a mathematical study of the discrepant motions of Uranus. His goal was to predict the position and size of the undiscovered Neptune. In 1846 Urbain-Jean Joseph Le Verrier of France independently started a similar study. Adams communicated his predictions to the Astronomer Royal, Sir George Airy, but was not taken seriously at first. In mid-1846, Sir John Herschel (William's son), familiar with the attempts of both Adams and Le Verrier to predict Neptune's position, said of the undiscovered planet, 'We see it as Columbus saw America from the shores of Spain. Its movements have been felt trembling along the far-reaching line of our analysis, with a certainty hardly inferior to that of ocular demonstration.' [27] Airy, now convinced that Adams' calculations were worthy of investigation, sent them to James Challis at Cambridge Observatory. Challis began his search shortly thereafter, using repeated observations of the predicted area of the sky to look for motions of the planet against the background of fixed stars.

Le Verrier also had difficulty convincing astronomers in his home country to engage in a serious search for the new planet. On 23 September, 1846, Le Verrier communicated his predictions to Berlin astronomer Johann Gottfried Galle, who had no trouble identifying Neptune that same night. One of the primary reasons for Galle's quick success was the fact that asteroid searches had prompted the Berlin Academy of Sciences to construct accurate star maps of the zodiacal [28] region. Galle and his student Heinrich Louis D'Arrest merely compared the section of sky with the unpublished star maps and found the interloper almost immediately. Because of the accuracy of their predictions, it is now generally accepted that both Adams and Le Verrier should be credited with Neptune's discovery.

As Bode's Law provided much of the impetus leading to the discovery of Neptune, it is ironic that Neptune's discovery resulted in the reduction of Bode's Law to a mere mathematical curiosity. Neptune was found to have a distance of 30.06 AU, much closer than the 38.8 AU predicted by Bode's Law. Even the later discovery of Pluto at 39.44 AU was insufficient to resurrect a law for which astronomers still find no physical basis.

NOTES AND REFERENCES

[1] Dreyer, J. L. E. (ed.) (1912) *The Scientific Papers of Sir William Herschel*, Royal Society and Royal Astronomical Society, **1**, 30.

[2] The brightest stars in the sky are designated 1st magnitude; the dimmest stars

visible with the unaided eye are designated 6th magnitude. Stars of 1st magnitude are very nearly 100 times as bright as 6th magnitude stars. Astronomers have formalized the magnitude scale such that two stars which differ in brightness by a factor of ten are assigned magnitudes which differ by precisely 2.5 magnitudes. Thus, a star of magnitude 1.0 is precisely ten times as bright as a star of magnitude 3.5 and 100 times as bright as a star of magnitude 6.0.

[3] Royal Astronomical Society manuscript Herschel W.2/1.2, 23; also quoted in Alexander, A. F. O'D. (1965) *The Planet Uranus*, American Elsevier Publishing Company, Inc., p. 26.

[4] RAS MSS Herschel W.2/1.2, 24.

[5] Dreyer, J. L. E. (ed.) (1912) *The Scientific Papers of Sir William Herschel*, Royal Society and Royal Astronomical Society, **1**, 30.

[6] RAS MSS Herschel W.1/13.M, 14.

[7] RAS MSS Herschel W.1/13.M, 14.

[8] Schaffer, S. (1981) Uranus and the establishment of Herschel's astronomy, *Journal for the History of Astronomy*, **XII**, 11–26.

[9] 'Lexell's Comet' was discovered in 1770, but there were insufficient measurements to establish its orbit, and it was 'lost' to astronomers.

[10] Perihelion is the point in the path of an object in (non-circular) orbit around the Sun where the distance to the Sun is at a minimum. The most distant point is known as aphelion.

[11] Dreyer, J. L. E. (ed.) (1912) *The Scientific Papers of Sir William Herschel*, Royal Society and Royal Astronomical Society, **1**, 30.

[12] Although bearing the date 1784, the *Berliner Astronomisches Jahrbuch* was published near the end of 1781. It contained data on the positions of bodies in the Solar System for three years in advance, hence the 1784 date.

[13] The ecliptic is the apparent path of the Sun among the background stars during the course of a year. It represents the plane in which the Earth revolves around the Sun and near which all of the other planets of the Solar System orbit the Sun.

[14] A circle is divided into 360 degrees, and each degree is divided into 3600 seconds of arc. As viewed from Earth, the apparent diameter of Uranus is slightly larger than 1/500th that of the Moon.

[15] The mean opposition distance of Uranus (2 720 000 km) is the mean, or average, yearly minimum distance from Earth to Uranus. Yearly minimum distances range from 2 581 900 to 2 858 100 km. Opposition refers to a configuration in which the Sun and Uranus (or another celestial body) are in opposite directions in the sky as viewed from Earth.

[16] RAS MSS Herschel W.1/12.M, 20.

[17] Houzeau, J. C. & Lancaster, A. (1882) *Bibliographie Generale de l'Astronomie*, **2**, col. 1456.

[18] Dreyer, J. L. E. (ed.) (1912) *The Scientific Papers of Sir William Herschel*, Royal Society and Royal Astronomical Society, **1**, 100.

[19] Dreyer, J. L. E. (ed.) (1912) *The Scientific Papers of Sir William Herschel*, Royal Society and Royal Astronomical Society, **1**, 100.

[20] Bode, J. E. (ed.) (1783) *Berliner Astronomisches Jahrbuch, 1786*, p. 258.

[21] An astronomical unit, abbreviated AU, is defined to be the mean distance from

the Earth to the Sun, which is equivalent to the semi-major axis of Earth's orbit, 149 597 900 km.

[22] Lubbock, C. (1933) *The Herschel Chronicle*, Cambridge University Press, p. 202.

[23] Herschel, W. (1802) Observations on the two lately discovered celestial bodies, *Philosophical Transactions*, **92**, 213–233.

[24] Bowell, E., Gehrels, T., Zellner, B. (1979) VII. Magnitudes, colors, types and adopted diameters of the asteroids. In: Gehrels, T. (ed.) *Asteroids*, University of Arizona Press, Tucson, p. 1111.

[25] Astronomers often chart the positions of 'stars' as a part of their observations. In the case of Uranus, one of these stars charted by pre-1781 astronomers later proved to be a planet. These pre-discovery sightings can be recognized because (1) they are close to the predicted position of the planet at that time, (2) there is no known star at that position, and (3) the sighted object has the appropriate brightness.

[26] Grosser, M. (1962) *The Discovery of Neptune*, Cambridge, Massachusetts.

[27] Herschel, J. (1846) letter on 'Le Verrier's planet', *Athenaeum*, 3 October 1846, p. 109.

[28] The zodiac is comprised of the 12 star constellations that lie along the ecliptic, the Sun's apparent annual path through the sky. The motions of the moon and the other planets as viewed from Earth lie within the zodiac.

BIBLIOGRAPHY

Alexander, A. F. O'D. (1965) *The Planet Uranus*, American Elsevier Publishing Company, Inc., New York, pp. 25–55.

Dreyer, J. L. E. (ed.) (1912) *The Scientific Papers of Sir William Herschel*, Royal Society and Royal Astronomical Society, Vol. 1.

Grosser, M. (1962) *The Discovery of Neptune*, Cambridge, Massachusetts.

Hoskin, M. (1986) 'Sir William Herschel Discovers Uranus', *The Planetary Report*, **VI**, No. 6, p. 3.

Hunt, G. (ed.) (1982) *Uranus and the Outer Planets*, Cambridge University Press, Cambridge, pp. 23–89.

Moore, P. (1983) 'William Herschel, Astronomer and Musician of 19 New King Street, Bath', published by P. M. E. Erwood, 95 Walton Road, Sidcup, Kent, in association with the William Herschel Society, 19 New King Street, Bath.

Ring, E. F. J. (1986) 'The Discovery of Infrared Radiation', *Thermology*, **1**, 174–180.

Ring, E. F. J. (1986) 'William Herschel, Pioneer of Thermology', published by the William Herschel Society, 19 New King Street, Bath.

2

Uranus's position in the Solar System

2.1 LAST OF THE NAKED-EYE PLANETS

The planet Uranus, with a magnitude of nearly 6.0, is close to the limit of visibility for the unaided eye. Because of this, it had undoubtedly been viewed by ancient observers of the heavens many times. However, until Herschel pointed his telescope at Uranus in March of 1781, no observer recognized it as a wanderer (planet) among the background stars. There are more than 9000 stars of visual magnitude 6.5 or brighter. Prior to 1800, charts of the constellations seldom included stars fainter than 3rd or 4th magnitude. Furthermore, the motion of Uranus was slow enough that it required seven years to traverse a single constellation (about 30°) in the zodiac. Saturn, by contrast, traveled that same distance across the sky in 2.5 years. Prior to the discovery of Uranus, Saturn was believed to be the most distant observable object in the Solar System. It is consequently not surprising that the seventh planet lay undiscovered for so many millennia.

Occurring so long after the dawn of astronomy, the discovery of Uranus did much to change scientific views about the heavens. Long-held theories about the size and nature of the Solar System had to be revised substantially. The Sun and its collection of planets would never again be primarily a celestial timetable for use in astrological predictions. Serious study of the night skies would never again be purely the realm of professionals. An amateur astronomer had discovered a major planet with a telescope made in his basement workshop. His discovery created a public awareness and excitement that would continue to grow until technology could permit mankind to view these distant worlds from close range. In a very real sense the discovery of Uranus tolled the death knell of the Dark Ages and gave birth to an age of exploration and reason.

2.2 TERRESTRIAL PLANETS AND GAS GIANT PLANETS

Herschel is noted for his telescope-making prowess and for discovering Uranus and at least four Solar System satellites. Perhaps of equal or greater significance were his adventures into comparative astronomy. He did not merely study individual planets, stars, or nebulae: he compared them to one another, making inferences about their relative ages or other characteristics. To his contemporaries such inferences were often beyond comprehension, even beyond the boundaries of astronomy as they understood it. Herschel abandoned the cautious attitudes of his fellow astronomers, and chose instead to speculate on origins, processes, and reasons for the differences he observed. His conclusions were not always correct, but a surprisingly large number are in harmony with those accepted by present-day astronomers. Because his methods and conclusions were often at odds with those of his colleagues, some of Herschel's most important contributions did not receive proper recognition until long after his death.

As the realization dawned that much could be learned by Herschel's methods, comparative planetology was born. When Solar System observers first became aware that Earth was a planet, they assumed that other planets must be like Earth. Gradually they discovered bits and pieces of information that showed some major differences between the planets. There were the obvious differences in their distances from the Sun and in the time it took them to orbit the Sun (see Fig. 2.1). It

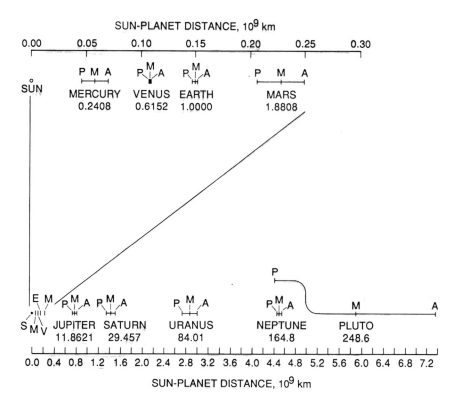

Fig. 2.1 — Relative distances from the Sun of the major planets. Minimum (P, perihelion), mean (M), and maximum (A, aphelion) distances are shown. Orbital periods in years are given under each planet name.

was also apparent that Jupiter, Saturn, Uranus, and Neptune were much larger than Earth (see Fig. 2.2). Some had natural satellites (moons) circling them, others had

Fig. 2.2 — Relative sizes of the major planets. (P-30676A)

none. From Newton's laws of motion, scientists could utilize measurements of the distances of the satellites from their planets and of their orbital periods to calculate the gravitational pull of each planet. That enabled them to determine the amount of mass contained within each planet. Jupiter, for example, has a volume more than 1300 times as large as Earth's, but it contains only 318 times as much mass as Earth. Its density (mass per unit volume) is less than one-quarter that of Earth, only 30% larger than the density of water. The contrast between Saturn's mass and its volume is even more startling: although more than 760 times Earth's volume, its mass is less than 100 Earth masses. Its density is so low that Saturn would actually float in water! The respective densities of Uranus and Neptune are about 20% and 60% greater than water. Certainly these worlds are not made of the same materials as Mercury, Venus, Earth, and Mars, whose densities range from 4 to 5.5 times the density of water. Tables 2.1 and 2.2 summarize some of the physical characteristics of the planets.

Table 2.1 — Physical characteristics of the terrestrial planets

	Mercury	Venus	Earth	Mars	Pluto
Mean distance from Sun (AU)	0.3871	0.7233	1.0000	1.5237	39.44
Orbital period (tropical years)	0.2408	0.6152	1.0000	1.8809	248.5
Inclination to Earth's orbit (°)	7.00	3.39	0.00	1.85	17.17
Equatorial diameter (Earth=1)	0.3824	0.9489	1.0000	0.5328	0.176
Volume (Earth=1)	0.0561	0.8573	1.0000	0.1508	0.006
Mass (Earth=1)	0.0558	0.8150	1.0000	0.1074	0.002
Mean density (g/cm³)	5.49	5.25	5.52	3.86	2.0
Rotation period (h)	1407	5832	23.9	24.6	153
Tilt, equator to orbit, (°)	0	2.6	23.4	24.0	96
Solid surface?	yes	yes	yes	yes	yes
Ring system?	no	no	no	no	no
Number of known satellites	0	0	1	2	1

Table 2.2 — Physical characteristics of the gas giant planets

	Jupiter	Saturn	Uranus	Neptune
Mean distance from Sun (AU)	5.2026	9.5547	19.2181	30.1096
Orbital period (tropical years)	11.862	29.458	84.014	164.793
Inclination to Earth's orbit (°)	1.30	2.49	0.77	1.77
Equatorial diameter (Earth=1)	11.209	9.449	4.007	3.87
Volume (Earth=1)	1321	764	63	57
Mass (Earth=1)	317.9	95.2	14.5	17.1
Mean density (g/cm³)	1.33	0.69	1.27	1.65
Rotation period (h)	9.92	10.66	17.24	16.11
Tilt, equator to orbit (°)	3.08	26.73	97.92	28.8
Solid surface?	no	no	no	no
Ring system?	yes	yes	yes	yes
Number of known satellites	16	17	15	8

Another startling fact became apparent to Solar System astronomers: Jupiter, Saturn, Uranus, and Neptune seem to have no solid surfaces. They are giant spheres of gas, held together by gravity. Their chemical makeup is more closely aligned with that of the Sun than with Earth. Collectively they are most often referred to as gas giant planets. Mercury, Venus, Earth and Mars are classified as terrestrial planets because of their solid surfaces.

Pluto is a special case. It is the most distant of the known planets, and the details of its size and density have been difficult to determine. It has a solid surface and is orbited by a small moon named Charon, but Pluto is smaller than Earth's Moon. Pluto's other characteristics (density only about twice that of water, and methane ice on its surface) are also in sharp contrast with the terrestrials of the inner Solar System. Except for its independent orbit and its moon, Charon, Pluto closely resembles a satellite of one of the gas giant planets. For many years, astronomers conjectured that it was an escaped moon of Neptune. However, since it possesses its own satellite and since no reasonable scenario for its escape from Neptune into its present orbit has been proposed, Pluto remains a classification enigma.

Uranus is clearly a gas giant. It is considerably smaller than Jupiter and Saturn, but its volume would still encompass 63 Earths, and it is nearly a twin in size to Neptune. It has several characteristics which make it unique among the gas giants (for example, the orientations of its rotation axis and its magnetic field). It nevertheless shares the gas giant characteristics of large size and absence of a solid surface. Its differences from the other gas giants are clues, which, if properly assembled, can help scientists to reconstruct its origins and past history.

2.3   THE ORIGINS OF URANUS

Being residents of Planet Earth, we have never observed the formation of a system of planets around a central star. We are becoming more familiar with our own Solar System in its present state. Processes and natural laws which affect the evolution of the Solar System are also beginning to be better understood. Astronomers have even seen evidence that other stars may be encircled by disks of material which may be primordial planetary systems. Scientific theories about the formation of the Solar System generally agree that the process requires thousands of millions of years. If these time scales are accurate, the entire history of the human family on Earth spans only a minute fraction of the age of a planetary system like the Solar System. One might suppose that with such a limited view, theorists would be unable to devise any rational hypotheses, and that their models of Solar System formation would be little more than conjecture.

Fortunately, the prospects for devising realistic models are not as bleak as they might seem. Whatever the formative processes, they must be consistent with the laws of nature. Science has not been infallible in discovering or interpreting these laws, a fact readily demonstrated by Aristotle's faulty deduction that Earth is at the center of the universe. Today the scientific method includes observing, hypothesizing, testing, revising hypotheses, and continued testing. This method substantially reduces (but does not eliminate) the possibility of error.

To be useful, hypothetical models of Solar System formation should also predict conditions consistent with those presently observed. When very little was known about the Solar System, there were fewer constraints on possible hypotheses. As scientific knowledge increased, hypotheses were either discarded, revised, or confirmed. A hypothesis becomes a theory when there is sufficient confidence in its ability to predict observed facts. A theory is recognized as a natural law when repeated confirmations have erased doubts about its validity.

Proposed hypotheses on Solar System formation should also be consistent with widely accepted cosmological [1] theories. It would be inappropriate, for example, to postulate a behavior for the Sun that violated the premises of stellar evolutionary theory for the myriads of other stars of the same type as the Sun. The Sun is classified as a G2 main sequence star, which means that it has a surface temperature of about 5800 K, a diameter of about 1 400 000 km, and an estimated age of about 5 000 000 000 years. The star Rigel Kentaurus (also known as alpha Centauri A) is a near twin to our Sun. Rigel Kentaurus is also one of four stars which collectively are the Sun's nearest stellar neighbors.

Two types of theories seem to fit the stated requirements: accretional theories and giant protoplanet theories. Some of the basic tenets of each of the two are outlined below. In discussing these theories, the author acknowledges heavy reliance on papers by Lewis [2] and by Podolak [3] and recommends that individuals interested in more details refer directly to one of these two excellent papers.

2.3.1 Accretional theories

Accretional theories of Solar System formation presuppose that the newly formed Sun was surrounded by a disk of both solid and gaseous material. There are differences of opinion about how the disk of material was formed. Perhaps as an enormous cloud of material inside our Milky Way Galaxy began to collapse owing to its own gravity (the most likely process in the birth of a star), the motions within the material were too rapid to permit complete collapse, and a spinning disk of material resulted. For purposes of the discussion here, the processes that occur after the formation of the disk are basically independent of disk formation processes.

Pressures and temperatures within the disk determined whether, at a given time, any particular constituent was solid or gaseous. Each of the planets began as solid material collected gravitationally to form a small core of rock. These small cores grew as inelastic collisions [4] with other rocky fragments occurred until most of the rocky material in the disk had been collected. At that point, the terrestrial planets had accreted (accumulated) nearly all of their present mass. By that time, gaseous material in the inner Solar System had been blown away by the very active solar wind flowing outward from the Sun. As the young Sun diminished in brightness, the surrounding disk cooled, and some of the gaseous materials in the outer Solar System began to condense into ices. The most abundant of these condensed gases were water (H_2O), ammonia (NH_3), and methane (CH_4). These new and more abundant solids were also slowly accreted by the rocky cores of the giant planets. Eventually these ice-covered rocky cores became massive enough to induce dynamic instabilities in the surrounding gases, composed primarily of hydrogen and helium. The result was that much of the hydrogen and helium was captured by the planets we now know as Jupiter, Saturn, Uranus, and Neptune.

Such theories can be made to successfully account for the formation of the inner planets and for some deviations from solar composition in the atmospheres of the outer planets, especially Uranus and Neptune. Accretion is inherently a relatively slow process. There is concern that the processes necessary to form Uranus and Neptune through accretion might require too much time. The hydrogen and helium gases in the nebular disk must not dissipate before the accretion process is completed, or the materials necessary to form the atmospheres of these two planets would not be available. Another obvious difficulty inherent in accretional theories is a reasonable explanation of why Jupiter and Saturn accumulated such large atmospheres compared to Uranus and Neptune. Theoreticians will continue to study these and other problems. Perhaps data from spacecraft and recent ground-based observations can be used to further refine accretion models of the origin of Uranus and the other planets of the Solar System.

2.3.2 Giant protoplanet theories

Let us suppose alternatively that the planets all started with a solar mixture of ingredients. Different scenarios have been suggested for the source of this material. In one, the gaseous material is stripped from a developing star encountered briefly by the Sun [5]. In another, the Solar System starts as a loose assemblage of gas 'floccules' randomly orbiting around the center of mass of the assemblage; collisions and transfer of momentum result in a central Sun and orbiting planets [6]. In still another scenario, the gravitationally collapsing proto-Sun leaves rings of gaseous material which break up and interact with each other [7].

In any of these cases, the planets start out with solar composition and undergo evolutionary changes to bring them to their present non-solar-composition state. The terrestrial planets lose all of their original atmospheres. Uranus loses a major part of its hydrogen and helium as well as most of its nitrogen. Although the giant protoplanet theories have difficulty explaining how these evolutionary changes come about, mechanisms have been proposed to overcome these and other apparent difficulties. There is, however, a more important reason why such theories are somewhat less favored than the accretional theories. Initiation of the collapse of gaseous protoplanets into the present-day gas giants occurs much more readily in the presence of the solid rocky planetary cores of the accretional theories.

2.3.3   Possible implications of the tilt of Uranus
One of the most unusual characteristics of Uranus is the 95° tilt of its rotation axis with respect to the pole of its orbit around the Sun. Although tiny Pluto's tilt is slightly larger, gravitational forces from its satellite Charon, which has a mass more than 10% that of Pluto, could account for the tilt. Pluto and Charon are tidally locked: they always keep their same faces toward each other. No similar tidal forces exist for Uranus. All of its moons and rings combined have less than 0.02% of the mass of Uranus: tidal forces from the moons and rings would be too small to tip the massive planet on its side.

The angle between the planes of the equator and the orbit of a planet is known as the planet's obliquity. The larger a planet's obliquity, the more pronounced are seasonal effects. Earth's obliquity is 23.4°; Mars, Saturn, and Neptune have obliquities only slightly larger. Each of these four planets have moderate seasonal effects due to the changing latitudes of solar illumination. Mercury, Venus, and Jupiter have such small obliquities that they exhibit almost no seasonal effects. The 98° obliquity of Uranus means that once each orbit around the Sun the northern hemisphere will remain almost entirely in darkness for a substantial period of time (as shown in Fig. 2.3), while the southern hemisphere basks in perpetual sunlight. Then, 42 Earth years later (half of Uranus's orbital period), the same conditions are repeated in the opposite hemispheres. The time it takes for the atmosphere to cool appreciably is many times longer than the 84-year Uranus orbit; the result is that the temperature at any latitude in the atmosphere is relatively unchanged with season. Mathematical calculations show that the extreme axial tilt causes the poles to receive more total solar energy per square meter during the 84-year orbital period than the equator, and polar temperatures were therefore expected to be a few degrees warmer than equatorial temperatures. The redistribution of heat from poles toward the equator was expected to cause winds in a direction opposite that of the rotation of the planet.

Seasonal effects on the airless Uranian satellites would be more substantial. Temperatures at the dark poles of the satellites could be extremely low, possibly

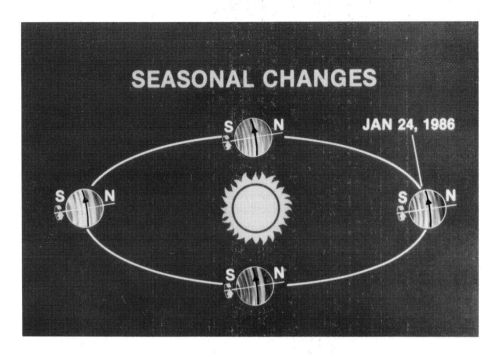

Fig. 2.3 — Seasonal variations in solar illumination during the 84-year period of Uranus caused
by the planet's 98° obliquity. (260-1772)

controlled by the outflow of heat from the interiors. In the absence of internal heat
sources, temperatures near absolute zero (−273°C) might occur.

What could have caused Uranus's high obliquity? There remain few, if any, clues
to aid in answering this question, and so theoreticians must make some *ad hoc*
assumptions that are little more than speculation. It is not impossible to imagine
conditions in Uranus's part of the early solar nebula which naturally led to the giant
planet's unorthodox orientation. Perhaps the collection of planetesimals (or the
swirling gases) that were to become Uranus already possessed the necessary angular
momentum. After all, the much larger Saturn has a 27° tilt, and outer neighbor
Neptune sports a 29° obliquity. A more popular explanation is that Uranus was born
with a small obliquity. Sometime during the late stages of its formation, before its
moons and rings were separate entities, it was struck by an Earth-sized icy body near
one of its poles. This collision caused an immediate and permanent shift in Uranus's
rotation axis, increased substantially the water content in the planet, mixed the gases
and liquids in the interior, and may account for the absence of an internal heat
source. It is possible that tiny fragments of this icy body remain as the present
satellites of Uranus. Until theoreticians have more information to constrain possible
models of Uranus's origin and history, this collisional scenario will remain a
reasonable working hypothesis.

NOTES AND REFERENCES

[1] Cosmology is a branch of astrophysics that deals with the origin, structure, and space–time relationships of the universe.

[2] Lewis, J. S. (1984) The origin and evolution of Uranus and Neptune. In: Bergstralh, J. T. (ed.) *Uranus and Neptune*, NASA Conference Publication 2330, National Aeronautics and Space Administration, Scientific and Technical Information Branch, pp. 3–24.

[3] Podolak, M. (1982) The origin of Uranus: compositional considerations. In: Hunt, G. (ed.) *Uranus and the Outer Planets*, Cambridge University Press, pp. 93–109.

[4] Inelastic collisions are collisions in which part of the energy of motion of the colliding particles changes to heat or another kind of radiation. In the context of accretion theory, the inelastic collisions usually result in the two colliding particles sticking together to form a single larger particle.

[5] Woolfson, M. M. (1978) The capture theory and the evolution of the solar system. In: Dermott, S. F. (ed.) *The Origin of the Solar System*, John Wiley & Sons, New York, pp. 179–217.

[6] McCrea, W. H. (1978) The formation of the solar system: a protoplanet theory. In: Dermott, S. F. (ed.) *The Origin of the Solar System*, John Wiley & Sons, New York, pp. 75–110.

[7] Cameron, A. G. W. (1978) Physics of the primitive solar nebula and of giant gaseous protoplanets. In: Gehrels, T. (ed.) *Protostars and Protoplanets*, University of Arizona Press, Tucson, pp. 453–487.

BIBLIOGRAPHY

Bergstralh, J. T. (ed.) (1984) *Uranus and Neptune*, NASA Conference Publication 2330, National Aeronautics and Space Administration, Scientific and Technical Information Branch, pp. 3–24, 377–423.

Hunt, G. (ed.) (1982) *Uranus and the Outer Planets*, Cambridge University Press, Cambridge, pp. 93–109.

3

Discovery of the five largest satellites

3.1 OBERON AND TITANIA

The two largest satellites of Uranus were not given formal names until 65 years after their 1787 discovery. By that time two more satellites had been discovered. Sir John Herschel, William's son, apparently was responsible for suggesting the names for the four satellites. The names he gave to the largest satellites were Oberon and Titania, the king and queen of the fairies in William Shakespeare's play *A Midsummer Night's Dream*. The smaller two were named Umbriel and Ariel, characters in Alexander Pope's poem *The Rape of the Lock*. Umbriel is 'a moody sprite whom Spleen supplies with sighs and tears'; Ariel is 'a sylph, the special guardian of Belinda'. Ariel is also the name of 'an airy, tricksy spirit, changing shape at will to serve Prospero, his master' in Shakespeare's Tempest.

3.1.1 Discovery by William Herschel

By January 1787, less than six years after Sir William Herschel discovered Uranus from his home in Bath, he was residing at Observatory House in Slough. Most of his research was done with his 18.2-in aperture 20-ft-long reflecting telescope. In a paper delivered to the Royal Society on February 15, 1787, he related the following:

> ... The 11th of January, ... in the course of my general review of the heavens, I selected a sweep which led to the Georgian planet; and, while it passed the meridian, I perceived near its disk, and within a few of its diameters, some very faint stars whose places I noted down with great care.
>
> The next day, when the planet returned to the meridian, I looked with a most scrutinizing eye for my small stars, and perceived that two of them were missing. Had I been less acquainted with optical deceptions, I should immediately have announced the existence of one or more satellites to our new planet; but it was necessary that I should have no doubts. The least haziness, otherwise imperceptible, may often obscure small stars; and I

judged, therefore, that nothing less than a series of observations ought to satisfy me, in a case of this importance. To this end I noticed all the small stars that were near the planet the 14th, 17th, 18th, and 24th of January, and the 4th and 5th of February; and though, at the end of this time, I had no longer any doubt of the existence of at least one satellite, I thought it right to defer this communication till I could have an opportunity of seeing it actually in motion. Accordingly I began to pursue this satellite on Feb. the 7th, about six o'clock in the evening, and kept it in view till three in the morning on Feb. the 8th; at which time, on account of the situation of my house, which intercepts a view of part of the ecliptic, I was obliged to give over the chase: and during those nine hours I saw this satellite faithfully attend its primary planet, and at the same time keep on, in its own course, by describing a considerable arch of its proper orbit. [1]

In the same paper, Herschel also told of tracing the orbit of an inner, faster satellite. He estimated the orbit periods to be about $13\frac{1}{2}$ days for the outer satellite (Oberon) and $8\frac{3}{4}$ days for the inner (Titania). A year later [2] he refined those periods to 13.462 days and 8.709 days, respectively, remarkably close to the presently accepted values of 13.463 days and 8.706 days.

Herschel's discovery of Oberon and Titania led him to redouble his efforts to discover additional satellites of Uranus, which he was convinced must exist. For more than 40 years, no other astronomer could compete with him in this search. In fact, it appears that for at least ten of those years, even Oberon and Titania were not seen in any telescope other than Herschel's own discovery instrument. In his final Uranus paper, he reported on his additional satellite searches on about 158 dates between 1787 and 1810. The task was not a very rewarding one: seeing conditions in England were often unsuitable, the objects for which he was searching were among the faintest objects visible with telescopes of that era, the number of background stars was large, and scattered light from Uranus was always a problem. Nevertheless, in 1798 Herschel reported to the Royal Society the discovery of four additional satellites. Two were supposedly exterior to Oberon's orbit, one between the orbits of Oberon and Titania, and the last one interior to Titania's orbit. The outer three satellites were never confirmed and are assumed to be spurious. The supposed inner satellite will be discussed later in this chapter (section 3.2.2).

Herschel's observations of Oberon and Titania led to other important discoveries about Uranus. From his measures of the distances and orbital periods of the satellites, he made an estimate of the mass of Uranus, calculating it to be 17.74 times as massive as Earth (the presently accepted value is about 14.54). Then, from his measurements of the size of Uranus, he estimated its volume as 80.49 times that of Earth and its density as 0.22 that of Earth. The respective present estimates for these values are 63.03 and 0.214. By watching the changing shape of the satellite orbits as viewed from Earth, he correctly concluded that they were orbiting in a retrograde (backwards) direction relative to the orbit of Uranus around the Sun. He surmised that Uranus was also 'tipped over', in sharp contrast with all the other planets whose rotations were known at that time. He observed a slight flattening of Uranus at its poles and expressed his opinion that Uranus, like Jupiter and Saturn, was rotating more rapidly than Earth. Uranus's rotation period is slightly less than three-quarters that of Earth.

3.1.2   Earth-based observations of Oberon

When one considers their remoteness and relatively small size, the satellites of Uranus are not easy targets for ground-based observations. Much of what was 'known' prior to Voyager 2's passage of Uranus in early 1986 was based on 'educated guesses' about their general characteristics. Only five satellites were known to exist (Miranda was discovered in 1948). Their orbital periods, orbital eccentricities (non-circular orbit shapes), and relative orbital inclinations (tilts) were known. It was assumed that they keep their same face toward Uranus, so that, like Earth's Moon, their rotation periods are equal to their orbital periods.

The apparent brightness (in stellar magnitude units) of each of the five 'classical' satellites was carefully measured, especially after photoelectric devices became available in the middle of the twentieth century. Oberon's apparent brightness at mean opposition [3] is 14.2 magnitudes, meaning that it is dimmer by a factor of almost 2000 than the dimmest naked-eye stars. If one were to move Oberon 1 AU [4] from both Earth and the Sun, the fully illuminated Oberon would have an apparent brightness of 1.52 ± 0.05 magnitudes [5] and would be one of the brighter 'stars' in the sky. Note that this brightness, designated by astronomers as Oberon's V(1,0), is for an impossible geometry, since an object can never be 1 AU from both Earth and the Sun and still be fully illuminated.

Natural surfaces also vary in brightness with changes in the solar phase angle (the angle between the Sun and the observer, as viewed from the target). The amount of variation is an indication of roughness of the surface. Unfortunately, Earth-bound astronomers can only observe phase angles between $0°$ and $3°$ for the satellites of Uranus, since Earth and the Sun are never more than $3°$ apart as viewed from the satellites. Nevertheless, Oberon, Titania, and Ariel all show significant brightness surges inside of $0.5°$ (see Fig. 3.1) [6]. This brightness surge is often called the 'opposition effect'.

As telescopes and photoelectric equipment continued to improve, the brightnesses of the satellites were determined for a spectrum of colors from the ultraviolet into the far infrared. By comparing such measurements with laboratory spectra and with the spectra of celestial bodies whose composition is known, astronomers can infer some of the main chemical constituents of the satellite surfaces. Oberon's spectrum shows it to be relatively gray in color (see Fig. 3.2), but with the absorption signature of water ice in the near infrared (Fig. 3.3): water ice is therefore assumed to be a main component of the surface of Oberon.

The far infrared brightness of a satellite is dependent on the size and surface temperature of the satellite. The higher the temperature and the larger the size, the greater the infrared brightness. If the surface light-scattering properties and surface temperatures can be independently estimated, the infrared brightness can be used to estimate the satellite size. For most satellites in the outer Solar System, a reasonable thermal model for the surface is one that behaves like beach sand. Beach sand has very low thermal inertia; that is, it has little memory of how much sunlight has fallen

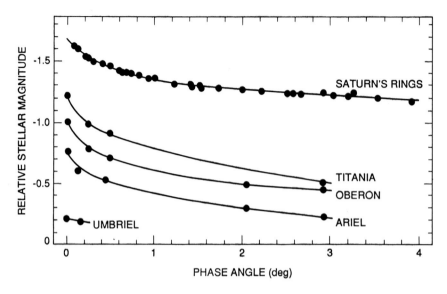

Fig. 3.1 — Near-infrared brightness surges ('opposition effect') of Ariel, Titania, and Oberon. Shown for comparison are Umbriel and Saturn's rings, whose brightness variations at small phase angle are less severe. The data contain arbitrary offsets to facilitate their intercomparison. Adapted from Brown [6].

on it in the past. Its temperature is determined by the amount of sunlight it is presently receiving. The Uranian satellites may be assumed to have the same light-scattering properties as the icy satellites of Saturn. This infrared-diameter method was used by Brown, Cruikshank, and Morrison [7] to estimate the diameters of the four outer satellites of Uranus. They found a diameter for Oberon of 1630 km, with an uncertainty of 140 km, with a corresponding albedo [8] of about 0.18±0.04.

Some attempts were made to model observed shifts in the positions of the satellites in their orbits and attribute the shifts to the gravitational attraction of the other satellites. Such measurements are extremely difficult, and small errors in the observations can lead to large errors in mass estimates for the satellites. For example, Oberon mass estimates by Veillet [9] were almost a factor of two higher than the later Voyager determinations. A combination of Veillet's mass estimate with the infrared diameter of 1630 km led to an unrealistically high density estimate for Oberon of 2.6±0.6 g/cm^3.

3.1.3   Earth-based observations of Titania

Titania is the brightest of the Uranian satellites. With a mean opposition stellar magnitude of 14.0, it is 20% brighter than Oberon, but almost a factor of 1600 fainter than the dimmest naked-eye stars. Titania's V(1,0) was found to be 1.27±0.14. Like Oberon, it brightens substantially as the phase angle drops below 0.5°.

There are small differences between the spectra of Oberon and Titania, but the dominant near-infrared spectral feature is still that of fine-grained water ice. The infrared diameter of Titania found by Brown, Cruikshank, and Morrison [7] was

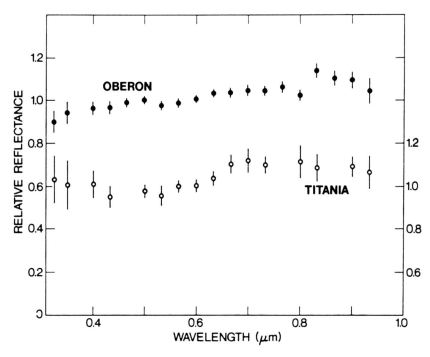

Fig. 3.2 — Reflectance spectra of Oberon and Titania from the ultraviolet (0.3 μm) to the near infrared (1.0 μm), each normalized to 1.0 at a wavelength of 0.56 μm (the wavelength of green light). The reflectance scale on the left is that for Oberon and that on the right for Titania. From Cruikshank, D. P. (1982) The Satellites of Uranus. In: Hunt, G. *Uranus and the Outer Planets*, p. 199.

1600±120 km. This value is slightly smaller than Oberon's infrared diameter. (Voyager data would later show that Oberon is actually slightly smaller than Titania.) The corresponding albedo of Titania is 0.23±0.04. The mass of Titania was also overestimated by Veillet [9], resulting in an erroneously high density of 2.7±0.6 g/cm³.

3.2 UMBRIEL AND ARIEL

For more than 60 years following Herschel's discovery of Oberon and Titania, astronomers continued to search for additional Uranian satellites. Most concentrated on trying to see the additional four already announced by Herschel, but no satellites with Herschel's calculated distances or periods were found. His son John was among the most avid of the searchers. John succeeded in confirming the existence of Oberon and Titania and in refining their orbit periods, but he never saw any of the additional satellites reported by his father.

The work of three other observers in the late 1840s is worthy of mention. Johann von Lamont was Director of the Royal Observatory at Bogenhausen (near Munich, now in West Germany). Englishman William Lassell, like Herschel, started as an

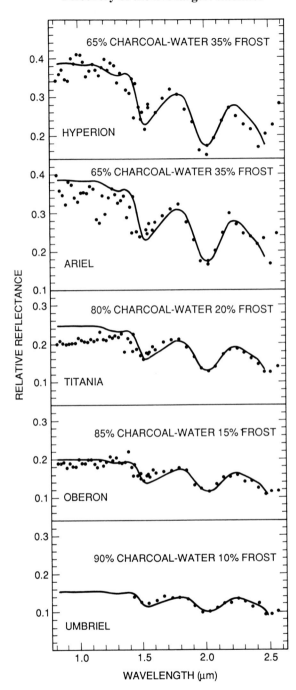

Fig. 3.3 — Reflectance spectra of Hyperion, Ariel, Titania, Oberon, and Umbriel overlaid with data constructed from a linear superposition of laboratory spectra for fine-grained water and charcoal (darkening agent). Note that the spectra match closely, confirming the existence of water ice on the surfaces of each of the Uranian satellites and on Saturn's Hyperion. From Brown [6].

amateur astronomer who made his own telescopes. Otto Wilhelm Struve was vice director and later director of the Pulkovo Observatory (near present-day Leningrad, Russia).

Lamont's main purpose in observing the satellites of Uranus was to use their motions to obtain a better estimate of the mass of the planet. He tried to make observations of all six of Herschel's announced satellites but claimed to have seen only three. He used measures of Oberon's and Titania's orbits for his Uranus mass estimates, but said he was only able to detect the outermost of Herschel's 'additional satellites' once [10]. No other observers have reported seeing this 'satellite', and it is likely that both Herschel and Lamont saw (different) background stars.

3.2.1   William Lassell's report

Neptune was first spotted by Galle on September 23, 1846, and English amateur astronomer William Lassell began observations of Neptune almost immediately. His homemade 24-in reflecting telescope was superior to others of his time in its rejection of scattered light near the planets. On October 10, 1846, less than a month after Neptune's discovery, he discovered Neptune's largest satellite, Triton. His search for additional satellites of Uranus had began earlier, but resumed in earnest in the autumn of 1847. Lassell later reported [11] that he had detected four satellites during that 1847 observing period, including Oberon and Titania (at that time still referred to as satellites II and I, respectively) and two others closer to Uranus. His observations of the inner satellites in 1847 were insufficient for him to determine an orbital period. He attempted to confirm his observations during 1848 and 1849, but apparently met with little success, possibly because of increasingly poor skies at his observing site in Liverpool, England, due to population and industrial growth in the area.

William Rutter Dawes was also an amateur astronomer and a friend of Lassell. He wrote an 1848 paper 'On the interior satellites of Uranus' [12], attempting to combine Lassell's 1847 data with some observations of Struve of a single inner satellite, made during the same year. But Lassell's and Struve's observations were not easily reconciled. In his paper, Dawes concluded that there must be three satellites interior to Titania's orbit. He estimated the orbital period of the outermost to be about 96 h, the innermost would be near 51 h, and the third would orbit Uranus at an intermediate rate. Umbriel's period is now known to be 99.46 h, Ariel's is 60.49, so Dawes's calculations were tantalizingly close to correct, even though his conclusions about the existence of three satellites were wrong.

Lassell's later observations clarified things. In a letter to the Royal Astronomical Society dated 3 November 1851, Lassell wrote:

> I am now able to announce to you my discovery of two new satellites of the planet Uranus. I first saw them on October 24th with a strong persuasion that they were really attendants on the planet, and obtained further observations of them on the 28th and 30th of October and also last night. The observations are all perfectly well satisfied with a period of revolution of almost exactly four days for the outermost, No. 2, and 2.5 days for the closest, No. 1. They are therefore both considerably within the nearest of the two bright satellites, and even within Sir William Herschel's first

satellite, to which he assigned a period of about 5 days 21 hours, but which satellite I have never yet been able to recognise. These satellites are very faint objects, probably much less than half the brightness of the conspicuous ones, and generally the nearest has appeared the brightest. All four were steadily seen at one view in the 20-foot equatorial with a magnifying power of 778 in the more tranquil moments of the atmosphere. The finest definition of the planet and freedom from all loose light in the field of view, is necessary for the scrutiny of these most minute and delicate objects. [13]

The Uranian satellites continued to be referred to by numbers until 1852, when Lassell adopted the names (Oberon, Titania, Umbriel, and Ariel) suggested by Sir John Herschel and recommended that the names be formally adopted by the Royal Astronomical Society. Earlier that year he had moved his 24-in telescope to Malta, where observing conditions were far better than those in Liverpool. There he continued his regular observations of the Uranus system. He conducted an extensive search for additional satellites exterior to Oberon, correctly concluding that there were no others within his detection limits. In a letter to the Reverend R. Sheepshanks, communicated to the Royal Astronomical Society in early 1853, he stated candidly:

Jan. 11. Surveyed Uranus with 1018 under a fine sky and admirable definition, the disk appearing perfectly round, and having a remarkably hard and sharp edge. On this occasion, as well as many others which I have not thought it necessary to particularise, I carefully scrutinised the neighborhood of the planet to the distance of 5′ from his centre, for the discovery of other satellites. In the course of this scrutiny I made many measurements and diagrams of the positions of small points of light, which all turned out to be stars; and I cannot now resist the conviction, amounting, indeed, in my own mind to certainty, that Uranus has no other satellites visible with my eye and optical means. In other words, I am fully persuaded that either he has no other satellites than these four, or if he has, they remain yet to be discovered. [14]

His continued observations also enabled Lassell in late 1853 to determine the orbital periods of Umbriel and Ariel to amazing precision: Umbriel's period was only 31 s different from present values, whereas Ariel's was accurate to 1/10 s! By 1862 Lassell began using a very large 48-in reflector of his own design, and it was evident that he had not yet given up the search for additional satellites of Uranus. It is curious that he seems to have been the only astronomer to observe the Uranian satellites during the 14-year period from 1851 to 1865.

Though some experts question whether Lassell should be credited with the discovery of Umbriel and Ariel, none question the fact that he was the first to observe them regularly and over short enough time spans to firmly establish their orbital periods. Rawlins [15] argues that Sir William Herschel should be credited with the discovery of Umbriel in 1801 and that Otto Struve should be credited with Ariel's discovery in 1847. The reasons for considering these possibilities are summarized in the following two sections.

3.2.2 Umbriel: discovery by William Herschel?

It now seems apparent that Sir William Herschel's 'announced satellites' between the orbits of Titania and Oberon and exterior to Oberon's orbit were stars mistakenly identified as satellites. But what of his satellite interior to Titania's orbit? He reported having observed this satellite on only four nights over an 11-year time span: 18 January 1790, 27 March 1794, 15 February 1798 and 17 April 1801. The orbital period he announced was 141.4 h, but with such scattered data, a reasonably accurate period determination would be almost impossible.

Herschel had said that the Uranian satellites generally disappeared from his view when they were within 22″ (1/80th of the Moon's diameter as seen from Earth). Ariel is never that far from Uranus in Earth-based views; Umbriel nearly reaches that distance during those brief periods each year when Earth is closest to Uranus and Umbriel is near its greatest angular distance from the planet. In other words, Herschel generally could not see satellites as close as Umbriel or Ariel. The implication is that occasionally he could discern objects closer to Uranus.

One of the first to suggest that Herschel had seen Umbriel was his son, Sir John Herschel. In the 1859 edition of his book, *Outlines of Astronomy*, the younger Herschel mentions the difficulties associated with viewing the larger satellites Oberon and Titania, and then said:

> . . . Umbriel (a much fainter object) was also very probably seen by Sir W. Herschel, and described by him as 'an interior satellite', but his observations were not sufficiently numerous and precise to place its existence . . . beyond question. It was rediscovered, however, by M. Otto Struve (October 8, 1847), and observed subsequently, on numerous occasions, by Mr. Lassell, to whom we also owe the first discovery of Ariel (September 14, 1847), as well as a fine series of observations and micrometrical measures of all four . . . [16]

In 1875, American astronomer Edward Singleton Holden presented a paper summarizing his analysis of Sir William Herschel's reported sightings of an 'interior satellite' [17]. To do this, he took the known orbital motions of Ariel and Umbriel and calculated their positions for each of the times reported by Herschel. Holden concluded that the first and fourth of Herschel's four reported sightings of the interior satellite were close to the predicted positions of Umbriel, and that Herschel should therefore be given credit for Umbriel's discovery. According to Holden's thesis, the second sighting was very likely of Ariel; the third was spurious.

In response to Holden's paper, Lassell pointed out [18] that after more than 70 years none of the additional satellites reported by Herschel had ever been verified. He added that Ariel and Umbriel were fainter and closer to the glare of Uranus than the outer satellites, and therefore beyond the means available to Herschel to detect them, especially when one considers Herschel's frequent difficulties in seeing the brighter and more distant Oberon and Titania. Most astronomers supported Lassell's viewpoint, including Holden's colleague and discoverer of the moons of Mars, Asaph Hall [19].

Rawlins [15] did an extensive reanalysis of the question, utilizing Herschel's own comments about his sightings. Ariel was undoubtedly hidden from Herschel's view by the glare from Uranus. Rawlins discarded Herschel's two earliest observations as being too vague in the descriptions of the directions and distances of the supposed satellite from the planet. The reported 1798 sighting he also discarded, firstly because it was on the wrong side of the planet to have been Ariel or Umbriel, and secondly because Herschel did not subsequently check the sky to see if a star was in the position. Rawlins concluded that the 17 April 1801 sighting was unquestionably of Umbriel, at a distance of only 18″ from Uranus. The direction and distance of Umbriel on that night match Herschel's description perfectly, and he carefully checked the star field on the night of 18 April 1801 to verify that no star was in the position occupied by Umbriel the previous night.

3.2.3   Ariel: discovery by Otto Struve?
Struve, like Lamont, was primarily interested in the satellites because of his desire to improve estimates of Uranus's mass. He made satisfactory measurements [20] of Oberon and Titania with his 15-in refractor, and on 13 nights in the fall of 1847 he observed an additional satellite interior to Titania's orbit. Strangely, Struve was only able to see this additional satellite when it was south of Uranus. He tentatively attributed his inability to see it north of Uranus as an indication that one side of the satellite was much darker than the other. From his rough measurements of its distance from Uranus, he estimated the orbital period to be between 3 and 4 days, probably 3.924 days. Struve felt he was seeing the same satellite earlier reported by Herschel, but that Herschel had erroneously assigned too large an orbital period.

Lassell's later comments on the 1847 observations [21] were that he and Struve had probably observed Umbriel on some occasions and Ariel on others. He thought this explained the incompatibility of the two data sets. Unfortunately, Lassell's explanation does not account for the fact that in 1847 Struve saw an inner satellite only south of Uranus, whereas Lassell saw one only north of the planet. Furthermore, Lassell later discarded Struve's observations (and Herschel's) as spurious instead of using them to refine the orbit periods of the satellites.

It is intriguing to note that nearly half of an 84-year Uranus orbit later, astronomers were having a certain amount of difficulty observing Ariel south of the planet, presumably because darker parts of its surface faced Earth [22]. Since Ariel always keeps the same face toward Uranus, in 1847 those same darker portions of Ariel's surface would face Earth when the satellite was north of the planet. Ariel is more than twice as bright as Umbriel, and were it not for the scattered-light glare from Uranus, it should be easier to detect. Rawlins [15] shows that Struve's 8 October 1847 observation corresponds very well with the predicted position for Ariel and that Struve's 1 November 1847 sighting was of Umbriel. None of Struve's later 1847–8 sightings of inner satellites match the positions of Ariel or Umbriel. Struve reported his sightings [20] to the Royal Astronomical Society in late 1847, nearly four full years before Lassell reported his 'discovery' of Ariel and Umbriel. Ironically, Lassell's friend and sometimes co-observer Dawes, mentioned in the last paragraph of his March 1848 paper [12] an observation made by Lassell on 27 September 1845 which almost perfectly matches the position of Ariel. Rawlins therefore argues that Lassell's 1845 observation is the first known (though second reported) sighting of Ariel.

One could debate long and heatedly about what constitutes 'discovery' of a planet or satellite, but it now serves no purpose, since all of the principals in the discovery of the four outer satellites of Uranus have long since died. Other than for historical reasons, it now seems of little consequence whether Lassell or Struve is called the discoverer of Ariel. The same also applies to Lassell and Herschel with regard to Umbriel's discovery. Almost all ventures of this nature are the results of the cooperative efforts of a large number of investigators. Perhaps we should adopt the motto, 'There is no limit to what a person can accomplish, if he (or she) doesn't care who gets the credit!'

3.2.4 Earth-based observations of Umbriel

Umbriel is the dimmest of the four outer Uranian satellites. Its mean opposition stellar magnitude is 15.3; the dimmest stars visible with the naked eye are more than 5000 times brighter! V(1,0) Umbriel's V(1,0) is 2.31±0.15 [23].

As with Oberon and Titania, the dominant near-infrared spectral feature is that of fine-grained water ice. The infrared diameter of Umbriel found by Brown, Cruikshank, and Morrison [7] was 1110±100 km; they also calculated an albedo of 0.19±0.04 for Umbriel. Both estimates are very close to those later determined by Voyager 2. The mass of Umbriel estimated by Veillet [9], was lower than Voyager estimates, but well within his stated uncertainties; combined with the infrared diameter, it results in a density of 1.4±0.6 g/cm^3.

3.2.5 Earth-based observations of Ariel

Ariel, except for its proximity to Uranus, would be almost as easy to spot in a telescope as Oberon and Titania. With its stellar magnitude of 14.4, it is more than twice as bright as Umbriel and only slightly fainter than the two largest satellites. Ariel's V(1,0) is 1.48±0.15 [23].

The near-infrared spectrum of Ariel again provides evidence for a surface dominated by fine-grained water ice, but the water-ice features are more pronounced for Ariel than for the outer three satellites of Uranus. Ariel's spectrum closely mimics that of Saturn's satellite, Hyperion. The infrared diameter of Ariel found by Brown, Cruikshank, and Morrison [7] was 1330±130 km. Their calculated albedo of 0.30±0.06 gave Ariel the distinction of having the most reflective surface in the Uranus system. Voyager determined that Ariel's surface is even more reflective than suspected from Earth-based observations, since the 1330-km diameter was about 15% too high. Veillet's [9] estimate of Ariel's mass was higher than later Voyager determinations, but if one includes the relatively large uncertainties of each, the match is reasonably good. Combination of Veillet's mass for Ariel with its infrared diameter results in an estimated density of 1.3±0.5 g/cm^3.

3.3 MIRANDA

'What is your name?'

'Miranda. — O my father,

I have broke your hest to say so!'
                    'Admired Miranda!
Indeed the top of admiration! worth
What's dearest to the world! ... you, O you,
So perfect and so peerless, are created
Of every creature's best!'

Shakespeare's *The Tempest*, Act III, Scene I

Lassell's conclusion in 1853 was that Uranus's 'retinue ... must be cut down to four attendants, until some astronomer arises rich enough to present him with some more.' [14] Little did he know that it would be very nearly a century before a 'richer' astronomer would come along. During that century, a ninth planet would be discovered; automobiles, airplanes, and rocketry would be invented; astronomical research would progress beyond Lassell's wildest dreams; and electronic computers and the space age would be within a decade of being born. Although the discovery that led to increasing Uranus's 'retinue' of satellites to five was less dramatic than the earlier Uranian discoveries, it is nevertheless an interesting story.

3.3.1 Discovery by Gerard P. Kuiper
On December 7th, 1905, in the little village of Harenskarpel, Netherlands, Gerard Peter Kuiper was born. Kuiper was destined to become one of the foremost Solar System astronomers of the twentieth century. After completing undergraduate and doctorate degrees at the University of Leiden he emigrated to the United States, where he joined the faculty of the University of Chicago and worked at the Yerkes Observatory in Williams Bay, Wisconsin.

Kuiper discovered carbon dioxide (now known to be the dominant constituent) in the atmosphere of Mars. He determined that Saturn's satellite Titan possesses an atmosphere containing methane and ammonia. He was also the first to propose the giant protoplanet theory for formation of the planets and their satellites (see section 2.3.2). His studies of Pluto led him to conclude that it is much smaller than the giant planets and that its rotation period is about 6.4 days. Of special significance for this chapter is Kuiper's discovery of two small satellites in the outer solar system: Neptune's Nereid and Uranus's Miranda.

With the improvement of photographic emulsions and techniques came the opportunity to use such photography as a powerful new tool in astronomical observations. In 1948, Kuiper was using the 82-in reflector of the McDonald Observatory near Fort Davis, Texas, to photographically observe Uranus's four outer satellites.

The fifth satellite of Uranus was first photographed on February 16, 1948, 2^h 55^m UT on a four-minute exposure of the Uranus system, taken at the Cassegrain focus of the 82-inch telescope (scale $1 \, mm = 7''.38$). This exposure was intended to provide data on the relative magnitudes of the four known satellites. The close companion to the planet was noticed at once but no opportunity to establish its nature occurred until March 1, 1948, when two control plates showed it to be a satellite and not a field star. [24]

Kuiper chose to name this fifth Uranian satellite Miranda, after a character in Shakespeare's *The Tempest*. Miranda's father, Prospero, told Miranda when she was younger, 'a cherubin thou wast that did preserve me'. Ariel was also a character in that play. Later in 1948, and again in 1950, Kuiper and co-worker D. L. Harris searched for more distant satellites of Uranus with negative results [25]. A continuation of this same survey of satellite systems netted Neptune's Nereid in 1949.

3.3.2 Earth-based observations of Miranda

Miranda is extremely difficult to see in a telescope. Its small size and closeness to Uranus combine to render most observational techniques useless. Persistent efforts to ferret out its secrets did provide ground-based astronomers with a few titbits of information prior to the Voyager encounter. Miranda's orbital period (and its rotational period, since it is in 'locked' rotation) was found to be 33.92 h. Its average distance from the planet is about 130 000 km; because of its eccentricity (non-circularity), that distance varies from slightly less than 128 000 km to more than 132 000 km. The plane of Miranda's orbit is not the same as Uranus's equatorial plane: 3.4° of inclination separate the two, so that during its 34-h orbit it slowly oscillates between 3.4° south latitude and 3.4° north latitude above the planet.

Miranda's mean opposition brightness is 16.5 stellar magnitudes; it is less than one-third the brightness of Umbriel, the dimmest of the outer four satellites. Extrapolation of that brightness to V(1,0) yields 3.79±0.17. [25].

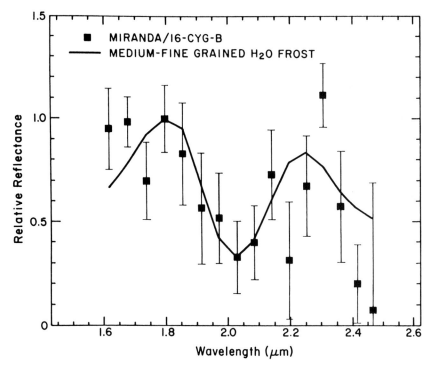

Fig. 3.4 — Reflectance spectrum of Miranda is compared to that of undarkened fine-grained water ice. Although the uncertainty in the Miranda data is large, the fit is good enough to confirm that water ice is the primary constituent of Miranda's surface. From Brown [6].

The near-infrared spectrum of Miranda (Fig. 3.4) is much less certain than those of the outer Uranian satellites, but clearly shows the absorption features of fine-grained water ice. The depth of that absorption led Brown and Clark [26] to conclude that the surface must be almost pure ice and therefore very reflective. Although a far-infrared measurement of Miranda's diameter did not prove feasible, Brown and Clark conservatively estimated that the surface albedo was 0.4 ± 0.3. From Miranda's $V(1,0)$ they then determined that the diameter would be 500 ± 225 km. They argued that the albedo was very likely higher than 0.3, so that the diameter was probably lower than 500 km. The actual diameter determined from Voyager imaging is 472 ± 6 km [27]. Veillet's [9] estimate of Miranda's mass was little more than an upper limit, so its combination with the 500-km diameter did not yield useful results for the mean density of Miranda. Brown [6] listed the density as approximately $3\,\text{g/cm}^3$.

NOTES AND REFERENCES

[1] Dreyer, J. L. E. (ed.) (1912) *The Scientific Papers of Sir William Herschel*, Royal Society and Royal Astronomical Society, **1**, 312–314.

[2] Dreyer, J. L. E. (ed.) (1912) *The Scientific Papers of Sir William Herschel*, Royal Society and Royal Astronomical Society, **1**, 317–326.

[3] Mean opposition refers to the average distance from Earth of the celestial body (Uranus in this case) when Earth is directly between the Sun and the celestial body.

[4] An AU, or astronomical unit, is the average distance between Earth and the Sun, or 149 597 900 km (92 955 800 miles).

[5] Degewij, J., Andersson, L. E., Zellner, B. (1980) Photometric properties of outer planetary satellites, *Icarus*, **44**, 520–540.

[6] Brown, R. H. (1984) Physical properties of the Uranian satellites. In: Bergstralh, J. T. (ed.) *Uranus and Neptune*, NASA Conference Publication 2330, National Aeronautics and Space Administration, Scientific and Technical Information Branch, pp. 437–461.

[7] Brown, R. H., Cruikshank, D. P., Morrison, D. (1982) Diameters and albedos of the satellites of Uranus, *Nature*, **300**, 423–425.

[8] Albedo is the technical name for the ratio of the amount of light reflected or scattered by a surface to that incident upon the surface. Unless otherwise specified, albedo will normally refer to geometric albedo (the fraction of incident light scattered back toward the light source) for visual colors (green to yellow). Fresh water ice or snow generally has an albedo near 1.00; the albedo of carbon black is near 0.05.

[9] Veillet, C. (1983) De l'observation et du mouvement des satellites d'Uranus. Ph.D. dissertation, University of Paris, Paris, France.

[10] Lamont, F. (1843) *Royal Astronomical Society, Monthly Notices*, **4**, 122.

[11] Lassell, W. (1848) Satellites of Uranus: observations by Mr. Lassell, *Royal Astronomical Society, Monthly Notices*, **8**, 43–44.

[12] Dawes, W. R. (1848) On the interior satellites of Uranus, *Royal Astronomical Society, Monthly Notices*, **8**, 135–137.

[13] Lassell, W. (1851) Extract of a letter from Mr. Lassell, dated Nov. 3, 1851, *Royal Astronomical Society, Monthly Notices*, **11**, 248.

[14] Lassell, W. (1853) Letter from Mr. Lassell to the Rev. R. Sheepshanks, *Royal Astronomical Society, Monthly Notices*, **13**, 147–151.

[15] Rawlins, D. (1981). Arguments were presented in a paper at IAU/RAS Colloquium No. 60, entitled 'Uranus and the Outer Planets', held at the University of Bath, England, from the 14th to the 16th of April, 1981. The same basic arguments are presented in: Rawlins, D. (1973/6) Discoveries of Ariel and Umbriel, *Astronomy & Space*, **3**, 26–40.

[16] Herschel, J. F. W. (1859) *Outlines of Astronomy*, 6th ed., Longman, Green, Longman, and Roberts, London, p. 368.

[17] Holden, E. S. (1874) *Royal Astronomical Society, Monthly Notices*, **35**, 16–22.

[18] Lassell, W. (1874) *Royal Astronomical Society, Monthly Notices*, **35**, 22–27.

[19] Hall, A. (1885) *Washington Observations — Publications of the United States Naval Observatory*, **28**, Appendix I, p. 5.

[20] Struve, M. O. (1847) Note on the satellites of Uranus, *Royal Astronomical Society, Monthly Notices*, **8**, 44–47; Struve, O. (1848) *Astronomische Nachrichten*, **26**, no. 623.

[21] Lassell, W. (1851) On the interior satellites of Uranus, *Royal Astronomical Society, Monthly Notices*, 12, 15–17.

[22] Rodgers, J. (1878) *American Journal of Science and Arts*, **15**, 195.

[23] Reitsema, H. J., Smith, B. A., Weistrop, D. E. (1978) Visual and near-infrared photometry of the Uranian satellites, *Bulletin of the American Astronomical Society*, **10**, 585 (abstract). Their data is also discussed in Cruikshank, D. P. (1982) The Satellites of Uranus. In: Hunt, G. (ed.) *Uranus and the Outer Planets*, Cambridge University Press, pp. 196–197.

[24] Kuiper, G. P. (1949) The fifth satellite of Uranus, *Publications of the Astronomical Society of the Pacific,* **61,** 129.

[25] Kuiper, G. P., Middlehurst, B. M. (1961) *Planets and Satellites* (Volume 3 of the *Solar System* series), University of Chicago Press, pp. 578, 587.

[26] Brown, R. H., Clark, R. N. (1984) Surface of Miranda: identification of water ice, *Icarus*, 58, 288–292.

[27] Davies, M., Colvin, T., Katayama, F., Thomas, P. (1987) The control networks of the satellites of Uranus, *Icarus*, **71**, 137–147.

BIBLIOGRAPHY

Alexander, A. F. O'D. (1965) *The Planet Uranus*, American Elsevier Publishing Company, Inc., New York, pp. 56–79, 104–127, 137–150, 291–297.

Bergstralh, J. T. (ed.) (1984) *Uranus and Neptune*, NASA Conference Publication 2330, National Aeronautics and Space Administration, Scientific and Technical Information Branch, pp. 437–480.

Hunt, G. (ed.) (1982) *Uranus and the Outer Planets*, Cambridge University Press, Cambridge, pp. 193–210.

4

The discovery of the rings of Uranus

4.1   DISTANT STARS AS ATMOSPHERIC PROBES

Most stars are extremely large objects. Our Sun, for example, has a volume large enough to encompass 1.3 million Earths. Yet all stars but the Sun are located at distances so remote from Earth that they appear as mere pinpoints of light even to astronomers equipped with the most powerful telescopes available. The smallest angle discernible from Earth-based telescopes depends both on the size of the telescope and on the atmospheric conditions. Typically, with perfect atmospheric conditions and a large telescope, features with angular sizes as small as 0.02″ might be resolved. The Hubble Space Telescope, which orbits Earth above its atmosphere and therefore is not degraded by atmospheric conditions, is designed to be able to resolve features down to an angular size of 0.005″. For the sake of comparison, the unaided eye has difficulty resolving angles smaller than 200″.

The 100 brightest stars in Earth's sky range in distance from 4.34 to 2450 light years [1], with a geometric mean of about 55 light years. The star Rigel Kentaurus is almost a twin to our Sun; at its distance of 4.34 light years its angular diameter is only about 0.007″. The star Betelgeuse, one of the largest stars known, has a diameter about 800 times that of our Sun; at its distance of about 650 light years it still subtends an angle of only 0.04″. Rigel Kentaurus and Betelgeuse are representative of the upper limit of star angular sizes. Angular sizes for typical stars used in stellar occultation measurements are factors of 100 to 1000 smaller than the range of sizes shown for Rigel Kentaurus and Betelgeuse. Almost all stars are therefore truly pinpoints of light far smaller than can be discerned from telescopes either orbiting the Earth or on its surface.

The planets of our Solar System, because of their relative nearness, appear as disks when viewed through telescopes. Jupiter, Saturn, Uranus, and Neptune, for example, can reach respective angular diameters as large as 50, 21, 4.1, and 2.4″ as viewed from Earth. Because of the orbital motions of the Earth and the other planets around the Sun, the planets appear to move with respect to the background stars.

Occasionally a planet will pass in front of a star and occult (block the light of) the star. Careful monitoring of the combined light of the planet and star in such circumstances can provide much information about the planet's atmosphere not obtainable from Earth in any other way. This technique is even more powerful if the event is simultaneously monitored by observers in several locations.

When the star is in view, the total light measured (usually by a sensitive photoelectric device) consists of light from the star plus light from the planet. After the star disappears behind the planet, only the light of the planet may be seen. Precise timings of the disappearance and reappearance of the additional starlight are obtained at several observatories, whose relative locations must also be precisely known. These timings are combined with a knowledge of the relative velocity of the planet with respect to the background star to determine chords of known length and relative location across the disk of the planet (see Fig. 4.1). If the measurements are

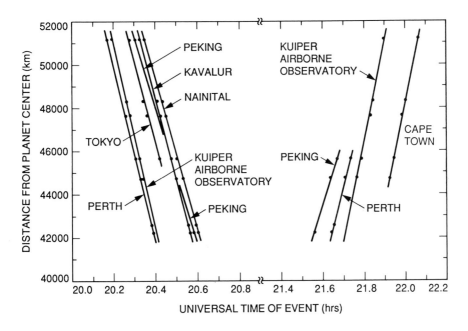

Fig. 4.1 — Diagram of the stellar occultation, showing approximate times, occultation chords, and recorded events for a widely observed occultation of the star SAO 158687 by Uranus and its rings on March 10, 1977. (Adapted from Bhattacharyya et al. [7]).

done carefully, they enable astronomers to obtain with high precision the instantaneous position of the planet, its diameter, and the oblateness (non-circularity) of its disk. If, in addition, the planet has an atmosphere, the starlight will disappear (or reappear) gradually. The rate of disappearance (or reappearance) can be used to determine the density (and hence the temperature) of the atmosphere at several different altitudes; it may also help determine the amount, altitude location, and nature of haze particles present in the upper atmosphere. These measurements will

be smeared over an altitude range in the atmosphere which results from a combination of the angular diameter of the star and the apparent altitude distance
traversed by the star during the time needed to acquire an individual data point. For a
star with an angular diameter of 0.001″ as viewed from Earth, an instantaneous
altitude in the atmosphere of Uranus spanned by the star's diameter is about 14 km.
Typical apparent velocities of Uranus relative to a background star are about 20 km/s
or less, but today's efficient photoelectric equipment usually permits collection of
many individual data points each second. Although the data are effectively smeared
over about 15 km, the oversampling enables data analysts to remove part of the
instrumental effects and smearing to obtain atmospheric information with altitude
resolutions substantially smaller than 15 km. Since direct imagery from Earth-bound
telescopes cannot resolve features smaller than several hundred kilometers, stellar
occultations provide a powerful means of probing distant planetary atmospheres.

4.2 THE 10 MARCH 1977 OCCULTATION OF SAO 158687

The orbits of the planets and the positions of background stars are known well
enough to predict the approximate times of stellar occultations well in advance of
their actual occurrence. One such occultation by Uranus was predicted for the night
of 10 March 1977. The star to be occulted was SAO 158687, a 9th magnitude star
somewhat redder in color than the Sun. Several teams of astronomers made plans to
observe the occultation from the vicinity of the Indian Ocean, where the event could
be seen well above the horizon in a dark sky.

Astronomers in western Australia, Japan, China, and India were too far north to
observe the star's disappearance behind Uranus's atmosphere: from their vantage
points, Uranus slipped just southward of the star (see Fig. 4.2). One of the teams had
arranged to use the Kuiper Airborne Observatory (KAO), a large airplane-mounted
telescope named posthumously for Miranda's discoverer, Gerard P. Kuiper. The
KAO team, headed by James Elliot, followed a flight path that took them into the
southern extremes of the Indian Ocean and well into the shadow of SAO 158687 cast
by Uranus. Another team, headed by Robert L. Millis, attempted the same
observations from the fixed location of the Perth Observatory in western Australia.
The two teams had worked together in a similar manner to observe an occultation of
Mars in early 1976.

Observations were made at red to near-infrared wavelengths where SAO 158687
was as bright as Uranus. In yellow light, Uranus is about 40 times as bright as SAO
158687, but at near-infrared wavelengths of 0.8 to 0.9 μm, Uranus is very nearly as
dim as the star. Observers in the KAO recorded the occultation events simultaneously at three wavelengths: 0.62, 0.73, and 0.85 μm. At Perth Observatory,
observations were obtained only at 0.85 μm.

The KAO flight plan was carefully laid out to permit the maximum possible
observing time within the fuel limitations of the airplane. The precise time at which
Uranus would occult SAO 158687 was not known, so both airborne and ground
observers were prepared and recording the data well before the predicted time of
occultation entry. The story of what happened aboard the KAO and at the Perth
Observatory is recorded in detail in an article written by Elliot, Dunham, and Millis
[2]. Both teams of astronomers observed periods when the starlight seemed to

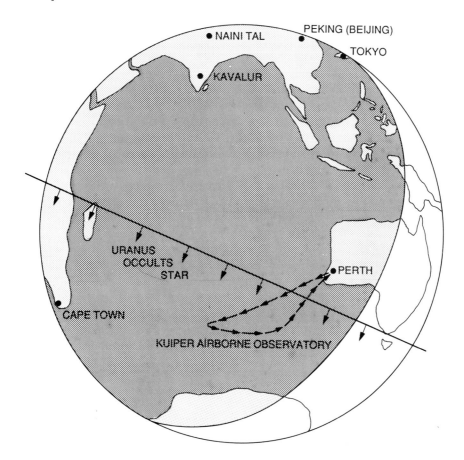

Fig. 4.2 —The approximate path of the northern edge of the 'shadow' of Uranus on Earth's surface during the March 10, 1977, occultation of the star SAO 158687. (Adapted from Elliot, Dunham, and Millis [2]).

disappear, long before the predicted entry into Uranus occultation. Each wondered whether the dips in starlight might be due to instrumental problems, or clouds, or alignment of the telescope. Careful checks revealed no problems, and the observations continued.

Observers in Perth did not see the occultation of SAO 158687 by Uranus, because their location was north of the shadow line. Furthermore, approaching sunrise prevented them from continuing their observations long enough to see any similar dips in starlight after Uranus had passed the star. Only the KAO observers were able to get a complete occultation record including the Uranus occultation and data spanning an hour on either side of Uranus's passage of the star. Partial occultations of SAO 158687 by the Uranus system were reported by seven other observing teams, including two teams in Perth, Australia, and one team in each of the following locations: Tokyo, Japan; Peking, China; Naini Tal and Kavalur, India; and Cape

Town, South Africa. All eight teams reported secondary events obviously separated from the occultation by the planet itself.

Most of the observers interpreted their results as evidence for one or more satellites in orbit around Uranus. Although rings had been suggested as an explanation for the data, astronomers initially chose to rule out any possibility of rings with widths as small as a few kilometers to several tens of kilometers. Before 1977, the only known example of planetary rings was the extended and complex ring system surrounding Saturn. As viewed from Earth, Saturn's rings appeared as vast 'sheets' of particles, and astronomers were comfortable with rings in that configuration. But Uranus was destined to challenge norms of the past.

The real or imagined abnormality of a system of narrow rings was in part based on dynamical considerations. It has long been known that the rate at which particles (or satellites) circle a planet varies with distance from the planet. Closer objects travel faster and circle the planet in less time than those farther away. In a similar fashion, particles within a ring should move faster at the inner edge of the ring than at the outer edge. This differential motion results in repetitive nudging of the particles by neighboring particles at slightly different distances from the planet. A ring of tiny particles will tend to spread out radially due to these soft collisions. If narrow rings really exist, one of two conditions must apply: (1) either the rings were too young to have undergone collisional spreading, or (2) they were confined to their narrow paths by some as yet unexplained force. Neither of these potential explanations seemed palatable in 1977.

4.3   FROM SKEPTICISM TO CERTAINTY

The complete occultation record obtained by the KAO team was the only single data set which could answer the question as to whether Uranus really had rings. Elliot and his team returned to the United States, arriving home the evening of March 13th. The following evening Elliot and his wife Elaine spread the recorded data across their living room floor. It was then that Elliot discovered that the five secondary occultation events recorded after the passage of the planet matched the five events recorded before Uranus occulted SAO 158687! Uranus indeed possessed a system of narrow rings.

Still somewhat skeptical, astronomers checked the data from the other observing sites. The Cape Town data showed six secondary events, five of which matched the ones reported by the KAO team; three of six events recorded by Millis and his team corresponded to rings seen from the KAO. Elliot and his colleagues published a paper announcing the discovery [3], calling the newly discovered rings by the Greek letters alpha, beta, gamma, delta, and epsilon, in order of increasing distance from the planet. The alpha and delta rings were missed by Millis and his team during periods when their telescope was being re-centered on Uranus. They reported seeing the epsilon ('1'), gamma ('2'), and beta ('3') rings and three others closer to Uranus, which they referred to as the '4', '5', and '6' rings [4]. In a later paper, the Perth, Cape Town, and KAO observations were combined [5] to provide a better measure of distance of each of the rings from the planet. The 4, 5, and 6 rings were identified in the KAO data, along with an additional ring between the beta and gamma rings, which the KAO team called the eta ring . No explanation was given for the omission of zeta (the sixth letter in the Greek alphabet) from the ring nomenclature scheme. The naming of the rings was somewhat haphazard at best: in order of increasing distance from the planet they have come to be known as 6, 5, 4, alpha, beta, eta, gamma, delta, and epsilon.

The Kavalur observers saw the obvious occultation by the epsilon ring, first reporting it as a new satellite of Uranus [6]. A later combined analysis of the Kavalur and the Naini Tal observations [7] identifies all nine rings reported by Elliot *et al.* (4, 5, and 6 were referred to by the more consistent names of theta, iota, and kappa). In addition, they reported several other possible ring features closer to Uranus than the rings reported by Elliot *et al.*, including an extended feature about 3000 km wide.

The sixth feature reported by the Cape Town observers was not seen at any other location. It corresponds to a distance well outside the other known rings. It is intriguing to note that the indicated distance is very near the orbital radius of Ophelia, one of the small moons later discovered by Voyager 2. No work has yet been done to see if tiny Ophelia was at the appropriate position in its orbit to occult SAO 158687 on March 10, 1977, as viewed from the Cape Town Observatory.

4.4   EARTH-BASED OBSERVATIONS OF THE RINGS

The ground-based evidence was by now overwhelmingly in favor of the existence of narrow Uranian rings. Whether they were unlikely or not, they most certainly did exist. It now remained for the theoreticians to explain how and why. To aid them in that task, much additional evidence was needed. Careful calculations of the motions of Uranus against the background stars showed that stellar occultations by the planet and its rings were relatively common in the years following 1977. Plans were made to observe those occultations with appropriately instrumented telescopes. Many astronomers also attempted to see the rings through direct imaging. Those attempts met with less success.

William Herschel thought he saw stubby rings around Uranus in 1787 and 1789. After several years of additional observations, he concluded in 1798 that his earlier observations were in error and that Uranus had no ring resembling the rings of Saturn. The Uranian rings, because of their narrowness, their proximity to the planet, and their low reflectivity, are far beyond the capabilities of Herschel's telescopes. They have not in fact been seen visually even with today's sophisticated telescopes.

Several Earth-based images of Uranus's rings do exist. The first was obtained by Matthews, Neugebauer, and Nicholson, [8] from the 5-m telescope at Palomar Mountain. At an infrared wavelength of 2.2 μm absorption by methane in the atmosphere of Uranus makes the planet somewhat darker than the rings. At 1.6 μm, the planet is much brighter than the rings. The authors scanned the planet at both wavelengths and then subtracted the 1.6-μm data from the 2.2-μm data in such a manner that the planet was made to disappear. The resulting images of Uranus's ring system, one of which is reproduced in Fig. 4.3, do not resolve individual rings, but serve primarily to verify the existence of the rings. The width of the band in Fig. 4.3 corresponds to the brightness of the rings. At its greatest breadth, the band spans a radial distance of about 50 000 km; the broadest ring is actually less than 100 km in width. The azimuthal non-uniform brightness of the band is in part a consequence of

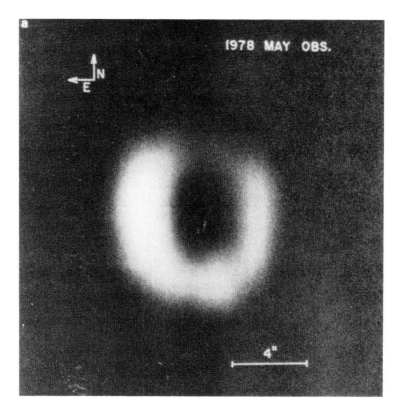

Fig. 4.3 — The first image of the rings of Uranus was obtained by differencing two infrared images, one at 2.2 μm wavelength and one at 1.6 μm wavelength. (From Matthews, Neugebauer, and Nicholson [8]).

variations in the outermost and widest of the rings, the epsilon ring. Ring epsilon is nearly 100 km in breadth at the bottom of the image and only 20 km wide at the top.

More direct imaging techniques have been used by other observers. Allen [9] used the 3.9-m Anglo-Australian Telescope to scan in a raster pattern at 2.2 μm to image the rings. Terrile and Smith [10] employed a special charge-coupled device (CCD) camera attached to the 2.5-m du Pont telescope at the Las Campanas Observatory near La Serena, Chile. Their observations covered 11 different wavelengths from 0.435 to 1.000 μm and were intended to concentrate on a search for cloud features in the atmosphere of Uranus. At 0.89 μm in the near infrared region of the spectrum (as at 2.2 μm), methane in the planet's atmosphere absorbs sunlight and makes Uranus appear very dark. Computer processing of an image of Uranus at 0.89 μm (Fig. 4.4) shows the ring system (dominated by the epsilon ring) as an apparent 'pedestal' surrounding the planet. Uranus and its five largest satellites also appear in the figure.

One of the major impediments to photographing the rings of Uranus is their extreme darkness. They typically reflect less than 5% of the sunlight that is incident upon them. Furthermore, sunlight at Uranus's distance from the Sun is weaker by a

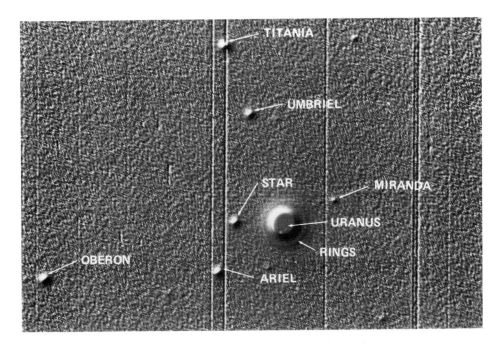

Fig. 4.4 — This CCD image of Uranus, its rings, and its satellites was obtained by Terrile and Smith [10], and was specially computer-processed to make the rings visible as a pedestal surrounding the planet. (P-27796)

factor of about 380 than it is at Earth's distance. The result is comparable to trying to photograph pieces of coal against a perfectly black background with light levels like those on Earth nearly an hour after sunset.

The vast majority of pre-Voyager data on the rings comes from the many stellar occultations telescopically observed at and since the 1977 discovery. At least 15 such occultations by Uranus's ring system were observed prior to the Voyager 2 encounter with Uranus in January, 1986. Many of these occultations were observed simultaneously from two or more sites. The resulting data provided literally hundreds of individual ring profiles. Summaries of the results of some or all of these occultation observations are contained in several excellent papers (Elliot [11]; Elliot and Nicholson [12]; French, Elliot, and Levine [13], and French *et al.* [14]). They provide information on the sizes, shapes, orientations, widths, and optical thicknesses of the nine 'classical' rings that was only marginally improved by the addition of Voyager encounter data in early 1986.

All of the rings except eta, gamma, and delta depart measurably from circularity. Widths of the epsilon, delta, beta, and alpha rings vary such that they are widest at their most distant excursions from the planet and narrowest when they are closest to the planet. The eta, gamma, delta, and epsilon rings are confined (within the measured uncertainties) to Uranus's equatorial plane; the five inner rings have small but measurable inclinations. Epsilon and the five inner rings (6, 5, 4, alpha, and beta) have orbits which precess (change the orientation in space of their elliptical orbits) uniformly with time. One of the more interesting results of the Earth-based studies of the rings is their tendency to behave almost like rigid bodies. For example, precession rates of the inner and outer edges of the epsilon ring would be expected on theoretical grounds to be slightly different, but they are identical.

Optical thickness is the term used by astronomers to describe how much of the light incident on a ring (or other target) passes through it. The less starlight that a ring passes, the more optically thick it is. When the incident light is reduced by a factor of e (=2.718), the object through which the light passes is said to have an optical thickness of 1. A reduction by e^2 (=7.389) corresponds to optical thickness 2. A perfectly opaque object has infinite optical thickness; a perfectly transparent object has zero optical thickness. In mathematical terms, optical thickness

$$t=\ln(I_0/I), \tag{1}$$

where I_0 is the intensity of the incident light, I is the intensity of the transmitted light, and ln means the natural logarithm of the ratio. An alternate method of expressing this relationship is

$$I=I_0 e^{-t}. \tag{2}$$

The angle of the incident starlight with respect to the plane of Uranus's rings varies with time as Uranus orbits the sun and (to a lesser degree) as Earth-bound observers of Uranus move in Earth's orbit. If the Uranian rings are not monolayers (that is, they are more than one ring particle in thickness), the observed optical thicknesses will also vary with viewing angle. If the angle between the viewing direction and a direction perpendicular to the ring plane is B, then the normal optical depth, t_n, can be determined from the observed optical depth, t, by the relationship

$$t_n=t \sin B. \tag{3}$$

Two other definitions which are useful in describing the rings of Uranus are equivalent width, E, defined as

$$E=W(1-I/I_0) \sin B, \tag{4}$$

and equivalent depth, A, defined as

$$A=Wt_n, \tag{5}$$

where W is the measured radial width of the ring.

Using the definitions for equivalent width and depth to describe the Uranus ring observations reveals something of their nature. If the rings were monolayers, there should be no variation of E around the rings. The four eccentric rings that show systematic variations of width with distance from the planet (alpha, beta, delta, and epsilon) all show variations of E with ring azimuth. These rings, and by analogy all of the Uranian rings, are not monolayers, but are many particles thick. Equivalent depth, A, provides a measure of the amount of material in a ring, and might be expected to be relatively constant around a ring. The ground-based observations were not conclusive on this point. Values of A around any given ring tend toward constancy, but the measured variations are larger than the estimated errors in the measurements. These variations in A could be due either to clumpiness (azimuthal non-uniformity in the numbers of ring particles) or to substantial radial variations in optical thickness across a given ring. Radial variations could invalidate the assumption that equivalent depth, defined by multiplying the measured width by an *average* optical thickness, is representative of the number of ring particles. The ground-based observations of the rings suggest that both clumpiness and radial non-uniformity contribute to the noted azimuthal variations in equivalent depth.

Radial optical depth variations in the epsilon ring are particularly apparent. The ring is optically thicker near its inner edge and halfway between its center and outer edge than near its center. This general profile persists even though the ring width varies from a minimum of 20 km near periapsis (the point in the orbit where the ring particles are nearest the planet) to a maximum of 96 km near apoapsis (the point where they are farthest away). The eta ring has a narrow component only a few kilometers wide and a broad optically thinner component extending several tens of kilometers outward. The alpha ring also has a 'double-dip' structure in its optical depth. Both the alpha and beta rings vary in width from about 5 km at periapsis to about 10 km at apoapsis. Typical ring occultation profiles for each of the nine rings are displayed in Fig. 4.5.

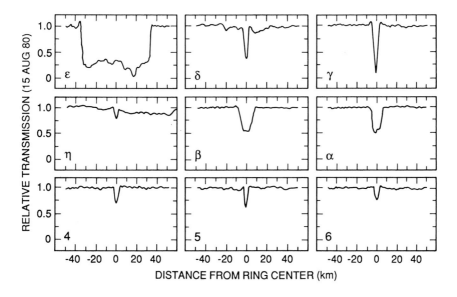

Fig. 4.5 — Typical occultation profiles for the nine Uranian rings, representing the fraction of starlight transmitted by the rings at various distances from the center of each ring. (Adapted from Elliot and Nicholson [12]).

The delta and gamma rings each have orbits which are primarily elliptical in shape, but which depart from normal elliptic orbit shapes by much larger amounts than allowed by uncertainties in the data. French, Kangas, and Elliot [15], analyzed the situation and concluded that the non-Keplerian orbit shapes might be attributable to gravitational perturbations from one or more undiscovered satellites orbiting Uranus.

The delta ring is shaped much like an ellipse with Uranus at its center (rather than at one of its two foci). Saturn's satellite Mimas causes a similar shape for the outer edge of the saturnian B ring. In the case of Saturn, Mimas orbits the planet in precisely twice the period required by ring particles at the outer edge of the B ring. Thus each of these ring particles are closest to Mimas at the same point in their orbit every second circuit of the planet. The repeated gravitational tugging by Mimas eventually distorts the B-ring edge into its centered elliptical shape. It was presumed that a similar mechanism was shaping Uranus's delta ring, although no known satellite existed at the appropriate orbital radius.

The gamma ring has an even stranger characteristic. This extremely narrow ring appears to 'breathe' as it circles Uranus. The period of these radial oscillations is very nearly the same as the orbital period of a particle in the gamma ring. It is possible that a small satellite or large ring particle within the gamma ring is responsible for this unusual behavior, but no direct ground-based detection of such a particle has occurred (nor, for that matter, has Voyager 2 detected any imbedded satellite in the gamma ring).

It was Goldreich and Tremaine [16], who in 1979 first proposed the theory for the formation and maintenance of Uranus's narrow rings. They attributed the narrow Uranian ring profiles to the gravitational effects of a number of undiscovered satellites. These satellites exerted just enough influence on the ring particles, according to the theory, to counteract the tendency for the rings to spread out into diffuse sheets of material. Without such influence, the Uranian rings might eventually dissipate outward into space or fall inward to become a part of the planet itself. This 'shepherding satellite' theory might have met much more resistance were it not for the obvious presence of the narrow rings of Uranus. Some purists argued that none of the proposed perturbing satellites had ever been seen, and that judgment on the validity of the theory should be withheld until observations verified their existence. They also pointed out that gravitational 'resonances' elsewhere in the Solar System (in Saturn's rings and in the asteroid belt between Mars and Jupiter) resulted in gaps rather than concentrations of material. The Cassini Division in Saturn's ring system was attributed to interaction with Saturn's satellite Mimas; the Kirkwood gaps in the asteroid belt were due to interactions with the planet Jupiter.

Vindication of the theory came as a result of Voyager 1 and 2 observations of Saturn's rings in 1980 and 1981. The Cassini Division was found to contain a large amount of material (Fig. 4.6), and there was only a narrow gap between the inner edge of the Cassini Division and the outer edge of the B ring (Fig. 4.7). Instead of creating a large gap in Saturn's rings, Mimas was responsible for shaping the outer edge of the B ring (Fig. 4.8) and creating in the process only a relatively narrow gap. The coup de grace came with discovery of two small shepherding satellites surrounding Saturn's narrow F ring (Fig. 4.9). There could now be little doubt that the shepherding mechanism proposed by Goldreich and Tremaine was operative in the Saturn ring system. By analogy, it seemed likely that shepherding satellites were responsible for the narrowness and sharp edges of Uranus's rings.

It should not be assumed that shaping forces operative in the Uranus ring system are well understood, even after the Voyager 2 encounter. Questions about the Saturn ring system, for example, were only partially answered by Voyager findings. Furthermore, the immense complexity of Saturn's ring system revealed by Voyager

Fig. 4.6 — The Cassini Division, once thought to be an empty gap within the rings of Saturn, is seen in diffusely transmitted sunlight in this Voyager 1 view of the underside of the rings. It appears brighter than the A ring (to the left) and the B ring (to the right), each of which are optically thicker and absorb more sunlight than the Cassini Division. (260–1129)

data has offered far more challenges for theoretical dynamicists than they anticipated. The same situation exists for Uranus's ring system. Although much more is known about the rings than ever before, the riddles posed by the observations will take decades or longer to solve.

4.5 ADDITIONAL INFORMATION ABOUT URANUS

One can often learn much about a master craftsman by studying his protégé. So it is that ground-based studies of the motions of the rings have revealed some things about the nature of Uranus that the planet itself was reluctant to disclose. Ring data discussed in this chapter are primarily the result of the mathematical fitting of ring models to the many occultation data points. A product of the solution for ring orbit sizes and orientations is the determination of Uranus's pole direction. The positive spin axis (which corresponds to Uranus's south pole, since it is south of the ecliptic plane near which most of the planets orbit the Sun) is near a right ascension of 76.60° and a declination of +15.1°. This point in the sky is near the border between the constellations of Orion and Taurus, about 8° east of the bright yellow star Aldebaran.

The shape of Uranus is very nearly that of an oblate spheroid: if one were to examine a cross-section through the poles, it would approximate an ellipse whose long axis lies in the equatorial plane. Because Uranus rotates relatively rapidly, its

Fig. 4.7 — Four views of the Huygens Gap at the outer edge of Saturn's B ring (bottom) and at the inner edge of the Cassini Division (top). These views are spaced at approximately equal intervals around the ring and illustrate the double 'bulge' of the B ring caused by its centered-ellipse shape. Note the normal elliptical shape of the narrow ringlet within the Huygens Gap. (260–1473)

polar radius is about 600 km less than its equatorial radius. This redistribution of mass toward the planet's equator effects the orientation of the elliptical and inclined ring orbits. For an elliptical orbit, the periapsis point precesses forward in the orbital plane by an amount which varies from ring to ring. The precession rate of the epsilon ring amounts to 1.36° per day; the 6 ring precesses at 2.76° per day. Values of the 'apsidal' precession rates for these and the four other rings with measurably eccentric orbits are given in Table 4.1. (The line of apsides connects periapsis and apoapsis points in an elliptical orbit; hence the terminology 'apsidal precession' used above to describe motion of the orbit periapsis point.)

A second effect, applicable for inclined orbits, is akin to the wobbling of a

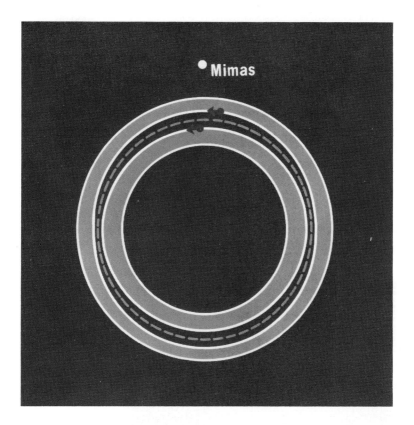

Fig. 4.8 — In a complex type of interaction, Saturn's satellite Mimas shapes the outer edge of the B ring. Ring particles at that distance circle Saturn in precisely one-half the time required for an orbit of Mimas. The resulting repeated gravitational pulling of Mimas effectively widens the Huygens Gap directly beneath it and on the opposite side of the planet. This widened shape circles the planet at the same rate as Mimas, even though the ring particles themselves move at a much faster rate. (260-1450)

spinning top. The line of nodes, defined by the intersection of the orbital plane with the plane of Uranus's equator, moves in the equatorial plane in a direction approximately opposite that of the orbital motion of the ring particles. As in the case of apsidal precession, this 'nodal regression' rate decreases with distance of the ring from the planet. The 6 ring node regresses at $-2.76°$ per day. Theory predicts a nodal regression rate for the epsilon ring of $-1.36°$ per day, but because the epsilon ring does not have a discernible inclination, this regression cannot be observed directly. Nodal regression rates for the other rings with measurable inclinations are given in Table 4.1.

The distribution of mass in the interior of a planet may be defined by a series of gravity harmonics, J_n, where $n=2, 3, 4$, etc., each of which would have zero value if the distribution were perfectly spherical. Within the gas giant planets, the distribution of mass must be nearly symmetric about the axis of rotation, and the odd-

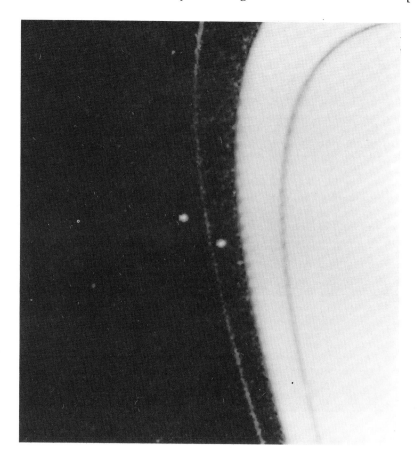

Fig. 4.9 — This Voyager 2 view of Saturn's F ring and its two shepherding satellites was the first clear confirmation that the shepherding mechanism proposed by Goldreich and Tremaine [16] was actually operating in nature. The image is somewhat smeared and the satellites (Prometheus and Pandora) appear much larger than they really are. Saturn's A ring and its Keeler Gap are seen at the right. (P-23911)

numbered harmonics are negligibly small. The most important harmonics are J_2, which defines the majority of equator-ward mass shift, and J_4, which defines mid-latitude deviations from the simple distribution described by the value of J_2. Apsidal precession and nodal regression rates of the Uranian rings are related to J_2 and J_4, and can be used to estimate the values of these low-order gravity harmonics. Pre-Voyager values for these gravitational constants given at the bottom of Table 4.1 are taken from a 1986 paper by French, Elliot, and Levine [17], as are all the other entries in Table 4.1.

Finally, the derived values of J_2 and J_4 can be combined with measurements of the optical oblateness (polar 'flattening') of Uranus, also obtained from ground-based stellar occultation data, to place limits on the radial distribution of mass within the

Table 4.1 — Characteristics of the rings of Uranus (pre-Voyager)

Ring Name	Semimajor Axis (km)[a]	Eccentricity (unitless)	Inclination (°)	Aps. precess. (°/day)	Nodal regr. (°/day)
6	41846±5	0.0010±0.0000	0.063±0.007	2.762±0.000	−2.758±0.001
5	42243±5	0.0019±0.0000	0.052±0.008	2.671±0.000	−2.665±0.001
4	42579±5	0.0011±0.0000	0.032±0.004	2.598±0.000	−2.592±0.003
alpha	44727±5	0.0008±0.0000	0.014±0.005	2.186±0.001	−2.183±0.007
beta	45669±5	0.0004±0.0000	0.005±0.003	2.030±0.001	−2.055±0.013
eta	47184±5	(0.0000±0.0000)	(0.002±0.003)	—	—
gamma	47632±5	(0.0001±0.0002)	(0.011±0.031)	—	—
delta	48306±5	(0.0000±0.0000)	(0.004±0.002)	—	—
epsilon	51156±5	0.0079±0.0000	(0.001±0.004)	1.363±0.000	—

$J_2 = 0.003346 \pm 0.000003$[b] $J_4 = -0.000032 \pm 0.000004$[b]

[a]For a mass (times gravitational constant) of $GM = 5.782223 \times 10^{21}$ cm^{3}g^1 s^{-2}.
[b]These Uranus harmonic coefficients of the gravity potential are referenced to a Uranus radius of 26 200 km.

planet. The mean density of Uranus is constrained by its total mass (determined from the orbital periods and orbit sizes of its satellites) and its dimensions (determined by direct visual observation as well as from stellar occultation measurements). But a planet with a large fraction of its mass in a concentrated rocky core overlain by a thick low-density atmosphere will have a smaller J_2 than one with uniform density throughout its interior. This is illustrated in Fig. 4.10, adapted from Elliot and

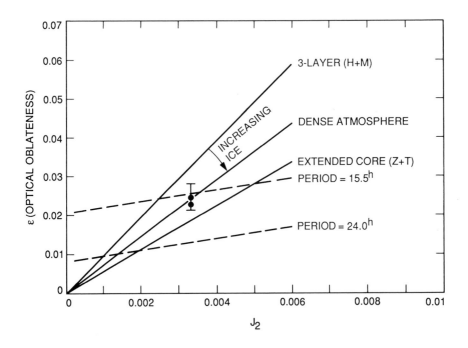

Fig. 4.10 — Graphs of this nature assist theoreticians in determining the nature of the interior of Uranus. The solid lines represent models with differing radial mass distributions within the planet. An optical oblateness of about 0.022 and gravity harmonic $J_2 = 0.00335$ implies a rotation period near 17 h and an internal structure somewhere between a model with no ice and one with uniform density throughout the interior.

Nicholson [18]. Note that, in addition to constraints imposed by the values and uncertainties in J_2 and the optical oblateness of Uranus, there are constraints imposed by the rotation period of Uranus, which was poorly known prior to the Voyager encounter. Any model of interior mass distribution must also be consistent with the determined value of J_4. With these constraints in mind, Hubbard [19] concluded that the best-fitting model had a rocky core constituting roughly 43% of the total mass of the planet, overlain by a relatively uniformly mixed layer of 'ices' (primarily water, ammonia, and methane) and gases (primarily hydrogen and helium). The ices would constitute about 38% of the total mass, and the gases the remaining 19%. Although a large variety of more complex models could fit the observed data, the general view of the interior presented by Hubbard remains basically unchanged by Voyager results (see Chapter 8).

NOTES AND REFERENCES

[1] A light year is the distance traveled by light in one year, or about 9 460 000 000 000 km. A light year is 63 240 AU, (where 1 AU=149 597 900 km, the mean distance between Earth and the Sun).

[2] Elliot J. L., Dunham, E., Millis, R. L. (1977) Discovering the rings of Uranus, *Sky and Telescope*, **53**, 412–416, 430.

[3] Elliot, J. L., Dunham, E. W., Mink, D. J. (1977) The rings of Uranus. *Nature*, **267**, 328–330.

[4] Millis, R. L., Wasserman, L. H., Birch, P. (1977) Detection of the rings of Uranus. *Nature*, **267**, 330–331.

[5] Elliot, J. L., Dunham, E. W., Wasserman, L. H., Millis, R. L., Churms, J. (1978) The radii of Uranian rings alpha, beta, gamma, delta, epsilon, eta, 4, 5 and 6 from their occultation of SAO 158687. *Astronomical Journal*, **83**, 980–992.

[6] Bhattacharyya, J. C., Kuppuswamy, K. (1977) A new satellite of Uranus. *Nature*, **267**, 331.

[7] Bhattacharyya, J. C., Bappu, M. K. V., Mohin, S., Mahra, H. S., Gupta, S. K. (1979) Extended ring system of Uranus. *The Moon and the Planets*, **21**, 393–404.

[8] Matthews, K., Neugebauer, G., Nicholson, P. D. (1982) Maps of the rings of Uranus at a wavelength of 2.2 microns. *Icarus*, **52**, 126–135.

[9] Allen, D. A. (1983) Infrared views of the giant planets. *Sky and Telescope*, **65**, 110–112. See also Color Plate 1 (p. 749) from Elliot, J. L., Nicholson, P. D. (1984) The rings of Uranus. In Greenberg, R., Brahic, A. (eds.) *Planetary Rings*, The University of Arizona Press, Tucson, pp. 25–72.

[10] Terrile, R. J., Smith, B. A. (1984) Cloud features on Uranus from CCD imaging. *Bulletin of the American Astronomical Society*, **16**, 657 (abstract only).

[11] Elliot, J. L. (1982) Rings of Uranus: a review of occultation events. In Hunt, G. (ed.) *Uranus and the Outer Planets*, Cambridge University Press, pp. 237–256.

[12] Elliot, J. L., Nicholson, P. D. (1984) The rings of Uranus. In Greenberg, R., Brahic, A. (eds.) *Planetary Rings*, The University of Arizona Press, Tucson, pp. 25–72.

[13] French, R. G., Elliot, J. L., Levine, S. E. (1986) Structure of the uranian rings. *Icarus*, **67**, 134–163.

[14] French, R. G., Elliot, J. L., French, L. M., Kangas, J. A., Meech, K. J., Ressler, M. E., Buie, M. W., Frogel, J. A., Holberg, J. B., Fuensalida, J. J., Joy, M. (1988) Uranian ring orbits from Earth-based and Voyager occultation observations. *Icarus*, **73**, 349–378.

[15] French, R. G., Kangas, J. A., Elliot, J. L. (1986) What perturbs the gamma and delta rings of Uranus? *Science* **231**, 480–483.

[16] Goldreich, P., Tremaine, S. (1979) Towards a theory for the uranian rings. *Nature*, **277**, 97–99.

[17] French, R. G., Elliot, J. L., Levine, S. E. (1986) Structure of the Uranian Rings. II. Ring Orbits and Widths. *Icarus*, **67**, 134–163.

[18] Elliot, J. L., Nicholson, P. D. (1984) The rings of Uranus. In Greenberg, R., Brahic, A. (eds.) *Planetary Rings*, The University of Arizona Press, Tucson, pp. 25–72.

[19] Hubbard, W. B. (1984) Interior structure of Uranus. In Bergstralh, J. T. (ed.) *Uranus and Neptune*, NASA Conference Publication 2330, National Aeronautics and Space Administration, Scientific and Technical Information Branch, pp. 291–325.

BIBLIOGRAPHY

Beatty, J. K., O'Leary, B., Chaikin, A. (1984) *The New Solar System*, 2nd Edition, Sky Publishing Corporation, Cambridge, Massachusetts, pp. 129–142.

Bergstralh, J. T. (1984) *Uranus and Neptune*, NASA Conference Publication 2330, National Aeronautics and Space Administration, Scientific and Technical Information Branch, Washington, D. C., pp. 291–325, 575–627.

Greenberg, R., Brahic, A. (eds.) (1984) *Planetary Rings*, The University of Arizona Press, Tucson, pp. 3–72, 589–659, 687–734.

Hunt, G. (ed.) (1982) *Uranus and the Outer Planets*, Cambridge University Press, pp. 111–124, 211–256.

5

Other pre-Voyager Uranus observations

5.1 THE INTERIOR OF URANUS

Studies of the interior structure of any planet are necessarily complicated by a paucity of relevant data. Even for the Earth, direct observations probe only a tiny fraction of the total radius of our planet, and much of our knowledge of inner Earth is obtained by building mathematical models and comparing the predictions of such models with conditions at Earth's surface.

Boundary conditions against which scientists may check possible models include several observable characteristics. The total mass of Earth (or of the remote planets) can be determined with fair precision by comparing the orbital periods of the satellites with the physical dimensions of those orbits. The size and shape of the planet, measured from its surface or from remote telescopic observations, can then be used to determine its volume and hence its overall bulk density (mass per unit volume). If satellite (or ring) orbits are elliptical or inclined to the planet's equator, the apsidal precession [1] rates of the elliptical orbits or the nodal regression [2] rates of the inclined orbits also provide clues about internal mass distribution. Reflected sunlight, infrared thermal radiation, and radiowaves given off by the planet contain information about the chemical composition of at least the outer layers of the atmosphere. Under certain conditions, such data can also reveal variations in composition with altitude and the atmospheric temperatures and pressures at those altitudes.

Armed with a knowledge of a planet's distance from the Sun and of the Sun's brightness at Earth's distance, one may calculate the amount of sunlight which reaches any planet in the Solar System. The difference between the amount of sunlight incident upon the planet and the amount of sunlight reflected back into space by the planet represents the solar energy absorbed by the planet. That solar energy heats the planet to a predictable temperature level. If the observed temperature of the planet (from infrared and radiowave data) is higher than that attributable to solar heating, the excess must come from heat sources within the planet, providing an additional constraint on interior models.

Any magnetic field around a planet generally arises from processes occurring in that planet's interior. Although it is often difficult to measure the strength and nature of a magnetic field from a remote observing site, one can often make crude estimates of those characteristics by observing the product of interactions within the field. One such interaction results in ultraviolet and visible auroral glows as electrons streaming outward from the Sun are channeled by the planet's magnetic field into the atmosphere near the magnetic poles. Another interaction typical of the gas giant planets gives rise to long-wavelength radiowaves whose intensity varies with a periodicity equal to the planet's rotation period. A third type of interaction is that of synchrotron radiation, caused by the motions of electrons accelerated by the magnetic field and sometimes visible in short-wavelength radio (microwave) emissions from the planet.

Rotation of a gas giant planet will cause the shape of the planet to depart from being precisely spherical. The resulting oblateness (polar flattening and equatorial bulging) is generally sufficient to be observed from Earth-based telescopes. The degree of distortion is dependent not only on the rotation rate, but also on the internal structure of the planet. If the rotation rate can also be measured (either by observing the periodicity in long-wavelength radiowave intensity or by observing the average rate of cloud motions across the disk of the planet), the planetary oblateness can serve as another constraint on interior models of the planet.

One additional constraint is generally applied to models of the gas giant interiors, namely that of consistency with accepted models for the origin and evolution of the giant planets. This constraint is less confining than others in that it constitutes a two-way street: realistic models of the giant planet interiors can also be used to test the validity of cosmogonic (origin of the 'cosmos') models.

5.1.1 Pre-Voyager models

Prior to the flood of data from the Voyager 2 encounter of Uranus in January of 1986, too little was known about Uranus to decide between several proposed models of the planetary interior. The mean density of the planet was known to within a small fraction of a percent, but the precise dimensions and shape were poorly determined. Furthermore, values for the low-order gravity harmonic coefficients were only beginning to be determined from observations of the rings, discovered only five months before the launches of the two Voyager spacecraft. Accurate determination of the planet's rotation period proved to be an almost hopeless task, and estimates of the rotation period ranged from less than 11 h to more than 24 h.

MacFarlane and Hubbard [3] and Hubbard [4] have described the state of interior modelling prior to the Voyager encounter. The discussions below follow mainly the latter paper.

Some of the main questions that interior models of Uranus attempt to answer are:

(1) What is the bulk chemical composition of Uranus? What respective fractions of Uranus's total mass are hydrogen, helium, methane, ammonia, water, and heavier rocky materials? How do each of these compare with the fractions of each presumed to exist in the primordial solar nebula?

(2) What is the chemical composition of the observable layers in the atmosphere of Uranus? What does this tell us about the internal composition of the planet?

(3) To what degree are the different components within the planet separated, and how does this compare with the other giant planets of the Solar System? What processes could lead to this chemical differentiation within Uranus?

(4) How much heat is coming from the interior of Uranus? What processes contribute to this heat flux? How are they related to the planet's formation processes? How are they related to the planet's atmospheric and interior dynamics? Do circulation patterns in the interior of Uranus give rise to a planetary magnetic field?

It is fortunate that at least partial answers to many of these questions were known for Jupiter and Saturn. The initial studies of these two planets by spacecraft were carried out via the Pioneer 10 and 11 spacecraft in the middle to late 1970s. These were supplemented by the Voyager 1 and 2 encounters with Jupiter in 1979 and with Saturn in 1980 and 1981. Extrapolation from these two planets, accounting for the greater distance and lower mass of Uranus, may provide some leverage for attacking an otherwise extremely complicated problem.

The relative abundance of elements in the early solar nebula has been deduced [5] on the basis of direct observation of the outer atmosphere of the sun and from chemical analyses of meteorites collected from Earth's surface. In order of relative abundance (in mass fraction), the ten most abundant elements in the early solar nebula are hydrogen (74.4%), helium (23.7%), oxygen (0.87%), carbon (0.39%), neon (0.19%), iron (0.14%), nitrogen (0.09%), silicon (0.08%), magnesium (0.07%), and sulfur (0.04%). The remaining 0.03% includes all the rest of the elements, none of which constitutes more than 0.01% of the total mass. Relative abundances are often stated as ratios of numbers of atoms of the given element to the number of hydrogen atoms. Stated in these terms, abundances are even more heavily weighted toward the lighter elements. For every 1 000 000 hydrogen atoms, there are 80 000 helium atoms, 739 oxygen atoms, 441 carbon atoms, 128 neon atoms, 91 nitrogen atoms, 39 magnesium atoms, 37 silicon atoms, 33 iron atoms, and 19 sulfur atoms.

Most Uranus model calculations make a simplifying assumption by grouping the elements into three categories: gas (G), ice (I), and rock (R). These groupings are according to the state of the matter in the solar nebula at the distance of Uranus, but just before Uranus was formed. The G-group contains those components that were gaseous then and remain gaseous now, primarily hydrogen and helium. The I-group contains those components whose condensation temperatures are close to the estimated existing temperatures at the onset of Uranus formation. These include water (H_2O, containing most of the oxygen), methane (CH_4, containing most of the carbon), and ammonia (NH_3, containing most of the nitrogen). Much of the sulfur might also be tied up in hydrogen sulfide (H_2S) 'ice'. The R-group comprises the remaining heavier elements and compounds, primarily those of magnesium, silicon, and iron.

A given model attempts to specify relative amounts of G, I, and R within Uranus. Although the relative amounts of each of these groups are not necessarily constrained to conform to solar abundances, most modelers assume that relative solar abundances are maintained within the I and R groups. The bulk density of Uranus is such that only a minor amount of the total planet can be composed of G-component.

Models must specify the relative amounts and radial distribution of I and R components.

If the I/R mass ratio is assumed to be consistent with solar abundance, it would have a value of about 3.5. This is the largest plausible I/R for any Uranus model which starts with a solar-abundance nebula. Any variation from this value would involve loss of the more volatile ices and a reduction of the I/R ratio. For example, if temperatures were such that most of the ammonia and methane were in the G-component at the time of Uranus's formation and the I component was primarily water, the bulk density of Uranus leads to an I/R ratio of 2.0. A suggestion by Lewis and Prinn, [6] was that carbon and nitrogen in the primordial nebula may have been tied up in carbon monoxide (CO) and nitrogen gas (N_2), and that reactions which combine the carbon and nitrogen with hydrogen may have been inhibited. In this process, more than half the oxygen would have been tied up in gaseous carbon monoxide, greatly reducing the oxygen available to form water. The resulting I/R would be 0.6, a practical lower limit to the ratio. Cosmochemical considerations and the mean density of Uranus thus serve to constrain the I/R ratio for Uranus to values between 0.6 and 3.5.

Additional constraints on the internal structure of Uranus may be provided by the present atmospheric relative abundances. In particular, Wallace [7] reported that the methane-to-hydrogen ratio in Uranus's atmosphere is 10 to 100 times the solar abundance of carbon. According to Gulkis, Janssen, and Olsen [8], on the other hand, the ammonia-to-hydrogen ratio is less than 1% the solar abundance of nitrogen! How can ammonia, which is less volatile than methane, be depleted by a factor of at least 100 while methane is substantially overabundant? These differences must say something about the processes and conditions under which Uranus was formed.

Another clue to internal composition is the ratio of deuterium (D) to normal hydrogen. Deuterium is a form of hydrogen which contains an extra neutron in the atomic core. It reacts to form water, methane, and ammonia more readily than it bonds to another hydrogen atom. Thus, if chemical equilibrium was attained during Uranus formation, a larger-than-normal amount of deuterium should have been trapped in the I component of Uranus in the form of deuterated water (HDO), deuterated methane (CH_3D), and deuterated ammonia (NH_2D). As Uranus cooled, there should have been free exchange between the deuterium in the I component and hydrogen in the thin G component, resulting in a marked atmospheric enhancement of deuterium relative to solar abundances. Such deuterium enhancement was not observed in the thick hydrogen atmospheres of Jupiter and Saturn. Rough measurements of the deuterium-to-hydrogen ratio from Earth [9,10] seemed to show no significant differences between Uranus and the larger giant planets.

If indeed the deuterium abundance is not enhanced over solar values, there are some possible explanations. Maybe deuteration of water, methane, and ammonia did not proceed to chemical equilibrium at the low temperatures associated with condensation of the I component. Or perhaps poor circulation within Uranus since its formation has not permitted deuterium to chemically equilibrate between the G and I components. A third possibility is that chemical equilibrium did occur with a much more extensive G component that the planet has since lost.

The helium-to-hydrogen ratio could similarly help resolve some of the unanswered questions about the origins and internal structure of Uranus. Present theories generally explain the thick hydrogen atmospheres of Jupiter and Saturn as captured atmospheres after the formation of the R-component and I-component cores. This would account for the similarity between the helium abundances on these planets and those of the pre-planet solar nebula. In the case of Uranus, it is possible that the hydrogen atmosphere came from the partial decomposition of methane rather than from a capture mechanism. Such an atmosphere would have a somewhat reduced abundance of helium. Measurement of the Uranian helium abundance from Earth-based observations is extremely difficult. The assumption of near-solar abundance of helium was briefly called into question just prior to the Voyager 2 encounter of Uranus. Orton's analysis of the spectrum of Uranus at radio wavelengths [11] seemed to imply that helium was grossly overabundant, a result that was bewildering. Voyager data were later to show that helium was present at solar abundance, to within the uncertainties of the measurements.

Models of Jupiter and Saturn include well-differentiated interiors. An inner core of molten R-component material was overlain by an outer core of molten I-component material. Because they constitute a relatively small percentage of the total mass of Jupiter and Saturn, little external difference in the gravitational harmonics would be seen between these two-layered core models and well-mixed cores. The extensive G-component layer of these two planets constitutes the majority of their mass.

A differentiated three-layered model of the type successfully employed for Jupiter and Saturn does not work well for Uranus. Too large a fraction of the mass is near the planet's center, and the resultant gravity harmonics would not match those deduced from observations of the rings. Zharkov and Trubitsyn [12] resolved this dilemma by assuming that the R and I components of Uranus are uniformly mixed. When adjusted for higher compressibility of water at high pressures, this two-layer model adequately represented the externally observed data except for the fourth-order gravity harmonic coefficient, J_4. This model is depicted schematically in Fig. 5.1.

The data (including J_4) are equally well represented by a three-layer model proposed by Hubbard and MacFarlane [13], in which the R component is separate from the I component. The R component is contained in a centrally condensed core with a mass of 6.2 Earth masses. The I and G components may also be separated, but if they are, the G component must have a composition which is nearly 50% methane and therefore contributes significantly to the gravity harmonics. It is perhaps more likely that the I and G components are well-mixed. In this 'dense atmosphere' model, the R core would contribute 43% of the planet's mass; the I and G components contribute about 38% and 19%, respectively. Fig. 5.2 is a schematic representation of the Hubbard and MacFarlane 3-layer model. Calculation of the temperatures and pressures at depth in the dense atmosphere model led to the conclusion that the I component was likely to be liquid rather than gaseous. Hence it was supposed that Uranus had a deep superheated ocean.

The two models presented are not unique in their ability to match the observational data and to fit well with cosmogonic models of Solar System formation. Perhaps the interior of Uranus is not layered, but consists of a mixture of materials whose relative abundances change smoothly with depth. More complex models with many

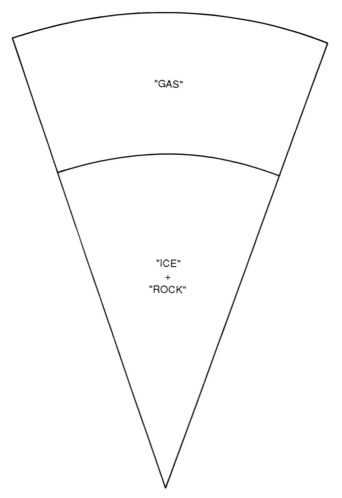

Fig. 5.1 — Uranus interior model of Zharkov and Trubitsyn. R and I components are assumed to be uniformly mixed. This model adequately represented the externally observed data except for the fourth-order gravity harmonic coefficient, J_4.

layers may also be possible. Even the Voyager 2 data from the Uranus encounter did not provide enough constraints to uniquely define the interior structure of this distant planetary neighbor.

5.1.2 Rotation period

From his observation of the motions of Titania and Oberon, Sir William Herschel correctly concluded in the early 1790s that the rotation axis of Uranus lay very nearly in the orbital plane of the planet. By the late 1790s he had further determined that the satellite orbits were retrograde (in a direction generally opposite to Uranus's orbital motion around the Sun) and that the planet very likely also rotated in a retrograde

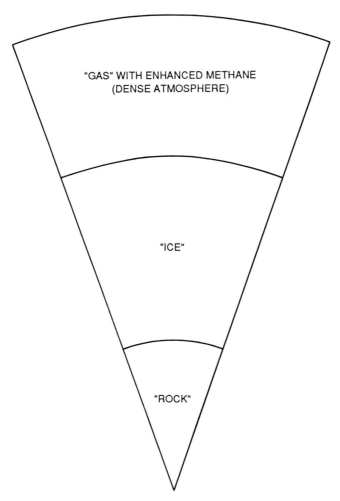

Fig. 5.2 — The 'dense atmosphere' model of Hubbard and MacFarlane. The I and G components were assumed to be well mixed over a separate rock core. Calculated temperatures and pressures implied that the I component was liquid and that Uranus had a deep superheated ocean.

sense. He had also noted a flattening of the poles, which he attributed to relatively rapid rotation of the planet.

More than a century passed before any real progress was made toward determining the length of a Uranian day. No visible features could be seen on the disk of the planet, so timing the passage of such features across the disk was not possible. In 1902, Deslandres [14] pointed out that sunlight reflected from the approaching limb of a rotating planet will show spectral features which are shifted to shorter (bluer) wavelengths than for the center of the disk. Similarly, spectral features seen in sunlight reflected from the receding limb will be shifted to longer (redder) wavelengths. These shifts are due to the Doppler effect [15]. Deslandres showed that if the

slit of a spectrograph were placed across the disk of Uranus, narrow spectral lines in
the reflected solar spectrum would be inclined by an amount determined by the
variation in radial velocity of Uranus along the spectrograph slit. The greatest tilt
would occur if the slit were placed along the equator of Uranus at a time when Earth
was passing through the equatorial plane of Uranus (see Fig. 5.3). Since that was not

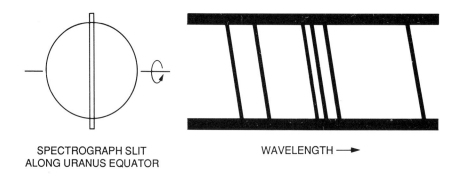

SPECTROGRAPH SLIT   WAVELENGTH ⟶
ALONG URANUS EQUATOR

Fig. 5.3—Uranus rotation results in tilting of spectral lines due to the Doppler effect, as shown
schematically in this figure. The greatest spectral tilt is expected when the spectrograph slit is
placed along the equator of Uranus at a time when Earth was passing through the equatorial
plane of Uranus.

due to happen until 1923, Deslandres had supposed that it would be necessary to wait
21 years to use this method to determine the rotation rate of Uranus.

Lowell and Slipher [16] halved the anticipated waiting period by using the
method of Deslandres in a paper published in 1912, in which they deduced a rotation
period of 10.8±0.3 h. Moore and Menzel [17] repeated the observations during the
late 1920s and derived a period of 10.84±0.16 h, remarkably close to the earlier
value. The good agreement between these two determinations of the rotation
period, along with supporting photometric data discussed below, misled astronomers into believing that the rotation period of Uranus was well determined. The
value of 10.84 h was widely quoted until 1977!

A different method was used by Campbell [18] in 1916 to 1918. He reasoned that
small variations in the brightness of Uranus as it rotated might enable him to use
photographic photometry to determine the rotation period. Although Campbell did
not publish his results until 1936, Pickering announced [19] in 1917 that Campbell
had found brightness variations of about 15% with a period of 10.82 h, in close
agreement with the spectroscopic value of Lowell and Slipher. Unfortunately,
Campbell's 1917 data did not fit well with his 1916 data. His explanation was that the
location of spots in the atmosphere had changed in the intervening time. Furthermore, his 1918 data showed no variation at all.

Reanalysis of the early spectroscopic data during the late 1970s showed that there
were major flaws in each, and that agreement between the published results was
grossly misleading. It is possible that a scientific 'bandwagon' effect occurred, where
later observers wanted to believe the early results of Lowell and Slipher and saw in

poor-quality data the confirmation for which they were looking. Whatever the reason for the seeming agreement, all three published rotation periods were known to be far too short, even before the receipt of definitive Voyager data in 1986.

Measurement of the Uranus rotation period from spectroscopic line tilts is extremely difficult, a fact that can be underscored by recent published results utilizing modern sophisticated spectroscopic equipment. Spectroscopic values for the rotation period published in 1980 range from 15 to 24 h [20,21,22]. After 1980, the Earth was again too far from the equatorial plane of Uranus for useful measurements.

Photoelectric brightness measurements were equally unsatisfying. Smith and Slavsky [23] tried observations over several years and concluded on the basis of maximum brightness variations of only 0.6% that the Uranus period was just under 24 h, suspiciously close to Earth's rotation period. A reanalysis of Campbell's 1916 data by Hayes and Belton [24] led to the conclusion that a period of 21.6 h best fit the data.

The most definitive pre-Voyager values for the rotation period were those derived by comparing the optical oblateness (polar flattening) of Uranus with the planet's second- and fourth-order gravity harmonics J_2 and J_4. The J_2 and J_4 harmonic coefficients describe deviations from a purely spherical mass distribution within Uranus. Positive values of J_2 represent a polar 'flattening' of the mass distribution; negative values of J_4 represent a mid-latitude deficit of mass. Brouwer and Clemence [25] derived a complex but straightforward mathematical relationship between the rotation period, the optical oblateness, and J_2 and J_4. Using J_2 and J_4 values from Table 4.1 and an optical oblateness of 0.024±0.003, French [26] derived a rotation period 15.6±1.4 h, retrograde. For an alternate value for the oblateness (0.022±0.001), derived from Stratoscope II imaging of Uranus, French derives a period of 16.6±0.5 h, retrograde. These two values represent the best estimates of the rotation period of Uranus prior to the Voyager encounter.

5.1.3 Polar flattening

Optical oblateness, or polar flattening, is defined as $1-(R_p/R_e)$, where R_p and R_e are the polar and equatorial radii of Uranus, respectively. As mentioned in the previous section, Sir William Herschel was the first to notice the polar flattening of Uranus. He suspected such flattening as early as 1783. Later in that decade he began to attribute the nonspherical appearance to the presence of rings, but again discarded that in favor of planetary oblateness in 1792. Although Herschel observed the nonsphericity of Uranus, he never managed to make quantitative measurements of the oblateness. The ideal time to have done so would have been in 1798, when the planet's equator was turned toward Earth-based observers, but Herschel seems to have made no observations of the planet itself after 1795.

The next good opportunity for side-on observations of Uranus's obliquity was near 1840. About that time, J. H. Mädler, director of the Dorpat Observatory in Latvia, attempted to measure the differences in the polar and equatorial diameters of Uranus. The fact that the oblatenesses he measured were about four times as large as presently accepted values serves to accentuate the difficulty of the measurements.

The next opportunity for vertical views of the equator came in 1882, and, thanks to a plea by A. Safarik of Prague for oblateness measurements of the planet, several astronomers with large telescopes at their disposal attempted the measurements. Of those who reported an oblateness (several could discern no flattening) all had large scatter in their data and all reported flattening two to four times as large as presently accepted values.

Little was added to knowledge of polar flattening in the 1924 time period. Not until relatively modern times has real progress been made toward determination of more realistic values for the Uranian oblateness. Earth was again near the Uranian equatorial plane in 1966. Only four years later, Danielson, Tomasko, and Savage [27] obtained high-resolution images of Uranus from a balloon-borne 91-cm-aperture telescope called Stratoscope II. They determined a Uranian oblateness of 0.01 ± 0.01. Franklin and his colleagues [28] reanalyzed the data in 1980, using techniques unavailable to the Stratoscope II investigators. They digitized the images and then processed them with a computer to find the best-fit ellipses to isophotes [29] near the edge of the disk image. On the basis of such fits to the best 23 images obtained by Stratoscope II, Franklin *et al.* set the oblateness of Uranus at 0.022, with a ±0.001 scatter in the determinations. Uranus displays noticeable limb darkening: the edges of the planetary disk are much dimmer than the center of the disk. It is assumed that the magnitude of this effect is not dependent on latitude; otherwise, isophotes might not represent the true optical shape of the planetary disk. For an equatorial radius of 25 500 km (at the 1-bar pressure level), an oblateness of 0.022 corresponds to a polar radius Uranus polar radius of about 24 940 km.

An independent method for relatively accurate determination of the polar flattening uses data from stellar occultations. The power of stellar occultation data for probing a planetary atmosphere and for ring searches was discussed in Chapter 4. Because of the finite size of a star, optical effects, and instrument response time, some adjustments to the occultation data must be made to utilize its full potential. The discovery of rings in 1977 provided an important tool in oblateness studies, namely, a precise method for determining the position of the center of the planet. Knowledge of the position and orientation of the planet can be combined with a single occultation chord across the planet to calculate the amount of polar flattening. Multiple chords across the planet during several occultations may then serve to reduce uncertainties in the calculated oblateness.

On the basis of stellar occultation measurements of the planet, the optical oblateness of Uranus was determined by Elliot *et al.* to be 0.024 ± 0.003. This coincides rather nicely (within the stated error bars) with the Stratoscope II results. It should be pointed out that these two results do not refer to the same level in the Uranian atmosphere. The Stratoscope II results provide an oblateness for the planet at a level near the cloud tops, where the pressure is close to 1 bar (about the same as Earth's sea-level atmospheric pressure). The stellar occultation measurements, on the other hand, refer to a level nearly 500 km higher in the atmosphere, where the pressure is only a millionth of a bar. It is not impossible that the oblateness of Uranus is slightly greater at the 1-bar level than at the 1-μbar level. For an equatorial radius of 26 100 km (at the 1-μbar level), an oblateness of 0.024 corresponds to a polar radius of 25 470 km.

5.1.4   Apparent lack of internal heat

Uranus has a mean distance from the Sun of 19.18 AU; Neptune's mean distance is 30.06 AU. Because of its greater distance, the intensity of the sunlight falling on Neptune is only 41% of that falling on Uranus. Yet the temperatures of the two planets are nearly identical. How can that be?

To fully answer the question one must first investigate the amount of solar energy actually absorbed by each planet. The ratio of the total solar radiation reflected by the planet to that incident upon it is known as the bolometric (all wavelengths) Bond (all directions) albedo, often designated as A^*. The fraction of incident sunlight actually absorbed by the planet is identical to the fraction not reflected back into space, namely $(1-A^*)$. It is this fraction which is responsible for solar heating of the planet. Any additional heat required to give the planet its observed temperature must originate in the interior of the planet.

The bolometric Bond albedo, A^*, of Uranus cannot easily be determined from Earth. Our atmosphere blocks many important wavelengths, especially in the ultraviolet and infrared regions of the solar spectrum. Light at wavelengths more readily transmitted by the atmosphere is also distorted (seeing effects) and selectively absorbed (transparency effects), each by time-varying amounts. Some of the problem can be overcome by using high-altitude observatories (like the Mauna Kea Observatory on the island of Hawaii), airplane-mounted telescopes (like the Kuiper Airborne Observatory), or balloon-borne telescopes (like Stratoscope II). Still greater altitudes can be achieved with sounding rockets (like the Aerobee or Scout rockets) or Earth-orbiting observatories (like the International Ultraviolet Explorer and the InfraRed Astronomy Satellite). Nearly all of these techniques have been used to collect information on the brightness of both the Sun and Uranus at a large variety of wavelengths.

One of the first to attempt detailed synthesis of data on the bolometric albedo of Uranus was Younkin [31] in 1970. Lockwood *et al.* [32] did their own analysis in 1983 and derived a value of 0.262 ± 0.009 for the bolometric albedo of Uranus. They also reanalyzed Younkin's data, finding a value for that earlier epoch of 0.228 ± 0.009 for the bolometric albedo. The earlier period represents data from a time when Uranus was presenting near-equatorial latitudes to the Earth and Sun; the later data correspond to a more nearly pole-on geometry. Taken together, the two data sets imply that the polar regions of Uranus may be about 15% brighter (bolometrically) than the equatorial regions. These values represent the best pre-Voyager bolometric albedo data obtained by Earth-based observers.

Determination of the light-scattering properties of the atmosphere of Uranus has even greater inherent difficulties. Because it is so distant from the Sun, Earthbound (including Earth-orbiting) observers and instruments always view a fully illuminated Uranus. The maximum angle between Sun and Earth as viewed from Uranus (i.e. the maximum phase angle) never exceeds 3°. From the vicinity of Earth it is impossible to view Uranus under a variety of lighting and viewing geometries and thereby deduce the Bond albedo. It is possible to watch for longitudinal brightness variations as the planet rotates, and seasonal or latitudinal changes as it revolves around the Sun, but only for a planetary disk that is fully illuminated.

The ratio of the Bond albedo, A (monochromatic) or A^* (bolometric), to the corresponding albedos at $0°$ phase angle (geometric albedos) is known as the phase integral, q (monochromatic) or q^* (bolometric). Before the advent of space probes to the outer planets, astronomers were forced to rely on theoretical models of light-scattering to estimate the value of q and q^*. Harris [33] tackled the problem of estimating the phase integral for the giant outer planets, deriving a value of 1.65 (for visual wavelengths). Combined with his estimate for the visual geometric albedo of Uranus (0.565), this yielded a visual Bond albedo, A_v, of 0.93. He pointed out that this was higher than for any other planet and that it implied that Uranus was only absorbing 7% of the visual sunlight falling on it. Harris's estimate was that Neptune absorbed 16% of the visual sunlight incident on it. If these ratios were accurate and had applied at other wavelengths as well, they might have explained the curious similarity in Uranus's and Neptune's temperatures.

As we have seen, however, the bolometric geometric albedos determined by Lockwood *et al.* [32] were found to be in the range 0.23 to 0.26. Younkin [31] estimated on the basis of very limited data that the bolometric phase integral was much closer to $q^*=1.25$. The bolometric Bond albedo would therefore lie near 0.30, and about 70% of the sunlight incident upon Uranus would go toward heating the planet.

The encounters of the Pioneer 10 and 11 spacecraft with Jupiter in December of 1973 and December of 1974 provided the first opportunities to actually measure the brightness of one of the gas giant planets at phase angles higher than those achievable from Earth. Tomasko, West, and Castillo [34] estimated on the basis of Pioneer data that the Jupiter bolometric phase integral is between 1.2 and 1.3, very close to the value earlier estimated by Younkin for Uranus. Wallace [35] argued on the basis of theoretical arguments and Pioneer data that the upper limit should be extended to 1.5. Lockwood *et al.* [32] used the observed maximum value of the monochromatic geometric albedo along with theoretical arguments to estimate that reasonable lower and upper limits for q^* were 1.33 and 1.64, respectively. Combined with the uncertainty in the bolometric geometric albedo, this range of values for q^* could set only rough limits on the bolometric Bond albedo: $0.30 \leqslant A^* \leqslant 0.44$ (for the 1981 time period). Thus, knowledge on the reflective properties of Uranus prior to the acquisition of Voyager data allowed for as much as 70% or as little as 56% of the incident solar energy to be absorbed by Uranus. The corresponding values for the bolometric Bond albedo for the 1961 epoch would be $0.26 \leqslant A^* \leqslant 0.39$.

The additional piece of data needed to answer the question of how much heat is coming from the interior of Uranus is an accurate measure of the amount of energy being emitted by the planet. Most of the solar energy used to heat the planet is in the form of ultraviolet, visible, and near-infrared light. Because of its low temperature, most of the energy re-radiated by Uranus is in the far infrared part of the spectrum. Earth-based observers again are limited by the transparency of Earth's atmosphere and by the small range of phase angles (less than $3°$) observable from Earth. The limitation in phase angle is not a severe problem in temperature determinations. This is because the giant planets with their deep atmospheres change temperatures at a rate much slower than a Uranus day or even a Uranus year. In other words, in contrast to terrestrial planets like Earth, Uranus's temperature at midnight is identical to its temperature at noontime, and its temperature in midsummer is identical to its temperature in midwinter.

Orton and Appleby [36] analyzed recent data on the brightness of Uranus at wavelengths from 4.5 μm to 3 mm and concluded that the total amount of energy being emitted by Uranus was equivalent to that of a blackbody [37] with a temperature of 58.3±2.0 K. Orton and Appleby further calculated on the basis of the bolometric Bond albedo estimates for 1961 and 1981 by Lockwood *et al.* [32] that the temperatures of Uranus (assuming solar heating only and no internal heat source) would be 57.0±0.8 K in 1961 and 55.8±1.0 in 1981. The 2.0–K uncertainty in the effective blackbody temperature of Uranus was sufficiently large to question whether Uranus had any discernible internal heat source. As Voyager approached Uranus, it was possible to say that Uranus's internal heat source (if one existed) could not be generating more than 23% as much energy as the solar energy input to the planet.

5.2   THE ATMOSPHERE OF URANUS

There were several early indications that the atmosphere of Uranus was markedly different from those of Jupiter and Saturn. Firstly, there was the difference in color noted by many early observers. Uranus was bluish in color; Jupiter and Saturn more orange and yellow. Secondly, there was a marked difference in size, with Uranus possessing a volume less than 10% that of Saturn or Jupiter. Thirdly, its overall density was nearly twice that of Saturn. Although its density was found to be comparable to that of Jupiter, astronomers recognized that its smaller size meant that Uranus had much lower pressures in its interior than Jupiter. The low-density elements that constituted a major fraction of Jupiter and Saturn must contribute a substantially smaller fraction of Uranus's mass.

5.2.1   Chemical composition

From the discussion above, it is clear that astronomers had some reasons to believe that the chemical composition of Uranus was 'strange' when compared to that of Jupiter and Saturn. The advent of the astronomical spectroscope [38] only served to accentuate those differences. Secchi [39] turned his spectroscope toward Uranus in 1869, not expecting to see anything other than the unaltered reflected light of the Sun. He was so surprised by what he saw that he had to check his instrument to be certain there was no malfunction. All of the colors near yellow (the dominant colors emitted by the Sun) seemed to be missing. '... If this spectrum is purely due to reflected sunlight (which might perhaps be questioned), it must undergo a considerable modification in the planet's atmosphere ...'.

Twenty years after Secchi's first spectral observations of Uranus, Keeler [40] demonstrated in 1889 that the darkened regions of the Uranian spectrum were due to absorption by some unknown constituent within the atmosphere of Uranus. He made a sketch (reproduced here as Fig 5.4) to show as faithfully as possible the appearance of the dark bands in the visible-light spectrum. There had been some speculation prior to Keeler's work that Uranus might be self-luminous, and that Uranus shone from its own light rather than from reflected sunlight.

The mystery of the cause of the dark bands in the spectrum of Uranus was not solved until 1932, when Wildt wrote a paper entitled, 'Methane in the atmospheres of the great planets' [41]. His conclusions, applicable to all four giant planets, were

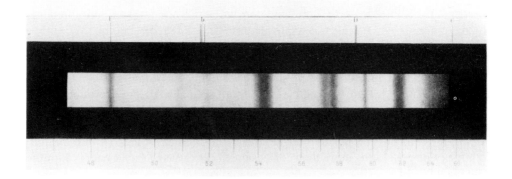

Fig 5.4 — Dark bands in the visible-light spectrum of Uranus, as sketched by Keeler, who demonstrated if 1889 that the darkened regions of the uranian spectrum were due to absorption, and not to gaps in emission.

based on theoretical considerations and extrapolations. The bands of methane (CH_4) that are seen in visible light are weak enough for Jupiter and Saturn that they were not prominent spectral features. A band at 0.886 μm in the infrared was the only feature in the spectra of Jupiter and Saturn that had been attributed to methane. Uranus's spectrum was so weak in the red and infrared that early observers could not obtain infrared spectra of the planet. Laboratory spectra of methane gas obtained up to that time were inconclusive, primarily because the methane path lengths (the amount of methane between the light source and the spectral detector) were too small. Wildt then used a formula derived by Dennison and Ingram [42] to explain the 0.886-μm infrared band to show that weaker absorptions should occur at 0.7258, 0.6191, and 0.5444 μm in the visible. The latter two of these are very close to the positions observed for the two darkest bands in Uranus's spectrum; the first was just beyond the long-wavelength limit of Uranian spectra available at that time.

Two things were necessary to confirm Wildt's conclusions. The first was to obtain spectra of Uranus that extended out to the 0.886-μm wavelength. The second was to obtain definitive laboratory spectra of methane at visible wavelengths and compare them with the Uranus spectrum. V. M. Slipher and S. Adel of the Lowell Observatory in Arizona accomplished both [43]. They used high-pressure tubing 5 cm in diameter and 12 m in length to hold the methane gas, which they then pressurized to 40 atmospheres (40 times Earth's sea-level atmospheric pressure). This provided an effective path length of 480 'meter-atmospheres' of methane. Glass windows at each end of the rigid tube enabled them to illuminate the gas at one end and place the spectrograph at the other end. They obtained a superb methane spectrum that covered wavelengths from violet to infrared.

Slipher and Adel earlier used the relatively high-altitude (to eliminate most of the absorption from Earth's atmosphere) telescope site at Lowell Observatory to obtain high-quality violet-through-infrared spectra of both Uranus and Neptune. They found that their laboratory spectrum of methane was almost a perfect match with each of the two planetary spectra. It had been proven that methane is a major component of the atmospheres of both Uranus and Neptune.

Hydrogen (H) and helium (He) had earlier been suggested as the major components of the Uranian atmosphere, but spectroscopic evidence for their existence was difficult to obtain. Since the solar spectrum was dominated by absorptions from these two elements, the reflected solar spectrum from Uranus would have to show substantial changes in those absorption lines to prove the existence of H and He. Some had suggested that a line of atomic hydrogen at 0.4861 μm was enhanced, but based on Slipher's methane results, Russell [44] showed that the atmosphere of Uranus was too cold to permit atomic hydrogen to exist in large enough amounts to be detectable from earth. He also showed that temperatures were too low to permit water vapor to exist in the visible atmosphere in measurable quantities.

Kuiper (discoverer of Miranda) utilized improved infrared photography techniques in 1948–9 to find a number of new absorption bands in the Uranus spectrum, which he and Phillips [45] then proceeded to show were not due to methane or any of its chemical byproducts, nor to ammonia, nor to other suggested molecular components of the atmosphere of Uranus. Herzberg [46] used a gas cell with a path length of 80 m filled with hydrogen at 100 atm pressure to obtain a spectrum, which he then showed reproduced the newly discovered Uranian spectral lines of Kuiper. Although atomic hydrogen (H) had been ruled out for the visible atmosphere, molecular hydrogen (H_2) was obviously present in large quantities. Astronomers soon came to realize that molecular hydrogen was the dominant constituent of the atmospheres of all the gas giant planets.

It was a natural step in logic to assume that the second most abundant gas was helium, although spectroscopic evidence for the amount would have to await the Voyager encounter. Courtin, Gautier and Lacombe [47] made the best pre-Voyager estimate from theoretical considerations that the ratio of the number of He molecules to the number of H_2 molecules was in the range $He/H_2 = 0.11 \pm 0.11$, or from 0 to 22%.

It should be emphasized that the atmospheric composition data given above refer to a level in the Uranian atmosphere that is above a reflecting cloud layer, probably composed of methane ice. The three main components of the atmosphere had been shown (or estimated, in the case of helium) to be molecular hydrogen, helium, and methane. Later observers were to detect deuterated hydrogen (HD) and trace amounts hydrogen sulfide (H_2S) and ethane (C_2H_6). By analogy with Jupiter and Saturn, ammonia (NH_3) was also presumed to be present. Gulkis, Janssen and Olsen [48] showed from the transparency of the atmosphere to 1-cm-wavelength microwaves that ammonia was probably less abundant than expected on the basis of the N/H ratio of the Sun. Pre-Voyager abundance estimates for the atmospheric gases above the cloud deck are given in Table 5.1, which has been adapted from data given by Hunten [49] and Atreya [50].

5.2.2 Temperatures

Uranian atmospheric temperatures vary substantially with altitude. Two general types of Earth-based observations have been used to infer temperatures at different levels in the atmosphere. The first of these is stellar occultation measurements, which

Table 5.1 — Pre-Voyager estimates of Uranian gas bundances

Gas	Uranian abundance	Solar abundance	Uranian/Solar
Molecular hydrogen (H_2)	500 to 700 km atm	H/H 1.000	—
Helium (He)	He/H=0.055±0.055	He/H=0.080	0.0 to 1.4
Methane (CH_4)	C/H=0.6 to $\geqslant 5\times10^{-3}$	C/H=4.4×10^{-4}	1.3 to $\geqslant$11
Deuterated hydrogen (HD)	D/H=0.2 to 4×10^{-4}	D/H 1.5×10^{-4}	0.1 to **3**
Ammonia (NH_3)	N/H$<10^{-6}$	N/H=1.3×10^{-4}	<.01
Hydrogen sulfide (H_2S)	Trace	S/H 1.7×10^{-5}	<1.0
Ethane (C_2H_6)	Trace		

provide information on temperatures relatively high in the atmosphere. The second is based on the observed intensity of Uranian radiation at various wavelengths in the infrared through microwave parts of the spectrum.

During a stellar occultation, changes in the apparent brightness of the star can be monitored as the atmosphere (of Uranus) passes in front of it. From such measurements and a general knowledge of the chemical composition (expressed as a mean molecular weight [51]) of the atmosphere, it is possible to deduce the variation of gas density with altitude. At high temperatures, gas density decreases slowly with altitude; at lower temperatures, the density decreases more rapidly with altitude. With some simplifying assumptions, one may then calculate the temperature and pressure of the gas as a function of altitude in the atmosphere.

The atmospheric pressure levels sensed by stellar occultations are near a microbar (0.000001 of Earth's sea-level atmospheric pressure). They correspond to an altitude of about 400 to 500 km above the 1-bar reference level (which is near the Uranian cloud tops). Temperatures (near the 1-μbar level) derived from stellar occultations by Uranus generally fall in the range from 120 K to 160 K [36]. Some variation with latitude has been observed, in the sense that temperatures are somewhat higher at near-equatorial latitudes. It has also been noted [52] that most profiles show a relative temperature maximum near 8 μbar, the cause of which is not well understood.

Determination of altitude–temperature profiles from infrared and microwave data is not as straightforward. Only data at wavelengths of 5 μm or longer (generally referred to as the thermal spectrum) are useful in this determination, since shorter wavelengths are dominated by reflected sunlight. An absorption or emission line or band in the thermal spectrum of Uranus is not formed at a single altitude within the atmosphere, but is the result of absorption over a range of altitudes. In order to reconstruct an altitude profile, one must construct a mathematical model of the atmosphere with a postulated temperature-versus-pressure profile and chemical composition. The model must also allow for the presence of haze layers or clouds of ice particles. With such a model, it is then possible in principle (though difficult in practice) to predict the amount of light that will escape Uranus at each point in the thermal spectrum.

The problem is simplified somewhat by the realization that at wavelengths longer than about 100 μm, the observed temperatures increase uniformly with wavelength and are representative of greater and greater depths within the atmosphere. For each of these longer wavelengths, one may determine an effective depth (and temperature) that is more nearly independent of small variations in the assumed chemical composition.

Deviations of the observed thermal spectrum from model predictions lead to adjustments of the model. When the fit is considered close enough, the model represents a best estimate by its author of the temperature–pressure profile and composition of the atmosphere. A number of pre-Voyager models are presented by Orton and Appleby [36]. They generally show a temperature minimum of 50 to 55 K near a pressure level of 0.1 bar (about 50 to 100 km above the 1-bar reference level), as shown in Fig. 5.5. Temperatures increase with depth below the 0.1-bar level, reaching 70 to 80 K at 1 bar. The rate of increase with depth below the 1-bar pressure level cannot be determined from Earth-based observations, but the 'lapse rate' (change in temperature per kilometer of depth) is generally assumed to be adiabatic [53].

Temperature-pressure relationships within the Uranian atmosphere deviate from those expected for a purely gaseous atmosphere if haze or cloud layers are present. The pre-Voyager estimates of these effects are discussed below.

5.2.3 Aerosols

Tiny liquid or solid particles suspended in a gas are termed 'aerosols' and can alter the temperature of the surrounding gaseous atmosphere. Smog is an example of an aerosol in Earth's atmosphere familiar to most urbanites. The nature and magnitude of the effects of such aerosol hazes in Uranus's atmosphere depend on how much sunlight they reflect, absorb, and transmit. Reflection of sunlight will reduce the amount of energy available for heating the lower atmosphere. Absorption will also reduce the transmitted sunlight, but additionally cause local heating as the particles transfer their absorbed heat to the surrounding gases. In general, it was expected that the aerosol particle concentrations in the upper atmosphere of Uranus would be small enough that such haze layers would have only a limited effect on the temperature profile.

The aerosols of importance in the atmosphere of Uranus are all chemical byproducts of methane (CH_4), primarily from interaction with ultraviolet light from the Sun ('photolysis') [54]. Ethane (C_2H_6) and acetylene (C_2H_2) both condense into ice particles at temperatures found just above the level of the temperature minimum (the 'tropopause') in Uranus's atmosphere. Methane gas is present in large enough quantities in the stratosphere (the first few hundred kilometers above the tropopause) that ethane and acetylene are produced by photolysis. Some of the ethane and acetylene then migrates slowly downward until the temperatures are low enough and the concentrations are high enough to form detectable ice hazes. Although the Earth-based estimates of the temperature–pressure profile are not well constrained

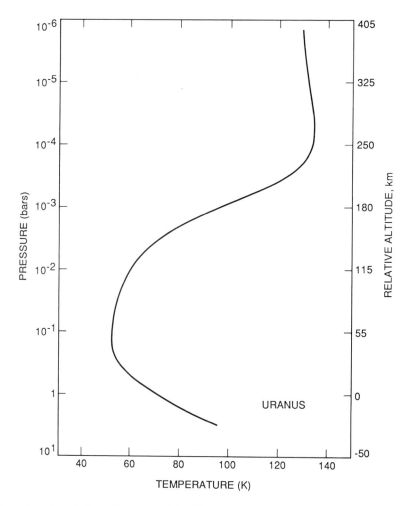

Fig. 5.5 — A typical pre-Voyager model of Uranus's atmospheric temperature at different pressure levels (and relative altitudes).

at these levels in the atmosphere, such condensation was expected to take place in the pressure range from 1 to 100 mbars (50 to 200 km above the 1-bar reference level).

5.2.4   Clouds and cloud motions
Clouds in the atmosphere of Uranus are formed by a process different from that which produces aerosols. Clouds occur at deeper levels within the atmosphere and are the result of condensation of the 'ice' constituents of the interior of the planet, namely methane (CH_4), ammonia (NH_3), water (H_2O), and possibly hydrogen sulfide (H_2S). Buoyant bubbles of warmer gas circulate upward from the interior of the planet, eventually reaching levels where temperatures are cold enough that

condensation of the icy component of the gas occurs. The heavier condensates reverse the upward flow and sink back to levels where the process is reversed and the condensates evaporate. An equilibrium cloud base is formed at the level in the atmosphere where the temperature is near the freezing point for each icy constituent.

In order of increasing depth below the Uranian troposphere, cloud layers of methane, ammonia, and water were expected to exist. The base of the methane cloud was expected to occur near a temperature of 85 K, somewhat below the 1-bar pressure level in the atmosphere. Ammonia would form a cloud base near a temperature of 145 K, at a pressure level near 8 bar. Water ice clouds would occur much deeper in the atmosphere, at pressure levels in excess of 100 bar.

The clouds of Jupiter and Saturn tend to organize themselves into bands and whirling storm-like systems (see Figs 5.6 and 5.7). Because of their higher atmos-

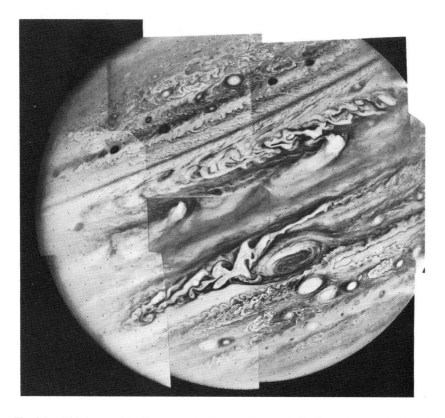

Fig. 5.6 — This image of Jupiter as seen by Voyager 2 dramatically shows the organization of atmospheric features into dark belts and brighter zones. Many individual storm systems and other discrete cloud features can be seen. (P-21147)

pheric temperatures, methane condensation does not occur, and the visible clouds are composed of ammonia ice particles. Earth-based telescopic observers were able to track discrete cloud features in the atmosphere of Jupiter. They noted that clouds

Fig. 5.7 — Voyager 2 showed that Saturn's clouds are also organized into belts and zones, but with far fewer visible storm systems than Jupiter. (P-23887)

at some latitudes moved faster across the disk than at other latitudes. The evidence for zonal winds was clear. Though cloud features were far more difficult to observe in Saturn's atmosphere, the occasional trackable features seemed to tell a similar tale for Saturn: prevailing winds differed markedly with latitude. Some observers reported band-like markings in the atmosphere of Uranus, but none succeeded in tracking discrete features across the disk to get an estimate of their rotation period. Either the clouds were too deep within the atmosphere to be observable from Earth, or the cloud cover was uniform and without much horizontal variation. Uranus's clouds seemed reluctant to provide any usable information on planetary rotation or differential (varying with latitude) wind speeds.

5.3   THE MAGNETOSPHERE OF URANUS

Magnetism and magnetic forces are commonplace in the our present-day lives. Small magnets are used to fasten notes to our refrigerators. Large and small magnets are an integral part of stereophonic speakers, telephones, hair dryers, electric drills, electric motors and generators, solenoid switches, computers, and countless other gadgets. Ultraviolet, visible, and infrared light, by which man discerns the nature of the universe around him, is a combination of electrical and magnetic waves. And almost everyone has at one time or another held a magnetic compass, which points northward because the Earth itself is an enormous spinning magnet. Earth's magnetic field protects us from the high-energy radiation streaming out from the Sun, makes long-distance radio communication possible, shapes the auroras (northern and southern lights seen at high latitudes), and creates the high-altitude belts of charged particles which surround our planet. The effects of the magnetic field on lava deposited on the ocean floors over eons of time is one of the most important pieces of evidence supporting the theory of plate tectonics (motion of continents) for the Earth. With all of its importance, Earth's magnetic field is not well understood by scientists. The Dynamo Theory suggests that our magnetic field is generated by the circulation of electrical currents in the molten iron core of the Earth. But the details of what keeps that dynamo running are sketchy at best.

The Sun and most of the planets of our Solar System also have magnetic fields. Detection and characterization of those magnetic fields from afar is almost impossible. It is usually the side-effects of a magnetic field that are most easily detected. The Sun's magnetic field gives rise to sunspots and solar flares and accelerates a high-energy stream of charged particles, called the solar wind, outward through the Solar System. The interaction of Jupiter's magnetic field with the solar wind particles generates low-frequency radio waves that can be detected from Earth. Occasionally auroral activity similar to that in Earth's aurora borealis might be viewed on another planet. But the most reliable method of detecting and measuring the magnetic field of another planet is to fly an appropriately instrumented spacecraft through that magnetic field.

In 1973, more than a decade before Voyager arrived at Uranus, the Pioneer 10 spacecraft passed through the Jupiter system and made definitive measurements of its magnetic field and associated radiation field. In 1979, Pioneer 11 made similar measurements in Saturn's magnetic field. The Mariner 10 spacecraft discovered in 1974 that Venus has no internal magnetic field and that Mercury has only a very weak magnetic field. A Soviet spacecraft to Mars also detected a weak magnetic field surrounding that planet. It was natural to assume that Uranus would also have a magnetic field.

5.3.1 The solar wind at Uranus's distance from the Sun

The outer boundary of the Sun's magnetic field (the heliosphere) extends well beyond the orbit of Neptune and perhaps well beyond the extended orbit of Pluto. The magnetic fields of the planets exist as magnetic bubbles (magnetospheres) within the larger heliosphere. The shapes and sizes of these planetary magnetospheres are greatly altered by the solar wind, that stream of charged particles (mainly electrons and protons) which flows outward from the Sun at hypersonic speeds. The solar wind is neither constant with time nor with distance from the Sun, but ebbs and flows depending on conditions near the outer boundaries of the Sun. The Sun itself undergoes an 11-year activity cycle, during which the observed daily number of sunspots grows to a maximum and then declines to a minimum. Near the time of minimum sunspot activity, the Sun's magnetic field actually reverses polarity: the north and south magnetic poles have the opposite orientation during any given 11-year activity cycle than in adjacent 11-year periods. All of these characteristics must be taken into account in order to more fully understand the magnetic field and charged particle environments near a planet.

Characteristics of the solar wind can be studied effectively only from interplanetary spacecraft. The Earth itself and most Earth-orbiting spacecraft are entirely within the Earth's magnetosphere, which excludes all but the highest-energy charged particles in the solar wind. Only the Pioneer 10 and 11 spacecraft preceded the two Voyager spacecraft into the outer Solar System beyond Mars. Barnes and Gazis [55] predicted on the basis of Pioneer data what the characteristics of the solar wind at Uranus would be at the time of the Voyager 2 encounter. The solar wind speed was predicted to be about 400 km/s, with short periods where departures of up to 50 km/s might occur. They also made predictions of typical values for the numbers of protons per cubic centimeter (0.025), the proton energy (expressed as a temperature of 10 000 K), and the dynamic pressure exerted by the solar wind on a Uranian magnetosphere (6×10^{-11} dynes/cm^2). Substantial variations from these values were expected. For example, the dynamic pressure, which helps determine the size of the Uranian magnetosphere, could vary by a factor of 5, resulting in 30% variations in the size of the magnetosphere.

Interaction of charged particles in the solar wind with the magnetosphere also results in an anti-sunward 'magnetotail' that behaves in some ways like an airport wind sock. The planetary field is stretched downwind until it eventually is indistinguishable from the solar wind (see Fig. 5.8). The planetary field also rotates with the planet, creating a twisted downstream appearance.

5.3.2 The magnetic Bode's law applied to Uranus
On the basis of magnetic field data obtained prior to 1975, Hill and Michel [56] proposed a 'magnetic Bode's law'. The purely empirical magnetic Bode's law relates the magnetic field strength of a planet to its rotational angular momentum (determined from the mass and spin rate of the planet). In contrast to Johann Bode's law of planetary distances, there exists a weak theoretical basis for such a relationship, even when one considers how little is known about how and why planetary magnetic fields are generated. The strength of Saturn's magnetic field was found to be close to the value predicted (see Fig. 5.9), and it seemed appropriate to use the magnetic Bode's law to predict approximate field strengths for a Uranian magnetic field. The field was predicted by Hill [57] to be that of a magnetic dipole approximately aligned with the rotation axis of Uranus and intermediate in strength between the magnetic fields of Saturn and Earth.

5.3.3 Excess ultraviolet radiation from Uranus
Data obtained with the Earth-orbiting International Ultraviolet Explorer (IUE) seemed to lend credence to Hill's prediction. Three different observing teams [58,59,60] reported seeing excess ultraviolet light from Uranus at two different wavelengths associated with hydrogen. These results were interpreted as an indication that Uranus might have active auroral regions. If such regions existed, they would constitute evidence that the planet possesses the magnetic field necessary to

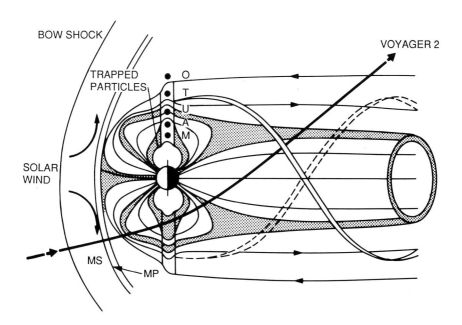

Fig 5.8 — A schematic representation of the pre-Voyager view of the Uranian magnetic field. Interaction with the solar wind reduces the sunward extent of the Uranian field and also results in an anti-sunward magnetotail.

generate such auroras. Since the south pole of Uranus was very nearly pointed at Earth, a polar-aligned magnetic field would produce an aurora which would be ideally situated for IUE observations. Cheng [61] suggested that the IUE data could mean that Uranus's magnetic field is very similar to that of Saturn, and that its interaction with the rings and satellites of Uranus could be responsible for darkening the surfaces of those bodies.

NOTES AND REFERENCES

[1] Apsidal precession refers to a slow turning of the orientation of an elliptical orbit. The 'line of apsides' is a line connecting the periapsis (closest) point of the orbit to the apoapsis (most distant) point. For the gas giant planets, the sense of this precession is prograde, or in the same general direction as the orbital motion of the ring particles or satellites.

[2] Nodal regression refers to an apparent retrograde motion of the points where an inclined orbit crosses the planet's equatorial plane. The 'line of nodes' is a line through the center of the planet which connects the two nodes, or crossing points. The sense of motion of the line of nodes is retrograde, or in a general direction opposite that of the orbital motion of ring particles or satellites. Its rate is measured in angular distance per unit time (e.g., degrees per day) in the planet's equatorial plane.

[3] MacFarlane, J. J., Hubbard, W. B. (1981) Internal structure of Uranus. In

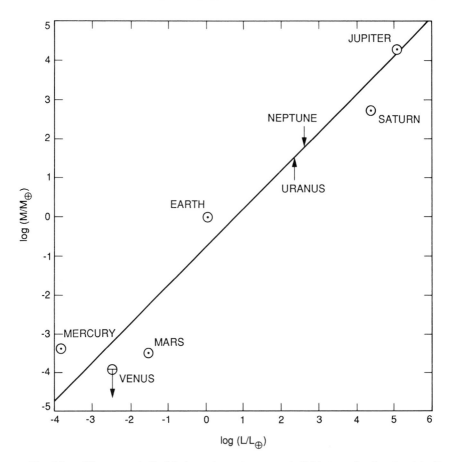

Fig. 5.9 — The magnetic Bode's law relates the magnetic field strength of a planet to its rotational angular momentum. Saturn's magnetic field was found by Pioneer 11 to be close to the value, and the magnetic Bode's law was used to predict approximate field strengths Voyager 2 might encounter at Uranus.

Hunt, G. (ed.) *Uranus and the outer planets*, Cambridge University Press, pp. 111–124.

[4] Hubbard, W. B. (1984) Interior structure of Uranus. In Bergstralh, J. (ed.) *Uranus and Neptune*, NASA Conference Publication 2330, National Aeronautics and Space Administration, Scientific and Technical Information Branch, pp. 291–325.

[5] Anders, E., Ebihara, M. (1982) Solar-system abundances of the elements. *Geochimica et Cosmochimica Acta*, **46**, 2363–2380.

[6] Lewis, J. S., Prinn, R. G. (1980) Kinetic inhibition of CO and N_2 reduction in the solar nebula. *Astrophysical Journal*, **238**, 357–364.

[7] Wallace, L. (1980) The structure of the Uranus atmosphere. *Icarus*, **43**, 231–259.

[8] Gulkis, S., Jafssen, M. A., Olsen, E. T. (1978) Evidence for depletion of ammonia in the Uranus atmosphere. *Icarus*, **34**, 10–19.

[9] Macy, W. Jr., Smith, W. H. (1978) Detection of HD on Saturn and Uranus and the D/H ratio. *Astrophysical Journal*, **222**, L73–L75.

[10] Trafton, L. M., Ramsay, D. A. (1980) The D/H ratio in the atmosphere of Uranus: detection of the $R_5(1)$ line of HD. *Icarus*, **41**, 423–429.

[11] Orton, G. S. (1986) Thermal spectrum of Uranus: implications for large helium abundance. *Science*, **231**, 836–840.

[12] Zharkov, V. N., Trubitsyn, V. P. (1978) *Physics of Planetary Interiors*, translated and edited by W. B. Hubbard, Pachart Publishing House, Tucson.

[13] Hubbard, W. B., MacFarlane, J. J. (1980) Structure and evolution of Uranus and Neptune. *Journal of Geophysical Research,* **85**, 225–234.

[14] Deslandres, H. (1902) *Comptes Rendus hebdomaires des Séances de l'Académie des Sciences,* **135**, 472.

[15] The Doppler effect was named for Christian J. Doppler, who showed that motion of a light-, radio-, or sound-wave source toward (or away from) an observer will cause the observed waves to be shifted to higher (or lower) frequencies. He also showed that the magnitude of the shift was directly proportional to the speed of the source relative to the observer. The Doppler effect is responsible for the apparent reduction in pitch in the whistle of a rapidly-moving train as it passes by a stationary listener standing at the railroad crossing.

[16] Lowell, P., Slipher, V. M. (1912) Spectroscopic discovery of the rotation period of Uranus. *Lowell Observatory Bulletin*, **2**, 17–19.

[17] Moore, J. H., Menzel, D. H. (1930) The rotation of Uranus. *Publications of the Astronomical Society of the Pacific*, **42**, 330–335.

[18] Campbell, L. (1936) The rotation of Uranus. *Harvard College Observatory Bulletin*, **904**, 32–35.

[19] Pickering, E. C. (1917) Variability of Uranus. *Harvard College Observatory Circular,* **200**.

[20] Münch, G., Hippelein, H. (1980) The effects of seeing on the reflected spectrum of Uranus and Neptune. *Astronomy and Astrophysics*, **81**, 189–197 (15-h Uranus rotation period, with uncertainties of +4 h, −2.6 h).

[21] Brown, R. A., Goody, R. M. (1980) The rotation of Uranus II. *Astrophysical Journal*, **235**, 1066–1070 (16.2-h Uranus rotation period with an uncertainty of ±0.3 h).

[22] Belton, M. J. S., Wallace, L., Hayes, S. H., Price, M. J. (1980) Neptune's rotation period: a correction and a speculation on the difference between photometric and spectrographic results. *Icarus*, **42**, 71–88.

[23] Smith, H. J., Slavsky, D. B. (1979) Rotation period of Uranus. *Bulletin of the American Astronomical Society*, **11**, 568 (abstract only). Slavsky, D. B., Smith, H. J. (1980) The rotational period of Uranus. *ibid.*, **12**, 704 (abstract only). Slavsky, D., Smith, H. J. (1981) Further evidence for rotation periods of Uranus and Neptune. *ibid.*, **13**, 733 (abstract only).

[24] Hayes, S. H., Belton, M. J. S. (1977) The rotational periods of Uranus and Neptune. *Icarus*, **32**, 383–401.

[25] Brouwer, D., Clemence, G. M. (1961) Orbits and masses of planets and satellites. In: Kuiper, G. P., Middlehurst, B. M. (eds.) *The Solar System*, **3**, University of Chicago Press, pp. 31–94.

[26] French, R. G. (1984) Oblateness of Uranus and Neptune. In Bergstralh, J. T. (ed.) *Uranus and Neptune*, NASA Conference Publication 2330, National Aeronautics and Space Administration, Scientific and Technical Information Branch, pp. 349–355.

[27] Danielson, R. E., Tomasko, M. G., Savage, B. D. (1972) High resolution imagery of Uranus obtained by Stratoscope II. *Astrophysical Journal*, **178**, 887–900.

[28] Franklin, F. A., Avis, C. C., Columbo, G., Shapiro, I. I. (1980) The geometric oblateness of Uranus. *Astrophysical Journal*, **236**, 1031–1034.

[29] An isophote is similar to a contour line in a topographic map: it traces the lines of constant brightness in an image.

[30] Elliot, J. L., French, R. G., Frogel, J. A., Elias, J. H., Mink, D. M., Liller, W. (1981) Orbits of nine uranian rings. *Astronomical Journal*, **86**, 444–455.

[31] Younkin, R. L. (1970) Spectrophotometry of the Moon, Mars, and Uranus. Doctoral Dissertation, University of California, Los Angeles.

[32] Lockwood, G. W., Lutz, B. L., Thompson, D. T., Warnock, A. (1983) The albedo of Uranus. *Astrophysical Journal*, **266**, 402–414.

[33] Harris, D. L. (1961) Photometry and colorimetry of planets and satellites. In Kuiper, G. P., Middlehurst, B. M. (eds.) *Planets and Satellites* (Volume III of a series on the Solar System), The University of Chicago Press, Chicago, pp. 272–342.

[34] Tomasko, M. G., West, R. A., Castillo, N. D. (1978) Photometry and polarimetry of Jupiter at large phase angles. *Icarus*, **33**, 558–592.

[35] Wallace, L. (1980) The structure of the Uranus atmosphere. *Icarus*, **43**, 538–544.

[36] Orton, G. S., Appleby, J. F. (1984) Temperature structures and infrared-derived properties of the atmospheres of Uranus and Neptune. In Bergstralh, J. T. (ed.) *Uranus and Neptune*, NASA Conference Publication 2330, National Aeronautics and Space Administration, Scientific and Technical Information Branch, pp. 89–155.

[37] A blackbody is an ideal body or surface that perfectly absorbs all radiant energy falling upon it, reflecting none. Such a body emits radiation whose energy is proportional to the fourth power of its absolute temperature. The constant of proportionality is called the Stefan–Boltzmann constant and is equal to 5.66956×10^{-5} erg cm^{-2} K^{-4} s^{-1}, where the temperature is specified in kelvins. A temperature of 0 K is equivalent to $-273.16°$ on the Celsius (centigrade) scale, and corresponds to absolute zero temperature. A kelvin is the same as a degree on the Celsius scale. A change of 1.0°C or 1 K is identical to a change of 1.8° in the Fahrenheit scale.

[38] A spectroscope splits the light from Uranus (or other source) into its component colors. Astronomical 'spectroscopists' use the brightness of Uranus (or other source) at different points in its 'spectrum' to deduce the relative abundances of the chemical constituents through which the light has passed. Ultraviolet, infrared, and radiowave spectra also provide information of a similar nature, including approximate temperatures at the levels within the atmosphere where the amount of light is being enhanced or diminished.

[39] Secchi, A. (1869) *Comptes Rendus hebdomaires des Sé ances de l'Académie des Sciences*, **68**, 761.

[40] Keeler, J. E. (1889) *Astronomische Nachrichten*, **122**, No. 2927, Col. 403.

[41] Wildt, R. (1932) Methane in the atmospheres of the great planets. *Naturwissenschaft*, **20**, 851.

[42] Dennison, D. M., Ingram, S. B. (1931) A new band in the absorption spectrum of methane gas. *Physical Review*, **36**, 1451–1459.

[43] Adel, A., Slipher, V. M. (1934) The identification of the methane bands in the solar spectra of the major planets. *Physical Review*, **46**, 240–241; (1934) The constitution of the atmospheres of the giant planets. *Physical Review*, **46**, 902–906.

[44] Russell, H. N. (1935) Atmospheres of the planets. *Science*, **81**, 1–9.

[45] Kuiper, G. P., Phillips, J. (1949) New absorptions in the Uranus atmosphere. *Astrophysical Journal*, **109**, 540–541.

[46] Herzberg, G. (1952) Spectroscopic evidence of molecular hydrogen in the atmospheres of Uranus and Neptune. *Astrophysical Journal*, **115**, 337–340.

[47] Courtin, R., Gautier, D., Lacombe, A. (1978) On the thermal structure of Uranus from infrared measurements. *Astronomy and Astrophysics*, **61**, 97–101.

[48] Gulkis, S., Janssen, M. A., Olsen, E. T. (1978) Evidence for the depletion of ammonia in the Uranus atmosphere. *Icarus*, **34**, 10–19.

[49] Hunten, D. M. (1984) Atmospheres of Uranus and Neptune. In Bergstralh, J. T. (ed.) *Uranus and Neptune*, NASA Conference Publication 2330, National Aeronautics and Space Administration, Scientific and Technical Information Branch, pp. 27–54.

[50] Atreya, S. K. (1984) Aeronomy. In Bergstralh, J. T. (ed.) *Uranus and Neptune*, NASA Conference Publication 2330, National Aeronautics and Space Administration, Scientific and Technical Information Branch, pp. 55–88.

[51] The molecular weight of a gas is roughly the sum of the numbers of protons and neutrons in the nuclei of its constituent atoms. A molecule of hydrogen gas (H_2), with one proton in the nucleus of each of its two hydrogen atoms, has a molecular weight of 2. Helium, with two protons and two neutrons in each atom, has a molecular weight of 4. A gas mixture of 90% hydrogen and 10% helium would then have a mean (average) molecular weight of $(2 \times 90\%) + (4 \times 10\%) = 2.2$. For a gas with 80% hydrogen and 20% helium, the molecular weight would be 2.4.

[52] Sicardy, B., Combes, M., Lecacheux, J., Bouchet, P., Brahic, A., Laques, P., Perrier, C., Vapillon, L., Zeau, Y. (1985) Variations of the stratospheric temperature profile along the limb of Uranus: results of the 22 April 1982 stellar occultation. *Icarus*, **64**, 88–106.

[53] If a volume of gas confined in an insulated container (one which allows no heat flow through its walls) is compressed, the absolute temperature of the gas will rise by an amount that is determined by the changes in the volume and the pressure of the gas. Furthermore, for a gas composed of 90% H_2 and 10% He, the increase in pressure is related to the decrease in volume by the relationship, $pV^{1.44} = $constant. For an ideal gas $p_i V_i T_f = p_f V_f T_i$, where i and f, respectively, stand for the initial and final values of the pressure (p), volume (V), and

absolute temperature (T). An adiabatic lapse rate is one for which the temperature of the gas increases with pressure by precisely the same amount as it would if it were confined in such an insulated container. Mathematical combination of the two above relationships yields $T_f = T_i (p_f/p_i)^{0.30}$. Thus, for an adiabatic lapse rate in a gas which is 90% H_2 and 10% He, the temperature approximately doubles for each decade increase in pressure.

[54] Atreya, S. K., Ponthreu, J. J. (1983) Photolysis of methane and the ionosphere of Uranus. *Planetary and Space Science*, **31**, 939–944.

[55] Barnes, A., Gazis, P. R. (1984) The solar wind at 20–30 AU. In Bergstralh, J. T. (ed.) *Uranus and Neptune*, NASA Conference Publication 2330, National Aeronautics and Space Administration, Scientific and Technical Information Branch, pp. 527–540.

[56] Hill, T. W., Michel, F. C. (1975) Planetary magnetospheres. *Reviews of Geophysics and Space Physics*, **13**, 967.

[57] Hill, T. W. (1984) Magnetospheric structures: Uranus and Neptune. In Bergstralh, J. T. (ed.) *Uranus and Neptune*, NASA Conference Publication 2330, National Aeronautics and Space Administration, Scientific and Technical Information Branch, pp. 497–525.

[58] Durrance, S. T., Moos, H. W. (1982) *Nature*, **299**, 428.

[59] Clarke, J. T. (1982) *Astrophysical Journal*, **299**, L105.

[60] Caldwell, J., Wegener, R., Owen, T., Combes, M., Encrenaz, T. (1983) Tentative confirmation of an aurora on Uranus. *Nature*, **303**, 310.

[61] Cheng, A. F. (1984) Magnetosphere, rings, and moons of Uranus. In Bergstralh, J. T., (ed.) *Uranus and Neptune*, NASA Conference Publication 2330, National Aeronautics and Space Administration, Scientific and Technical Information Branch, pp. 541–556.

BIBLIOGRAPHY

Alexander, A. F. O'D. (1965) *The Planet Uranus*, American Elsevier Publishing Company, Inc., New York, pp. 151–161, 218–224, 259–271, 278–290.

Beatty, J. K., O'Leary, B., Chaikin, A. (1984) *The New Solar System*, 2nd Edition, Sky Publishing Corporation, Cambridge, Massachusetts, pp. 23–32, 169–172.

Bergstralh, J. T. (1984) *Uranus and Neptune*, NASA Conference Publication 2330, National Aeronautics and Space Administration, Scientific and Technical Information Branch, Washington, D. C., pp. 27–278, 497–572.

Hunt, G. (ed.) (1982) *Uranus and the Outer Planets*, Cambridge University Press, pp. 125–191.

6

The saga of Voyager 2

6.1 CALIFORNIA INSTITUTE OF TECHNOLOGY'S JET PROPULSION LABORATORY

The extraordinary successes of the unmanned Voyager missions to the outer planets cannot be attributed to fortunate happenstance. The two hardy robot explorers were (and continue to be) the outward expression of the human craving to 'see what there is to see and know what there is to know'. It was that same desire which drove ancient astronomers to tirelessly chart the apparent motions of the Sun, the Moon, and the planets among the background stars. It led Galileo Galilei to point his crude telescope toward the heavens. Sir William Herschel would never have discovered Uranus without the insatiable drive to know and understand.

The advent of the Space Age brought with it the possibility of utilizing exploration tools not previously available. Astronomers, frustrated with the observational difficulties associated with Earth-based observations of other Solar System objects, began to realize the potential of a new type of remote sensing. Up to then, astronomical remote sensing was restricted to situations in which both the observer and his instrumentation were distant from the object being observed. Now for the first time in history an observer, while himself remaining remote from the target of his observations, could send his instrumentation to the immediate vicinity of those remote worlds.

Early missions were designed for our closest planetary neighbors: the Moon, Venus, and Mars. The goal has always been to stretch space exploration to the limits of its capability. The only limits seemed to be (1) the available monetary resources that could be put to use in such endeavors, and (2) the extended lengths of time needed for planning, for technology development, for construction and testing of the spacecraft, and for spacecraft travel to the planets. Finding a sufficient number of individuals with the necessary interest, knowledge, patience, and ingenuity has never been a problem.

The Jet Propulsion Laboratory (JPL) in Pasadena, California, has roots which go back to 1936 to basic research in rocketry and jet propulsion at the Guggenheim Aeronautical Laboratory of the California Institute of Technology (GALCIT). GALCIT was directed by Caltech professor Theodore von Karman, Hungarian-born pioneer of aeronautical engineering and aerodynamics. Frank J. Malina and a group of interested GALCIT students conducted the first test firings of liquid rocket engines in 1938 near the present site of JPL. The success of their research brought financial rewards in the form of a National Academy of Sciences contract to develop rockets for jet-assisted takeoff (JATO) for Army Air Corps planes.

From 1941 until 1958 JPL operated under the jurisdiction of the United States Army, developing rockets with ever greater thrust capability and reliability. It was also during this time that JPL scientists and engineers developed the capability to control the rocket instrumentation through radio commands and to receive data back by means of a radiowave-borne telemetry stream. Following the successful launches of the Soviet Union's Sputnik 1 and 2, JPL and the Army Ballistic Missile Agency were authorized to proceed to build and launch an Earth-orbiting satellite. Just 83 days later, on January 31, 1958, Explorer 1 was launched into orbit, carrying with it equipment to measure radiation levels in space. The equipment was designed by James A. Van Allen, chairman of the Physics Department at the University of Iowa. It was Explorer 1 which discovered the famous Van Allen radiation belts which encircle the Earth. Just 54 days later Explorer 3 was launched (Explorer 2 suffered a launch failure). A successful launch of Explorer 4 followed on July 26, 1958.

Tracking of these Earth-orbiting spacecraft was becoming a problem, so a remote site, away from manmade radio interference, was found at Camp Irwin, 80 km north of Barstow, California. In a relatively short time, JPL engineers constructed a steerable parabolic tracking antenna with a diameter of 26 m. The site, which now has three larger antennas used for tracking interplanetary spacecraft like Voyager, has come to be known as the Goldstone Deep Space Communications Complex (GDSCC). Similar tracking antennas have been built near Canberra, Australia (CDSCC), and near Madrid, Spain (MDSCC).

In late 1958 the United States government created its civilian space agency, now known as the National Aeronautics and Space Administration (NASA). Jurisdiction over the work being done by JPL was transferred from the Army to NASA shortly thereafter. JPL has been intimately involved with the unmanned spacecraft exploration of the Solar System since that time.

Pioneer 3 was launched on December 6, 1958. It was intended to escape Earth's gravity and orbit the Sun. Instead, its booster rocket shut off early, and 38 h after launch it plunged back into Earth's atmosphere. During its short flight, however, it discovered a second radiation belt higher than the one discovered by Explorer 1. Pioneer 4 was successfully launched into solar orbit three months later. Explorers 7 and 8 were launched into Earth orbit in 1959 and 1960, respectively.

Techniques used to communicate with spacecraft also aided telephone communications on Earth. In 1960 and 1961, the first two-way transcontinental telephone call to use the Moon as a radio reflector connected JPL to Bell Laboratories in New Jersey. A similar call used Earth-orbiting Echo 1 as the passive reflector, and the Moon was used to connect JPL to Woomera, Australia. In 1961 and 1962, the Goldstone 26-m tracking station was used to bounce radar off the surface of Venus. These experiments provided a double bonus: a vastly improved value for the length of the astronomical unit (Earth's mean distance from the Sun), and the first measure of Venus's rotation period (244.3 days, retrograde!). In 1962 and 1963, successful radar bounces off the surfaces of Mars and Mercury were achieved.

Rangers 1, 3, 4, and 5 began the Moon probe series in 1961 and 1962. All four were 'successfully' launched, although none achieved their objectives, either because they did not achieve the proper trajectory, or because of failure of the data collection electronics. An electronics failure also prevented Ranger 6 in 1964 from getting pictures of the Moon from close range, but Rangers 7, 8, and 9 successfully achieved that goal in 1964 and 1965. Many of the Ranger 9 pictures were seen 'live' on commercial television.

In late 1962, Mariner 2 became the first spacecraft to successfully encounter another planet. It revealed an extremely hot surface beneath the thick clouds of Venus. Mariner 3 was launched toward Mars in 1964, but failed to achieve a Mars trajectory and went into orbit around the Sun. Mariner 4 was successfully launched toward Mars, sending back pictures of that planet in 1965. Mariner 5 carried a more extensive instrument complement past Venus in 1967.

A series of Surveyor spacecraft were designed to land on the Moon's surface during 1966, 1967, and 1968. Surveyors 1, 3, 5, 6, and 7 successfully landed and sent back pictures and other data about the Moon's soil. Surveyor 2 and 4 survived their launches, but Surveyor 2 began tumbling uncontrollably and crashed into the Moon's surface. Surveyor 4 failed during final descent to the Moon's surface. Successful launches and missions of Lunar Orbiters 1 through 5 (managed by NASA's Langley Research Center in Hampton, Virginia) provided high-resolution mapping of the surface of the Moon in 1966 and 1967.

6.2 A 'GRAND TOUR' OF THE OUTER PLANETS PROPOSED

While tracking these various interplanetary spacecraft, navigation specialists noted that the velocity of a spacecraft (in Sun-centered coordinates) would change as it passed a planet. If the spacecraft passed a planet over its 'orbitally receding' side, it would actually gain energy from the planet and its velocity would increase. This concept has come to be known as 'gravity assist'. In 1966, Gary Flandro, a JPL and Caltech scientist, showed how gravity assist could be used to send a single spacecraft to several of the outer planets in succession, substantially reducing flight time to the more distant planets. 'Of particular interest,' he wrote, 'is the 1978 Earth–Jupiter –Saturn–Uranus–Neptune 'grand tour' opportunity which would make possible close-up observation of all planets of the outer solar system (with the exception of Pluto) in a single flight [1]. The idea was an exciting one, and JPL advanced mission planners began almost immediately to study such a mission in detail.

Scientific enthusiasm for such a mission was apparent. Two separate studies funded by the National Academy of Sciences strongly supported such a mission. One study, chaired by the same James Van Allen who designed Explorer 1's scientific instrumentation, set out the specific scientific goals for such a mission. The report also noted that 'professional resources for full utilization of the outer-solar-system mission opportunities in the 1970s and 1980s are amply available within the scientific community, and there is a widespread eagerness to participate in such missions.' The second study, headed by F. S. Johnson, was equally enthusiastic:

An extensive study of the outer solar system is recognized by us to be one of the major objectives of space science in this decade [1970s]. This endeavor is made particularly exciting by the rare opportunity to explore several planets and satellites in one mission using long-lived spacecraft and existing propulsion systems. We recommend that . . . spacecraft be developed and used in Grand Tour missions for the exploration of the outer planets in a series of four launches in the late 1970s.

JPL proposed such a mission to NASA in 1970. The mission was to include two spacecraft to be launched in 1976 and 1977 to Jupiter, Saturn, and Pluto, followed by two spacecraft launches to Jupiter, Uranus, and Neptune in 1979. NASA's 'Grand Tour' plans were received enthusiastically by the United States Congress and funding was recommended, but due to fiscal limitations several other NASA proposals were less favorably received. Faced with a difficult decision, NASA asked JPL to propose a version of the Grand Tour that might meet many of the same objectives but at a lesser total cost.

Planetary exploration successes in the inner Solar System continued. Mariner 6 and 7 had successful flyby reconnaissance missions of Mars in mid-1969, providing more extensive and higher-resolution coverage of the southern hemisphere and south polar cap of Mars than had Mariner 4. Mariner 8 suffered a launch failure, but its twin, Mariner 9, was successfully inserted into orbit around Mars in late 1971 and completed extensive mapping of the entire surfaces of Mars and its two moons over the next year. The gravity-assist concept was used in 1973 and 1974 to send Mariner 10 first to Venus, followed by three successive passes of Mercury, sending back pictures of these two planets. The ambitious Viking missions succeeded in placing two sophisticated spacecraft into orbit around Mars in 1976, each of which deployed successful landers to the Martian surface. The Viking mission was initially managed by NASA's Langley Research Center; management of the program was transferred to JPL after completion of the Viking primary mission objectives.

Ames Research Center at Moffett Field, California, managed the Pioneer program for its three successful planetary missions. Pioneer 10 was launched toward the outer Solar System in early 1972; Pioneer 11 followed a year later. These two spin-stabilized trailblazers for the later Voyagers showed that spacecraft could safely pass through the asteroid belt between Mars and Jupiter. They also did extensive studies of the radiation environments at Jupiter and (for Pioneer 11 only) at Saturn and sent back pictures and infrared data on these two planets that were of higher quality than any obtained from Earth. Pioneer 12 was launched toward Venus in mid-1978; Pioneer 13 was launched about $2\frac{1}{2}$ months later. Pioneer 12 was inserted into orbit around Venus on December 4, 1978, and used radar to map a surface otherwise perpetually hidden by clouds. Pioneer 13 dropped one large and three small instrumented probes into the Venus atmosphere.

The Soviet Union has also conducted a large number of unmanned space missions to Venus, Mars, and the Moon. Of particular import are the Soviet studies of Venus by means of flyby and orbital missions, instrumented soft landers, and atmospheric balloon-borne experiments. In 1986, the Soviet Union, Japan, and the European

Space Agency conducted definitive, though brief (due to the rapid flyby speeds), studies of Halley's Comet.

6.3 SCALING DOWN TO A VOYAGER MISSION TO FIT NASA'S BUDGET

In response to requests from NASA Headquarters for a less expensive version of the Grand Tour, JPL proposed a two-spacecraft Mariner Jupiter–Saturn (MJS) mission. The spacecraft were to be designed specifically for scientific investigation of the two planets only, with no additional instrumentation or other requirements for Uranus, Neptune, or Pluto observations. The spacecraft components were to be reliable enough to provide a high probability of successful spacecraft operation for four years, the period from launch through the Saturn encounters. The cost for such a mission was estimated to be about $250 million, only one-third of the estimated cost of the earlier, more ambitious Grand Tour proposal. NASA and the United States Congress approved funding for MJS, and the project was officially started July 1, 1972. Harris M. Schurmeier of JPL was appointed Project Manager for the new project; Caltech physicist Edward C. Stone was appointed as Project Scientist. From 31 scientific investigation proposals submitted to NASA, nine were selected with their associated Principal Investigators (PIs) and Co-Investigators (Co-Is). Each of these nine teams of scientists would provide the scientific instrumentation needed for their investigations, which would then be integrated into the overall spacecraft design. An additional 19 scientists were selected from 46 individual proposers to participate in imaging or radio science investigations for which the instrumentation was to be provided by NASA. A Team Leader (TL) was chosen for each of these two investigations. Stone and the 11 PIs and TLs constituted the Science Steering Group (SSG). The first meeting of the SSG was in December 1972 at JPL.

The initial selection of instruments included particulate material detectors. On December 3, 1973, just 17 months after the MJS project start date, Pioneer 10 encountered Jupiter. It had detected 67 hits on its micrometeoroid detector during the flight, ten of these hits near Jupiter [2]. But the rate of such impacts showed no increase during passage through the asteroid belt between Mars and Jupiter. By mid-1974, Pioneer 11 had also passed through the asteroid belt without detecting any increased numbers of particles. Pioneer trajectory specialists had also discovered that Pioneer 11 could utilize gravity assist at Jupiter to direct the spacecraft to the opposite side of the Sun where it could make a pre-Voyager encounter of Saturn (see Fig. 6.1). Since instrument selection for MJS had been made at least partially contingent upon the scientific findings of Pioneers 10 and 11, a decision was made to drop the particulate material detector investigation.

Pioneers 10 and 11 also served to emphasize the severity of the Jupiter radiation environment. By early 1975, a decision had been made to 'harden' the spacecraft and science instruments to make them less vulnerable to severe radiation effects. NASA also agreed to authorize inclusion of a plasma wave investigation that would be diagnostic of the radiation environment.

The MJS spacecraft design was developing into something unlike any prior Mariner spacecraft because of the longer flight times, the greater distances from the Sun, and the harsher environments it would have to survive to complete its mission.

PIONEER SATURN VOYAGE

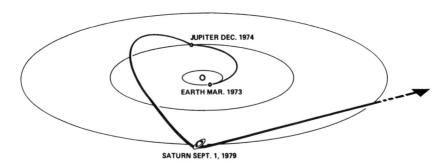

Fig. 6.1 — Trajectory of the Pioneer 11 spacecraft from Earth to Saturn.

It was decided to give recognition of that fact by calling the spacecraft Voyager. The name became official, and the older MJS designation was dropped by mid-1977. It is interesting to note that an early version of the Viking Mission to Mars was known as Voyager, but that designation was not used after 1974.

Voyager management, with the assistance of its Mission Analysis and Engineering team, provided a rich array of potential trajectories for consideration by the SSG. Close flybys of Jupiter's large moons, Io, Europa, Ganymede, and Callisto, and of Saturn's largest moon, Titan, were considered extremely important. Another vital consideration at Saturn was provision for a flight path that would permit the spacecraft as viewed from Earth to traverse the full radial extent of the rings: such a trajectory would allow the Radio Science team to obtain a valuable occultation scan of the rings. Initially, it seemed that the Titan and ring requirements were mutually exclusive. It was then discovered that a single spacecraft, with a properly timed arrival, could perform both experiments, leaving open the geometry of the flyby of the second spacecraft. The logical choice seemed to be a Saturn flyby trajectory which allowed a continuation on to Uranus, once the successful completion of the Saturn objectives of Voyager 1 was assured.

All but one of the science instruments could be used without revision for observations of Uranus. The sole exception was the Infrared Interferometer Spectrometry and Radiometer (IRIS). IRIS could adequately measure the thermal radiation from Uranus, but would have poor sensitivity at near infrared wavelengths, due to the much lower intensities of reflected sunlight at the distance of Uranus. In July of 1976, NASA gave approval for building a more sensitive Modified IRIS (MIRIS), with the condition that completion of its construction and testing not delay the scheduled Voyager launch dates. In the same letter, approval for consideration of a continuation of the second Voyager spacecraft on to Uranus was granted. The actual choice of the Voyager 2 Saturn flyby trajectory would remain dependent on successful Voyager 1 Saturn and Titan encounters.

The science objectives of the basic mission, outlined by the SSG in December 1972, are:

(a) to conduct comparative studies of the Jupiter and Saturn systems: (1) the environment, atmosphere, surface, and body characteristics of the planet; (2) one or more of their satellites; (3) the nature of the rings of Saturn.
(b) to perform investigations in the interplanetary and interstellar media.

With the conditional approval for a Uranus mission, these scientific objectives for an extended mission were added:

(c) to extend interplanetary and interstellar media investigations well beyond the orbit of Saturn;
(d) to extend these investigations to the Uranus systems if conditions should permit implementation of the Uranus targeting option.

The following types of information would be investigated at each of the bodies: (1) physical properties, dynamics, and compositions of atmospheres; (2) surface features; (3) thermal regimes and energy balances; (4) charged particles and electromagnetic environments; (5) periods of rotation, radii, figures, and other body properties; and (6) gravitational fields.

The Voyager Project Plan, dated April 1, 1977, further states [3]:

> The spacecraft will continue to escape from the solar system toward the solar apex, and communications could be maintained as long as the spacecraft continues to function. If the spacecraft continues to function past Saturn encounter, an extended mission could be conducted in anticipation of penetrating the boundary between the solar wind and the interstellar medium, allowing measurements to be made of interstellar fields and particles unmodulated by the solar plasma. Also, the second spacecraft to arrive at Saturn could be targeted for a gravity-assist aiming point which would permit an encounter with Uranus in January of 1986. Objectives at Uranus would be similar to those at Jupiter and Saturn, including its physical properties, atmosphere and methane bands, satellites, and surrounding environment.

6.4 THE LAUNCHES OF VOYAGERS 1 AND 2

The two spacecraft were scheduled to be launched from the same launch pad at Cape Canaveral, Florida. The launch window opened in late August and continued through mid-September of 1977. Voyager 2 would follow a slower trajectory so that passage outside the visible rings of Saturn would still permit sufficient gravitational bending of the flight path to direct the spacecraft on to Uranus. Voyager management also wanted to avoid passing Voyager 2 as deeply through the intense Jupiter radiation environment as Voyager 1, and the slower trajectory accommodated that desire.

Three spacecraft (including a spare) took the long and arduous trip via specially equipped moving vans from JPL in Pasadena, California, to Cape Canaveral, Florida, arriving in June of 1977. At the Cape, the spacecraft underwent extensive testing, including final integration with the Titan IIIE/Centaur launch vehicle (Fig. 6.2).

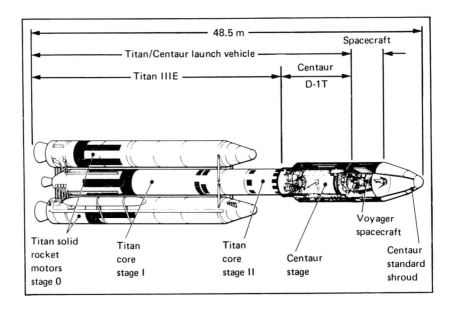

Fig. 6.2 — Schematic of the Titan IIIE/Centaur launch vehicle.

It was during the last few weeks before launch that a late addition was made to each spacecraft. Imaging Team member Carl Sagan had been requested in December of 1976 by Project Manager John R. Casani to organize an effort to place a message on each of the two spacecraft. The message was intended for intelligent beings from other worlds outside our Solar System who might someday find one of the Voyager spacecraft. Sagan accepted the challenge, selected five team members, and with help from a large number of other individuals, his team prepared a two-sided copper record. The record contains 118 digitized images of typical scenes on Earth, greetings in 54 languages, a variety of music, and recordings of some of the sounds of Earth. The history and contents of the record are described in detail in the book, *Murmurs of Earth* [4]. Two identical records were made and encased in a polished aluminum housing, gold in color, with instructions on the use of the records etched on the housing. One encased record with accompanying stylus, cartridge, and mounting bracket was affixed to the exterior of each of the two Voyager spacecraft (Fig. 6.3). It is anticipated that these sights and sounds of Earth will remain decodable for millions, perhaps even billions of years into the future.

While the spacecraft were being prepared for launch, the infrared investigations team and the Texas Instruments contractors were struggling to complete fabrication and testing of two MIRIS instruments in time to include them in place of the IRIS instruments already on the spacecraft. The launch window for Voyager 2 opened on

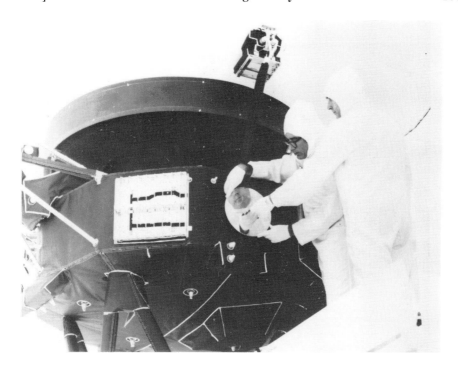

Fig. 6.3 — Voyager record being mounted on the spacecraft. (P-19411A)

August 20, 1977. The MIRIS instruments were completed, but testing disclosed some problems that had to be corrected before they could be shipped to the Cape. Since Voyager 2 was the spacecraft destined to have the opportunity to go on to Uranus, it was especially important that a MIRIS be included on it. Nothing occurred to delay the launch, and Voyager 2 began its journey to the stars on that first day of the launch window, sans MIRIS. The launch was perfect (Fig. 6.4), though later analysis of the engineering data showed that the Titan rocket had provided Voyager 2 a much rougher ride than anticipated. There was still hope that MIRIS might be completed in time for the launch of Voyager 1. The launching pad was refurbished, and Voyager 1 was mounted in the nose cone of the Centaur booster. On September 5, 1977, just 16 days after the launch of its twin, Voyager 1 lifted off atop the giant Titan/Centaur rocket. The launch was picture perfect and much smoother than had been the Voyager 2 launch. In spite of a herculean effort, MIRIS had missed both of its appointments with destiny.

6.5 PROBLEMS ALONG THE WAY

Dear Bride and Groom: Now that you are married I can promise you that your troubles are at an end. *Which* end they are at you will very soon find out!

The launch environment is generally the most hazardous part of a spacecraft's

Fig. 6.4 — Voyager 2 launch from Cape Canaveral on August 20, 1977. (373-7511A)

journey. More spacecraft are lost in the first five minutes of flight than in any individual one-month period thereafter. For a mission with the complexity and longevity of Voyager, however, there is a high probability that some electronic or mechanical components will fail sometime during the postlaunch period. It is a tribute to Voyager designers and engineers that none of the problems encountered up to the present time have been fatal to spacecraft operation or to the achievement of major mission science objectives.

Because of the enormous distances from which the Voyager spacecraft would be required to communicate its problems to Earth, radio signals travelling at the speed of light (299 792 km/s) would eventually take hours to reach Earth (see Fig. 6.5 for a diagram of trajectories of Voyagers 1 and 2 through the Solar System). Assuming

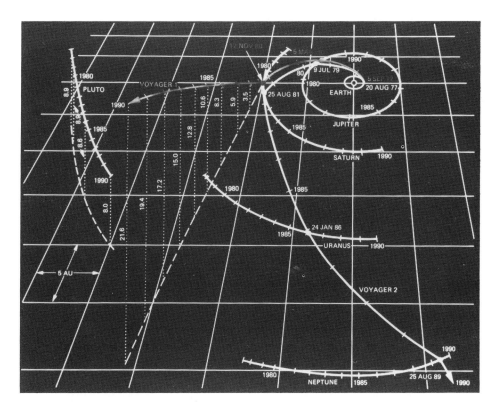

Fig. 6.5 — Trajectories of Voyagers 1 and 2 through the Solar System. (JPL-10687A)

immediate analysis and generation of corrective commands by ground personnel were possible, it would still take a like amount of time for those commands to reach the spacecraft. Some problems could develop beyond the point of recovery in the required 'round trip light time'. For that reason, automatic responses to certain types of anomalous spacecraft behavior were built into the spacecraft computers, so that by the time news of the problems reached Earth, corrective action would have been taken.

Although the idea is sound, pre-programmed responses by a nonreasoning robot often led to difficulties. Spacecraft acceleration at the time Voyager 2 separated from its final propulsive rocket was larger than anticipated by Voyager's computers, which initiated a number of corrective actions. As a result, the antenna was slightly mispointed, and several commands sent to it by ground controllers were ignored by Voyager 2. That, combined with the unexpected commands generated by the spacecraft computer, gave spacecraft engineers cause for concern. Once they understood what had happened, normal operation was restored.

During the launch period, several long appendages are either folded alongside the spacecraft or else wound into small containers waiting for commands which would cause them to be released and unfurled. The two largest appendages are opposing booms which support the power supplies (radioisotope thermoelectric generators, or RTGs) on one side and instrumentation for seven of the eleven science investigations on the other side of the spacecraft body (Fig. 6.6). The RTG

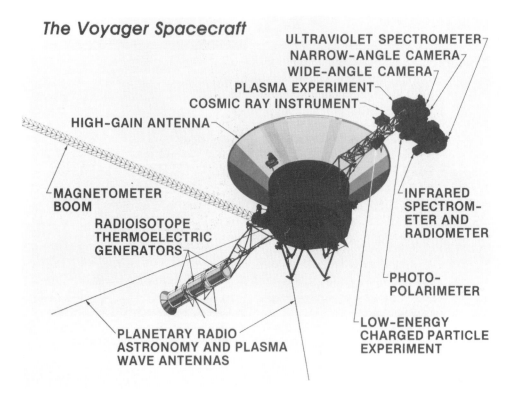

Fig. 6.6 — The two largest appendages on the Voyager spacecraft are the RTG boom, which supports the Radioisotope Thermoelectric Generators, and the science boom, which supports instrumentation for 7 of the 11 scientific investigations. (P-29473A)

boom deployed successfully, but data from the spacecraft indicated that the science boom was not fully deployed and latched. Careful analysis later led to the conclusion that the science boom was fully extended and probably latched, but that the electronic sensor which signaled latching had given an erroneous signal.

Several of Voyager's critical components exist in duplicate pairs on each spacecraft. There was, for example, a backup radio receiver to intercept commands from Earth in case the primary receiver failed. Since a failed receiver could not react to commands from Earth, Voyager's computers were programmed to restart a timer each time the radio sensed a command receipt. If seven days elapsed without a received command, the computer was to assume that its primary receiver had failed and switch to its backup receiver. By early April of 1978, command activity on Voyager 2 had diminished considerably, and a period of seven days elapsed without any commands being sent to the spacecraft. Responding correctly to its programmed instructions, the computers switched to the backup receiver. Voyager personnel, somewhat chagrined, generated the commands to switch back to the primary receiver and transmitted them to the spacecraft. When Voyager 2 at first failed to respond to commands from Earth, an additional problem was discovered. Each receiver was equipped with electronic circuitry designed to automatically detect the frequency of the incoming radio signal and tune the receiver to that frequency. A critical part of that circuitry was a tracking loop capacitor (TLC). Voyager 2's TLC had apparently failed in flight, so that the backup receiver would respond only to commands which arrived at precisely the right frequency. The most probable failure mechanism was electrical shorting caused by the migration of tiny conducting particles into the space between the capacitor plates. Continual operation of the primary receiver had prevented the buildup of such particles in its TLC.

Calculation of the appropriate radio frequency to transmit is not a simple task. In order to assure that the appropriate frequency arrives at Voyager 2's receiver, one must account for the motion of the transmitting tracking station on Earth due to rotation of the Earth, motion of the Earth around the center of mass of the Earth–Moon system, orbital motion of the Earth around the Sun, and the velocity of the spacecraft relative to the Sun. An additional complication arises from the fact that the radio frequency of the receiver changes with the temperature of the receiver. A temperature change of only $\frac{1}{4}$°C ($\frac{1}{2}$°F) shifts the receiver frequency by 96 Hz (cycles per second), half of its total bandwidth. Each time the electrical load being used by the spacecraft is changed by as little as 2 W, the receiver's 'best lock frequency' (BLF) must be monitored for at least 24 h before reliable commanding of the spacecraft can continue.

The command to switch back to Voyager's primary receiver was transmitted at a range of frequencies, and this time the spacecraft responded. The response was not very satisfying, however, for during the switching process, a short developed in the primary receiver which blew both of the redundant fuses. Another seven days had to pass before Voyager 2's computers switched to the backup receiver.

Several responses to the TLC failure in the backup receiver have been developed. Procedures are followed to regularly determine the BLF of the receiver. Elaborate computer programs assist Deep Space Network (DSN) personnel to determine the precise frequency to be transmitted so that commands will arrive at the spacecraft at a frequency very close to the BLF. When power loading on the spacecraft must be changed, a 'command moratorium' of 24 to 72 h is observed, its duration depending on the nature and magnitude of the power change. Since no usable receiver for backup purposes now exists, the computers are reprogrammed after each planetary encounter to execute a minimal set of observational sequences during the period surrounding close approach to the next planet. Fortunately, the receiver is still operating, and none of these Back-up Mission Loads has been used. The lack of automatic frequency tracking by the receiver did not result in any significant science loss during the Jupiter, Saturn, Uranus, or Neptune encounters of Voyager 2.

Electronic sensors on the spacecraft and its science instruments regularly monitor temperatures, voltages, and mechanical positions of movable parts. The readouts of those sensors are sometimes transmitted in a low-rate engineering data stream, sometimes in the science data stream, and sometimes in both. The low-rate engineering stream is multiplexed in such a way that some of the engineering readouts are sampled every 12 s, while others are sampled at 2-min, 3-min, 6-min, 12-min intervals, or not at all. An electronics failure of one of Voyager 2's 'tree switches' late in 1977 rendered 72 of several hundred engineering channels inaccessible. Fortunately, none of the lost engineering data are critical to the success of the Voyager missions.

The science instruments themselves were not without problems. The infrared instrument successfully jettisoned its telescope cover shortly after launch, but by early December of 1977 its sensitivity had degraded markedly. The problem was traced to crystallization of some of the bonding material which held the mirrors internal to its spectrometer. The crystallization caused a misalignment or warping of the mirrors which led in turn to the loss of sensitivity. The instrument was furnished with a 'Flash-Off Heater' (FOH) designed to heat the external primary telescope mirror and promote rapid evaporation of any impurities deposited during the launch period. After trying a number of other things which had no effect on the problem, the PI elected to turn the FOH on. The heating miraculously restored almost all of the prelaunch sensitivity. Apparently the heat reversed the crystallization process, allowing the bonding material to relax to its former shape. The instrument must be cold to operate properly, but prior to the 1989 Neptune encounter it was the practice to keep the instrument warm between planetary encounters in order to minimize the rate of sensitivity degradation.

Inside the photopolarimeter (a very sensitive light meter) are three command-able wheels. The first wheel contains a series of colored filters; the second contains drilled holes of various sizes to control the angular view of the instrument; the third has polarizing filters with a variety of orientations. The photopolarimeter was designed to be able to use any combination of positions of the three wheels. Excessive use of the wheels prior to Jupiter encounter, combined with radiation damage during the Jupiter flyby, caused a complete failure of the Voyager 1 instrument and degradation of the Voyager 2 instrument. Of the eight wheel positions available for each wheel at launch, only three filter positions, four aperture positions, and four polarization analysis positions were usable after the Jupiter encounter.

A host of other problems have occurred on each of the spacecraft. Most of the remaining problems either occurred only on Voyager 1 or subsequent to the start of the Voyager 2 Jupiter encounter period or were less significant than those chronicled in this section. Radiation damage will be discussed in connection with the Jupiter encounter discussion; scan platform seizure will be discussed in the Saturn section; computer memory loss will occupy a portion of the next section.

6.6   THE VOYAGER TEAM

The accomplishments of Voyagers 1 and 2 are due primarily to the ingenuity and dedication of a large number of people. The Science Steering Group (SSG) has already been mentioned. It consists of a Chairman, Team Leaders (TLs) for the Imaging Science (ISS) and Radio Science (RSS) investigations, and Principal Investigators (PIs) for the nine other science investigations (see Fig. 6.7). Edward C. Stone, who is also the Project Scientist, has ably filled the role of SSG Chairman since the inception of the Voyager Project in 1972. Bradford A. Smith of the University of

Fig. 6.7 — Members of the Voyager Science Steering Group and other Voyager Project Management personnel at JPL during the Uranus encounter. Pictured, from left to right, are Von Eshleman (former RSS TL), Pieter deVries (FSO Mgr), Ellis Miner (APS), Laurence Soderblom (Asst ISS TL), Lonne Lane (PPS PI), Bradford Smith (ISS TL), Rudolph Hanel (IRIS PI), Frederick Scarf (PWS PI), Herbert Bridge (PLS PI), 'Tom' Krimigis (LECP PI), Edward Stone (Proj Sci and CRS PI), Lyle Broadfoot (UVS PI), Leonard Tyler (RSS TL), George Textor (Mission Dir), Norman Ness (MAG PI), James Warwick (PRA PI), and Richard Laeser (Proj Mgr). (P-29700)

Arizona's Lunar and Planetary Laboratory has been ISS TL, with Laurence Soderblom of the United States Geological Survey in Flagstaff, Arizona, as his deputy. Von R. Eshleman led the RSS team until 1978, when G. Leonard Tyler assumed the role of RSS TL. Both are associated with Stanford University's Center for Radar Astronomy in California.

The cosmic ray investigation (CRS) was led by Rochus E. Vogt of Caltech until 1984, when Stone was given the responsibility of CRS PI in addition to his roles as SSG Chairman and Project Scientist. Rudolph A. Hanel guided the Infrared Interferometer Spectrometry investigation (IRIS) until after the Uranus encounter. Barney J. Conrath became IRIS PI in 1987. Both are scientists at NASA's Goddard Space Flight Center in Maryland. Low Energy Charged Particle (LECP) PI is S. M. (Tom) Krimigis of Johns Hopkins University's Applied Physics Laboratory in

Maryland. Norman F. Ness, PI for the Magnetometer (MAG) investigation was initially associated with NASA's Goddard Space Flight Center, but now heads the Bartol Research Institute at the University of Delaware. The Plasma (PLS) investigation was headed by Herbert S. Bridge until after the Uranus encounter, when John W. Belcher was named PLS PI. Both are associated with Massachusetts Institute of Technology. Excessive problems with the Photopolarimetry investigation (PPS) led to the dismissal of Charles F. Lillie of the University of Colorado's Laboratory for Atmospheric and Space Physics as PI in 1978. Charles W. Hord (also of LASP) was named interim PI until the appointment of JPL's Arthur L. Lane in 1979. James W. Warwick of Radiophysics, Incorporated, in Colorado has been the sole PI for the Planetary Radio Astronomy investigation (PRA). The Plasma Wave investigation (PWS) was led by Frederick L. Scarf of TRW Defense and Space Systems in California until his untimely death in 1988; Donald A. Gurnett, of the University of Iowa was appointed PWS PI in the same year. A. Lyle Broadfoot, initially associated with the University of Southern California, is now a part of the University of Arizona's Lunar and Planetary Laboratory. He is PI for the Ultraviolet Spectrometry investigation (UVS).

Eight different men have served as Voyager Project Manager, beginning with Harris M. Schurmeier, followed (in order) by John R. Casani, Robert J. Parks, Raymond L. Heacock, Esker K. Davis, Richard P. Laeser, Norman R. Haynes, and George P. Textor. The Project Manager oversees the full-time effort at JPL to adequately plan, implement, and monitor the activities of the two Voyager spacecraft. The Project Manager also serve as liaison with NASA Headquarters personnel in matters of policy and funding.

The Mission Director works under the Project Manager to manage the detailed activities of the Voyager Flight Team. Patrick Rygh, Richard P. Laeser, George P. Textor, and Richard Rudd served successively as Mission Director. Long-range overview planning and establishment of guidelines and constraints for operation of the spacecraft are under the direction of the Mission Planning Office (MPO) Manager. Ralph F. Miles, Jr, held that responsibility very early in the pre-launch time period. Charles E. Kohlhase has been MPO Manager since two years before the Voyager launches. A Ground Data System Engineering Office, managed by Allan L. Sacks during the Saturn–Uranus Cruise and Uranus Encounter time periods, has been in operation when there were problems associated with the multitude of computer programs needed to handle the spacecraft. These programs are of two general types: those associated with generating commands to be sent to the spacecraft ('uplink'), and those associated with data being returned by the spacecraft ('downlink').

The remainder of the Voyager Project organization at JPL is contained in three organizations: the Flight Science Office (FSO), the Flight Engineering Office (FEO), and the Flight Operations Office (FOO), as depicted in Fig. 6.8.

The FSO is managed by J. Pieter deVries. James E. Long and Charles H. Stembridge preceded deVries as FSO Managers. There are four teams in the FSO (Fig. 6.9). The Project Scientist (PS) and Assistant Project Scientist (APS) are also an integral part of the FSO Staff.

The Science Investigation Support Team (SIS) handles liaison with each of the 11 investigation teams and assures that encounter science, encounter preparations, and cruise science requirements are appropriately planned and implemented. The

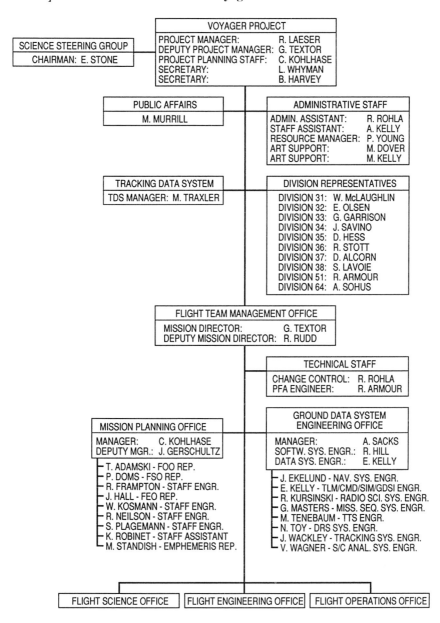

Fig. 6.8 — Voyager Project organization chart during the Uranus encounter period.

individuals who take the lead for each investigation are the respective Experiment Representatives (ERs) and Assistant ERs. The APS has the responsibility to coordinate those efforts for the encounter and encounter preparation observations. The APS also arbitrates between conflicting requirements and assures that the overall science observation plans will return the maximum available science within

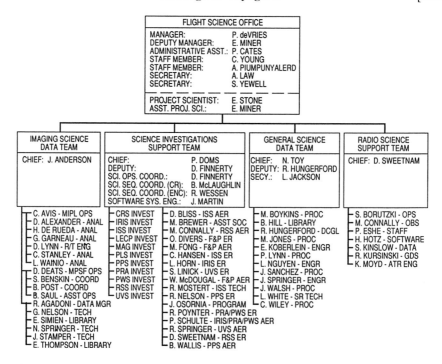

Fig. 6.9 — Voyager Flight Science Office organization chart during the Uranus encounter period.

the capabilities of the spacecraft and ground support systems. The SIS Team Chief has the responsibility to see that the work of SIS is efficiently performed. He and others of his staff make certain that both intrateam and interteam efforts are properly coordinated. The Radio Science Support Team (RSST) coordinates efforts to provide appropriate instrumentation and procedures for the collection of Radio Science data at the DSN tracking stations. The Imaging Science Data Team (ISDT) coordinates the massive amounts of processing and cataloging of imaging data. Preliminary data processing and sorting for the other nine investigations is done by the General Science Data Team (GSDT).

The FEO was managed by William I. McLaughlin for the Uranus encounter time period. Francis Sturms, Jr., Edward L. McKinley, and Saterios (Sam) Dallas preceded McLaughlin in that responsibility, and Lanny J. Miller occupied that post during the Neptune preparations and encounter. FEO consists of three teams, the Sequence Team (SEQ), the Navigation Team (NAV), and the Spacecraft Team (SCT), plus an Advanced Software Development Group (ASD), as shown in Fig. 6.10.

SEQ assists SIS in the early development of the spacecraft science and engineering sequence of events. Once that process reaches a certain level of development, SEQ takes the lead and is assisted by other elements of the Voyager Project until the commands are fully prepared and ready to transmit to the spacecraft for execution. SEQ also validates these sequences to ensure that they are consistent with operational, hardware, and software constraints.

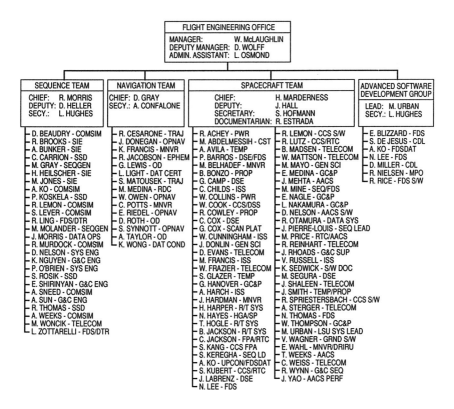

Fig. 6.10 — Voyager Flight Engineering Office organization chart during the Uranus encounter period.

NAV is charged with the responsibility of determining the precise trajectory being followed by the spacecraft, both relative to Earth and relative to the target planet and its satellites, and with making required adjustments to the trajectory by designing corrective propulsive maneuvers. NAV also provides information to the rest of the Voyager Project about the range of possible trajectories, their characteristics, and the associated delivery and knowledge uncertainties at each epoch where decisions must be made.

The tools NAV uses for these tasks include Earth-based observations of the planets and their satellites, radio tracking of the spacecraft to determine both range and velocity, and two other more esoteric processes. Delta Differential One-way Ranging (DDOR) utilizes simultaneous comparison of the spacecraft frequency and the frequency of a celestial radio source from two widely separated tracking stations to provide an independent determination of the spacecraft distance from Earth. Optical Navigation (OPNAV) uses the high-resolution ISS camera to image the planet and/or its satellites against a background of stars whose directions in the sky have been accurately determined from Earth. This provides an accurate direction and an improved range from the spacecraft to the planet as well as an improved orbit for each of the satellites imaged. For the distant outer planets, the position of Voyager relative to Earth and Sun is generally much better known than the position of the planet Voyager is approaching, so OPNAV provides a powerful and proven technique for determination of the spacecraft position relative to the planet.

SCT is the largest of the Voyager teams. These are the individuals who are charged with the responsibility of knowing the electronic and mechanical structure, health, capabilities, and limitations of each of the science and engineering subsystems. It is this knowledge base and the expertise of SCT members that has led to enormous improvements in spacecraft capability utilization during the course of the mission. The SCT is also largely responsible for innovative and insightful ways to overcome the limitations imposed by spacecraft hardware failures and the continually increasing distances of the spacecraft from Sun and Earth.

There are three pairs of computers aboard each Voyager spacecraft. The two Attitude and Articulation Control Subsystem (AACS) computers control the attitude of the spacecraft, keeping the antenna pointed at Earth by use of Sun and star sensors and an array of attitude control jets. These computers have been reprogrammed several times in flight by SCT personnel to improve the stability of the spacecraft and the pointing accuracy of the science instruments. The AACS also controls the steerable scan platform which points the ISS, PPS, UVS, and IRIS instruments. The two Flight Data Subsystem (FDS) computers provide science instrument timing and control functions; they also format the science and engineering data for transmission to Earth. FDS programming is the responsibility of the ASD, which works closely with SCT in this task. Portions of FDS memory on each spacecraft are no longer usable, but clever use of the remaining memory locations have minimized the impact of this memory loss. In the event one of the two FDS computers fails completely, a 'Single Processor Program' was developed for Uranus (and another was developed for Neptune) by ASD to minimize science data loss from such a failure. The two Computer Command Subsystem (CCS) computers contain the second-by-second timed command sequences to be executed by the spacecraft in performing its observations and calibrations. They also control access to the FDS and AACS computers and provide the logic for the automatic spacecraft fault responses mentioned earlier.

The FOO Manager during the Uranus encounter period was Douglas G. Griffith. He was preceded by Michael W. Devirian and Raymond J. Amorose, and was succeeded by Terrence P. Adamski. FOO is directly responsible for communications with the spacecraft and consists of four teams (Fig. 6.11): the Mission Control Team (MCT), the Mission Control and Computing Center (MCCC), the Deep Space Network Operations Control Team (NOCT), and the Data Management Team (DMT).

MCT directs and controls real-time Voyager operations, including transmission of commands to the spacecraft, monitoring the overall status of the spacecraft and the DSN tracking stations, and initial collection and dissemination of data from the spacecraft. MCCC provides the working interface between the Voyager Project and the various computer facilities used by, but not directly a part of, the Voyager Project for both uplink and downlink products. NOCT interfaces with DSN operational elements to ensure that appropriate DSN tracking support is provided and to monitor the status of radiometric (tracking) and telemetry (science and engineering

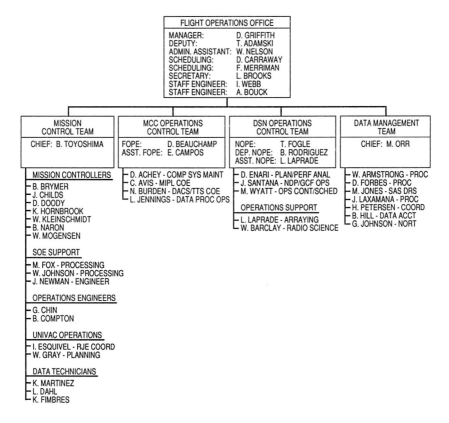

Fig. 6.11 — Voyager Flight Operations Office organization chart during the Uranus encounter period.

data) streams coming from the spacecraft. The DMT prepares the Experiment Data Records, which are magnetic tape records of spacecraft data for each of the individual investigations. The DMT also provides routine graphical displays of minimally processed science and engineering data for 'quick-look' analysis.

These, then, are the people of Voyager, exclusive of the science investigation team members at various institutions in the United States, England, France, and Germany. Since those science team members are all listed as co-authors in the preliminary science reports for the Uranus encounter [5], I will not list all of their names here. Their individual contributions to the success of the Voyager 2 Uranus encounter are gratefully acknowledged.

6.7 THE EYES AND EARS OF VOYAGER

Many parts of the human body contribute to our sensory perception of the world around us. The heart and brain play major roles in that process. Touch, taste, and smell are also important senses, but are limited to perception of things within our immediate vicinity. Because of their ability to detect waves generated by processes occurring at remote distances from our bodies, the eyes and ears play a unique role in our sensory perception. So it is with the Voyager spacecraft. The science instruments served as our 'eyes and ears' for conditions extant at the remote distance of Uranus, enabling us to 'see' things never before seen by humans. It is likely that future generations may have the opportunity to travel to the vicinities of the major planets to see such marvels personally, but until then robot exploration will do much to temporarily sate our intellectual appetites for knowledge about these distant worlds.

The location of the instrumentation for the 11 scientific investigations on the Voyager spacecraft is shown in Fig. 6.12. Voyagers 1 and 2 are basically identical, except as they have been altered by in-flight hardware failures or by differences in the contents of their computers. Where differences exist, the description here will be of Voyager 2, since it is the spacecraft that was sent to Uranus. All 11 science investigations on Voyager 2 were operative throughout the period of the mid-1989 Neptune encounter.

The most prominent feature of the spacecraft is its 3.66-m diameter parabolic antenna, which serves both to receive commands from Earth and to transmit science and engineering data back to Earth. The RSS investigation [6] utilizes the X-band and S-band transmitted carriers (the radiowave signals onto which the modulated telemetry data are superimposed). When using the internal ultrastable oscillator as a frequency standard, X-band and S-band signals are transmitted simultaneously. The X-band frequency is 8420.4 MHz (wavelength 3.560 cm); the S-band frequency is 2296.5 MHz (wavelength 13.054 cm). Voyager 2's S-band receiving frequency is 2113.3 MHz. When Voyager's radio system is 'in lock', the transmitted frequencies are determined by the received frequency, and are closer to 8415.0 MHz and 2295.0 MHz, respectively. The transmitted frequencies are in the ratio 11/3 (X-band/S-band). Much of the necessary instrumentation for the RSS investigation is co-located with the DSN tracking stations. The primary RSS science objectives at Uranus were:

(1) to study by means of radio occultations the atmosphere and ionosphere of Uranus, including atmospheric turbulence;
(2) to study by means of occultations and scattering experiments the rings of Uranus, including radial structure, particle size distribution, and radio optical depth; and
(3) to study the gravitational fields of Uranus and its satellites, including mass and density determinations and higher-order gravity harmonics.

The ISS [7] is a modified version of the slow-scan vidicon camera designs that were used in earlier Mariner flights (Fig. 6.13). The system consists of two cameras, a high-resolution narrow-angle (NA) camera and a more sensitive low-resolution wide-angle (WA) camera. The NA camera has a focal length of 1499.125 mm and a field-of-view size of 7.40 milliradians (0.424°) on a side. The WA camera focal length is 201.568 mm; it has a field of view 55.31 milliradians (3.169°) square. The output of each camera is an array of 800 lines of video by 800 picture elements per line, or 640 000 'pixels' per image. Pixel brightness level is represented by an eight-digit binary number (i.e. the brightness can be any decimal number between 0 and 255). Exposures times at Uranus could vary from 0.005 s to 15.360 s; they could also be integer multiples of 96 s longer. The available filters are listed in Table 6.1. The frame

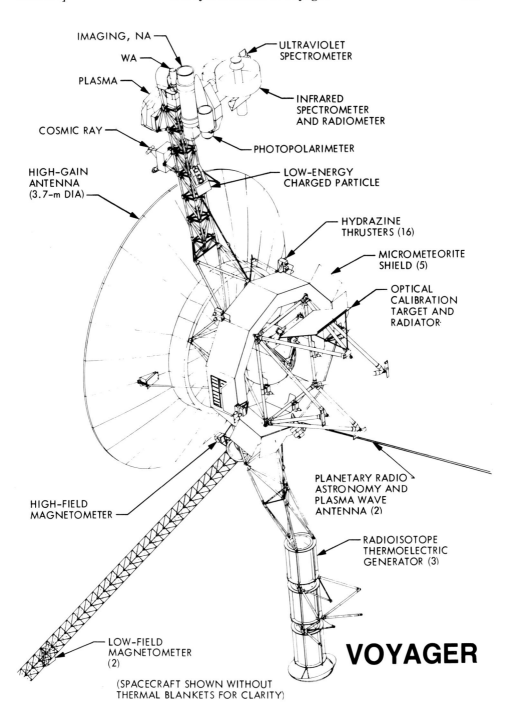

Fig. 6.12 — Diagram of the Voyager spacecraft, showing the location of the science instrumentation for the 11 scientific investigations. (P-19409)

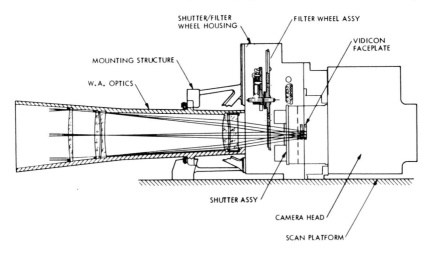

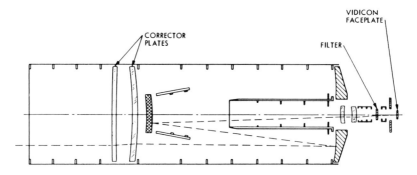

Fig. 6.13 — Diagram of the Voyager Imaging Science Subsystem slow-scan vidicon camera optics. The WA camera is shown above and the NA camera below.

readout time was as short as 48 s (for recorded images) or as long as 480 s (for nonrecorded images). Most of the nonrecorded images were passed through a data-compression routine in the FDS to reduce the total number of transmitted bits by a factor of about 3 without losing any information. ISS science objectives at Uranus were:

(1) to observe and characterize the circulation of the Uranus atmosphere, provide limits on atmospheric composition, and determine wind velocities in the regions observed;
(2) to map the radial and azimuthal distribution of material in the ring plane and search for new rings;
(3) to obtain global multispectral coverage of all satellites, establish their rotation rates and spin axis orientations, study their surface morphologies, and search for undiscovered satellites; and
(4) to provide support images to assist IRIS, PPS, and UVS in their data reduction.

   The IRIS [8] combines a Michelson interferometer for measurements in the mid-

Table 6.1 — Imaging spectral filters

Filter number and name	Characteristics and brief description	Effective wavelength (μm)	Filter factor
NA camera			
(#0) Clear	Broadband	0.497	1.0
(#1) Violet	Wideband, centered at 0.400 μm	0.416	7.4
(#2) Blue	Wideband, centered at 0.480 μm	0.479	3.5
(#3) Orange	Cut-on at 0.570 μm	0.591	7.0
(#4) Clear	Broadband	0.497	1.0
(#5) Green	Cut-on at 0.530 μm	0.566	3.3
(#6) Green	Cut-on at 0.530 μm	0.566	3.3
(#7) Ultraviolet	Wideband, centered at 0.325 μm	0.346	46
WA camera			
(#0) Methane, J/S/T	Narrowband, centered at 0.619 μm	0.6184	60
(#1) Blue	Wideband, centered at 0.480 μm	0.476	3.1
(#2) Clear	Broadband	0.470	1.0
(#3) Violet	Wideband, centered at 0.400 μm	0.426	7.2
(#4) Sodium-D	Narrowband, centered at 0.589 μm	0.589	250
(#5) Green	Cut-on at 0.530 μm	0.560	3.5
(#6) Methane, U	Narrowband, centered at 0.541 μm	0.541	40
(#7) Orange	Cut-on at 0.590 μm	0.605	15

infrared region and a radiometer for measurements in the visible and near-infrared range (see Fig. 6.14). The two share a 50-cm diameter reflecting telescope and have a 0.25° (4.4 milliradian) circular field of view, boresighted with the imaging cameras. IRIS requires 48 s (including mirror flyback time) to obtain a spectrum from 180 to 2500 cm^{-1}, corresponding to wavelengths from 55 to 4 μm. The radiometer is sampled every 6 s and covers from 2 to 0.33 μm. The primary IRIS objectives at Uranus were:

(1) to determine the atmospheric vertical thermal structure, which aids modeling of atmospheric dynamics;
(2) to measure the abundances of hydrogen and helium as a check on theories regarding their ratio in the primitive solar nebula; and
(3) to determine the balance of energy radiated to that absorbed from the Sun to help investigate planetary origin, evolution, and internal processes.

The UVS [9] covers the wavelength range of 0.05 to 0.17 μm with 0.001-μm resolution. There are no moving parts: spectral coverage is obtained by use of a reflective diffraction grating which spreads the light and focuses it onto an array of 128 adjacent detectors. Brightness levels for each channel range from 0 to 1023. A spectral scan is normally completed in 3.84 s; in a special occultation data mode, each scan is completed in 0.32 s. No lenses are used to define the pointing direction: 13 identical aperture plates are placed at precalculated distances from each other such that off-axis light is effectively rejected. The field of view thus defined is 0.1° by 0.86° (1.7 by 15 milliradians) and is boresighted with the ISS. A smaller aperture is offset by 20° by use of a small mirror. The field of view for this solar occultation port is 0.25° by 0.86° (4.4 by 15 milliradians). At UVS wavelengths between 0.05 and 0.12 μm,

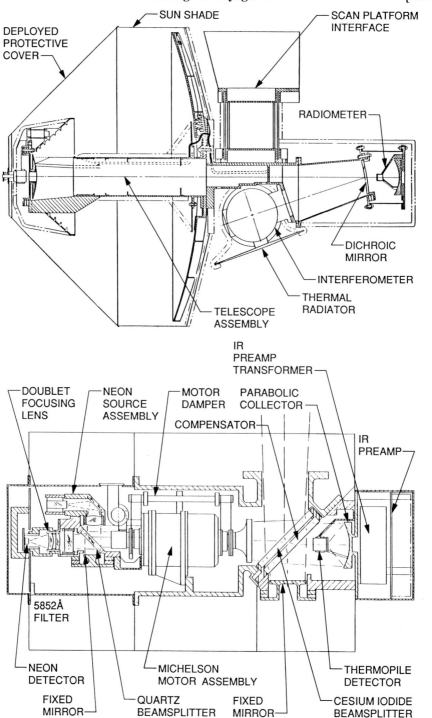

Fig. 6.14 — The Infrared Interferometer Spectrometer and Radiometer consists of two Michelson interferometers and a wideband visible and near infrared radiometer.

Earth's atmosphere is opaque; these wavelengths are covered by no other spacecraft-borne ultraviolet spectrometers. The UVS is therefore a useful tool for studies of stars and distant galaxies. The primary science objectives for UVS at Uranus were:

(1) to determine the scattering properties of the lower atmosphere of Uranus;
(2) to determine the distribution of atmospheric constituents with altitude;
(3) to determine the extent and distribution of hydrogen coronae of Uranus and its satellites;
(4) to investigate night airglow and auroral activity;
(5) to determine the ultraviolet scattering properties and optical depths of the rings; and
(6) to search for emissions from the rings and from any ring 'atmosphere'.

The PPS [10] instrument was partly described near the end of section 6.5. It consists of a sensitive photoelectric photometer mounted behind a 15.2-cm diameter f/1.4 Cassegrain telescope (see Fig. 6.15). It has a four-position aperture wheel which allows for fields of view of 0.12° (2.1 milliradians), 0.33° (5.8 milliradians), 1.0° (17 milliradians), and 3.5° (61 milliradians). It can access the three filter wheel positions and four polarizing analyzer positions shown in Table 6.2. It also has two sensitivity levels separated by a factor of about 50, J-mode on for high light levels, and J-mode off for low light levels. Each sample is represented as a brightness level from 0 to 1023. The PPS sampling rate is dependent on the FDS data mode, normally returning a sample every 0.6 s, but sampling 100 times per second in the special occultation data mode. The PPS scientific objectives for Uranus were:

(1) to determine the vertical distribution of cloud particles (atmospheric aerosols) down to an optical depth of unity;
(2) to determine the scattering and polarizing characteristics of the cloud particles and ring particles to obtain information on size, shape, and probable composition;
(3) to determine the Uranian atmospheric optical depth as a function of altitude, and to search for satellite atmospheres;
(4) to determine the Bond albedos of Uranus, its satellites, and its rings; and
(5) to use stellar occultation measurements to determine ring optical depths, radial distributions, and particle sizes.

The PRA [11] utilizes two vertically mounted 10-m electric antenna elements to monitor radio waves in two frequency bands (see Fig. 6.16). These antennas were extended by ground command after separation from the launch rocket. The low frequency PRA band covers from 1.2 kHz (250-km wavelength) to 1228 kHz (0.24-km wavelength) at intervals of 19.2 kHz and a bandwidth of 1 kHz. The high frequency band covers from 1.228 MHz (240-m wavelength) to 40.5504 MHz (7.4-m wavelength) at intervals of 0.3072 MHz and a bandwidth of 0.2 MHz. In its basic mode, the PRA scans the total of 198 frequencies in 6 s, representing each frequency level as a logarithmic intensity of 0 to 255. On occasion it may also sample one or more frequencies at much higher rates. At its highest rate (6400 bits every 0.06 s), PRA data output is identical to the highest ISS data output and must be recorded for more leisurely playback at a later time. The PRA cannot operate simultaneously in

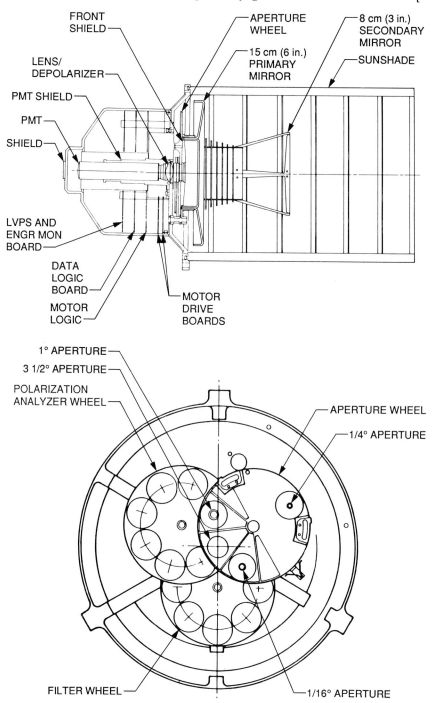

FRONT SHIELD

APERTURE WHEEL

8 cm (3 in.) SECONDARY MIRROR

LENS/ DEPOLARIZER

15 cm (6 in.) PRIMARY MIRROR

SUNSHADE

PMT SHIELD

PMT

SHIELD

LVPS AND ENGR MON BOARD

DATA LOGIC BOARD

MOTOR LOGIC

MOTOR DRIVE BOARDS

1° APERTURE

3 1/2° APERTURE

POLARIZATION ANALYZER WHEEL

APERTURE WHEEL

1/4° APERTURE

FILTER WHEEL

1/16° APERTURE

Fig. 6.15 — The Photopolarimeter is a shielded photoelectric photometer with three motor-driven wheels to control the size of the field of view, the filter passband, and the polarizer orientation.

Table 6.2 — Photopolarimeter apertures, filters, and polarizers

	Configuration	Purpose	Number and comments
Aperture	0.12°, 0.33°, 1.0°, 3.5°	Determines field of view of instrument	(#0) 0.33°, (#1) 1.0° (#2) 3.5°, (#4) 0.12°
Filter	(#0) 0.590 μm (#4) 0.265 μm (#6) 0.750 μm	Photometry	Only these three filters positions are now available.
Analyzer	(#2) 60° (#3) 120° (#6) 135° (#7) 45°	Polarization	Only these four analyzer positions are now available.

this high data rate and in its lower data rate mode. The PRA science objectives for the Uranus encounter were:

(1) to locate and explain radio emissions from Uranus at wavelengths ranging from tens of meters (decametric) to kilometers (kilometric);
(2) to detect radio emissions indicative of a Uranian magnetic field, and to determine the rotation rate of such a field;
(3) to describe planetary radio emissions and their relationship to the Uranian satellites;
(4) to measure plasma resonances near Uranus; and
(5) to detect lightning in the Uranian atmosphere.

The PWS [12] uses the same two 10-m antennas used by PRA, but connected in such a way that they act as a single 7-m dipole antenna (see Fig. 6.16). The PWS consists of two instruments: a spectrum analyzer, which operates continuously, and a wideband waveform receiver, which operates for brief periods on command. The spectrum analyzer samples frequencies of 0.0100, 0.0178, 0.0311, 0.0562, 0.100, 0.178, 0.311, 0.562, 1.00, 1.78, 3.11, 5.62, 10.0, 17.8, 31.1, and 56.2 kHz every 4 s with logarithmic intensity levels of 0 to 255. The frequency bandwidth of each frequency channel is 15% of the sampled frequency. The wideband waveform receiver is used to provide occasional views of the full frequency range from 0.050 to 10 kHz with higher time resolution. It is sampled 1600 times each 0.06 s with logarithmic intensity levels of 0 to 15. The PWS science objectives at Uranus were:

(1) to determine the role of interactions between charged particles and wave motions in various parts of the magnetic field of Uranus;
(2) to determine what mechanisms control the interactions of the rotating magnetic field of Uranus with its satellites;
(3) to determine the strengths and frequencies of variation of electric fields generated by flows of interacting charged particles (i.e. by 'plasmas') within the Uranian magnetic field;

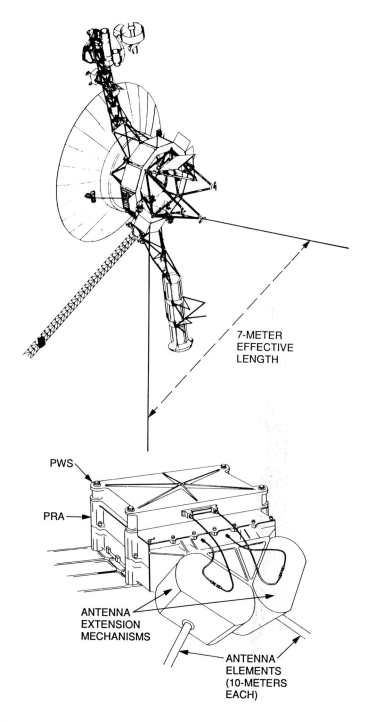

7-METER
EFFECTIVE
LENGTH

PWS

PRA

ANTENNA
EXTENSION
MECHANISMS

ANTENNA
ELEMENTS
(10-METERS
EACH)

Fig. 6.16 — Instrumentation packages for the Planetary Radio Astronomy and Plasma Wave
investigations are mounted together and utilize the same pair of 10-m antennas.

(4) to map variations in plasma density along the spacecraft trajectory; and

(5) to monitor low-frequency radio waves generated by atmospheric lightning discharges.

The MAG [13] sensors consist of two high-field magnetometers (HFMs) mounted on the spacecraft body and two low-field magnetometers (LFMs) located, respectively, at distances of 7.4 m and 13.0 m along a special magnetometer boom that was unfurled after separation of Voyager from the launch vehicle (see Fig. 6.17).

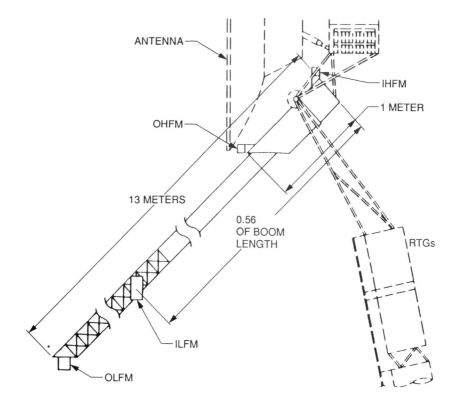

Fig. 6.17 — Four magnetometers are mounted at the base of and along the 13-m magnetometer boom.

The separation of the LFMs from the spacecraft body was necessary because of the small magnetic fields generated by electrical currents flowing within the science and engineering components of the spacecraft. Each magnetometer measures the ambient magnetic field strengths along three mutually perpendicular axes. The HFMs have two ranges: ±0.5 Oe (oersted) and ±20 Oe. For comparison, Earth's surface magnetic field strength varies from about 0.29 Oe near the equator to about 0.68 Oe near the South magnetic pole. Gain changes occur automatically when signals reach levels within 256 data numbers of the low end or the high end of the 0 to

±2047 data number range. There are eight separate LFM ranges: $\pm 8.8 \times 10^{-5}$Oe, $\pm 26 \times 10^{-5}$Oe, $\pm 79 \times 10^{-5}$Oe, $\pm 240 \times 10^{-5}$Oe, $\pm 700 \times 10^{-5}$Oe, $\pm 2100 \times 10^{-5}$Oe, $\pm 6400 \times 10^{-5}$Oe, and $\pm 50000 \times 10^{-5}$Oe. Note that the highest range for the LFMs is very nearly identical to the lowest range for the HFMs. The total data rate of the MAG during the Uranus encounter period was 750 bits/s. The scientific goals of the MAG at Uranus were:

(1) to measure and analytically represent the Uranian magnetic field;
(2) to determine the magnetosphere structure of Uranus;
(3) to investigate the basic physical mechanisms and processes involved both in interactions between the solar wind and the magnetosphere and in internal magnetosphere dynamics;
(4) to investigate the interactions of the satellites of Uranus with its magnetosphere; and
(5) to study the solar wind in the vicinity of Uranus.

The PLS [14], LECP [15], and CRS [16] all study electrons and other charged particles (ions of elements from hydrogen to iron). All utilize detectors which can only sense particles which strike the detectors. Each investigation is designed to discriminate between particles with different energies; each also has some ability to determine the incoming directions of such particles. The scientific goals of the three instruments at Uranus were:

(1) to study energetic particles in the solar wind near Uranus;
(2) to study the composition, sources, and other characteristics of charged particles with the Uranian magnetosphere;
(3) to study the interactions of these charged particles with rings and satellites; and
(4) to search for evidence of galactic cosmic ray particles.

The PLS consists of two Faraday cup plasma detectors, one pointed in the general direction of Earth and the other at right angles to a line from the spacecraft to Earth (Fig. 6.18). The Earth-facing sensor has three separate apertures, which in combination permit determination of the plasma velocity, density, and pressure. This cluster has a combined field of view that covers the entire Earth-facing hemisphere; the overlap between the three is a cone of half angle 45°. Each of the four (including the side-looking sensor) fields of view has an approximately conical field of view with half angle of 60°. Only the side-looking sensor is capable of detecting electrons. The PLS senses singly charged particle energies of from 10 to 5950 eV, where an eV is the kinetic energy acquired by an electron when being accelerated across an electrical potential of 1 V, an energy of about 1.602×10^{-12} erg. The instrument has four operating modes: M (=medium and high resolution ion mode), L (=low resolution ion mode), E1 (=high resolution electron mode), and E2 (=low resolution electron mode). The PLS cycled through the four modes every 12 min during the Uranus encounter, providing a data output rate of only four samples (32 bits) per second.

The LECP consists of two arrays of detectors on a rotating platform (see Fig. 6.19). The first array, with eight solid state detectors, is the Low Energy Magnetospheric Particle Analyzer (LEMPA). The detectors in the LEMPA are designed for detection of particles with energies as low as 10 to 15 keV, discrimination between ions and electrons, high sensitivity, and a large range of energies. The second array,

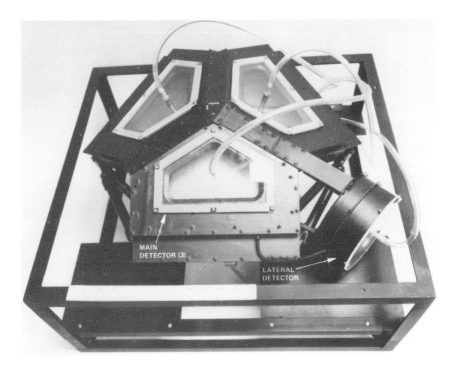

Fig. 6.18 — The Plasma Subsystem is mounted on the spacecraft science boom. (260-786B)

with seven solid-state detectors and an anti-coincidence shield of eight active detectors, is the Low Energy Particle Telescope (LEPT). The smallest of the Voyager 2 detectors (called 'D1c') failed shortly before the Saturn encounter. LEPT detectors are designed to measure the distributions of charge and energy of 0.1 MeV/ nucleon to 500 MeV/nucleon ions. The data rate output of the LECP was 600 bits/s during the Uranus encounter. The LECP telescope normally steps one 45° step each 48 s during an encounter. Within 24 h of closest approach, the LECP stepping occurred in a 12-min cyclic fashion, such that two eight-position cycles were completed in 1.6 min, followed by 10.4 min of no stepping. This minimized electronic interference with other instruments caused by the LECP stepping motor.

The CRS consists of three particle telescope systems. The High Energy Telescope (HET) system consists of two double-ended telescopes, each with 15 solid-state particle detectors and seven anti-coincidence detectors. The HETs measure the energy spectra of electrons and all elements from hydrogen to iron over a broad range of energies. They can also discriminate between isotopes [17] of hydrogen through oxygen. The four Low Energy Telescopes (LETs) each have four surface barrier detectors designed to determine the three-dimensional flow patterns of cosmic ray particles and to extend element discrimination to lower energies than detectable by the HETs. The Electron Telescope (TET) consists of eight solid-state detectors and six tungsten absorbers, all surrounded by a grid of six solid-state anti-

Fig. 6.19 — The Low Energy Charged Particle instrument consists of two particle telescope systems mounted on a rotating platform. (260-781B)

coincidence detectors. TET measures the energy spectrum of electrons between 5 and 110 MeV. The combination of HET and LET measures ions in the energy range from 1 to 500 MeV/nucleon. The data rate of the CRS is 260 bits/s. Both CRS and LECP completely sample their respective arrays of detectors each 96 s.

6.8 'JUPITER THE GIANT'

Voyager 1 reached its closest approach with Jupiter on March 5, 1979; Voyager 2 reached that milestone on July 9, 1979. The paths of the spacecraft through the Jupiter system are shown in Figs 6.20 and 6.21. Each spacecraft received a substantial gravity assist from Jupiter, which increased their respective velocities sufficiently to enable each to completely escape the Sun's gravity and eventually enter interstellar space (see Fig. 6.22). Effects due to the intense radiation field surrounding Jupiter were noted on both spacecraft. The Voyager 1 PPS, already partially crippled before the encounter, failed completely and has been unusable since that time. The UVS on each spacecraft returned few useful data near Jupiter closest approach, but each recovered shortly after the spacecraft exited the most intense radiation period. The clocks internal to the FDS computers on Voyager 1 were 'reset' 40 times within a several-hour period, causing them to lose synchronization with the CCS computer clocks. Some of the ISS, PRA, and PWS data was degraded as a result, but no permanent damage was done to the computers or these three instruments. The radio

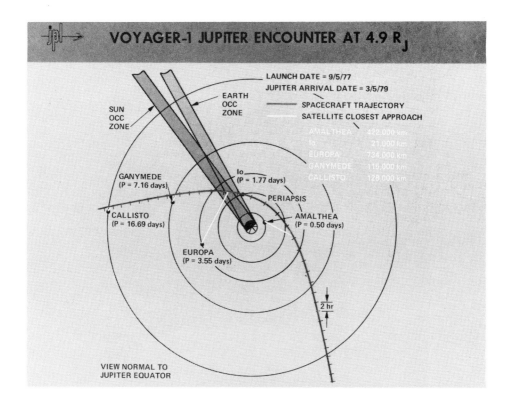

Fig. 6.20 — A trajectory-plane view of the path of Voyager 1 through the Jupiter system. Voyager 1's closest approach to Jupiter occurred on March 5, 1979. (260-284B)

transmitter frequencies on both spacecraft were slightly shifted by radiation effects, but these shifts have had no effect on subsequent operations. The optics in the Canopus Star Trackers (CSTs) on both spacecraft were darkened, resulting in an effective 13% reduction in sensitivity. The CST on Voyager 1 additionally lost some pointing capability, resulting in a reduction in the number of usable post-Jupiter reference stars.

In spite of the above difficulties, the Voyager encounters with Jupiter were an unqualified success. Formal Voyager papers on the Jupiter results are still being published nearly 10 years after the encounters. Dedicated issues or large groupings of scientific papers resulting from the Voyager encounters appeared in *Science, Nature, Geophysical Research Letters*, and in the *Journal of Geophysical Research* [18]. An excellent narrative containing the highlights of the scientific findings was given by Morrison and Samz [19]. A very brief summary of the major scientific findings is given below:

Jupiter atmosphere

(1) The atmosphere of Jupiter above the cloudtops is composed almost exclusively of hydrogen and helium. There are 26 ± 5 g of helium for every 100 g of hydrogen.

Voyager VOYAGER 2 TRAJECTORY

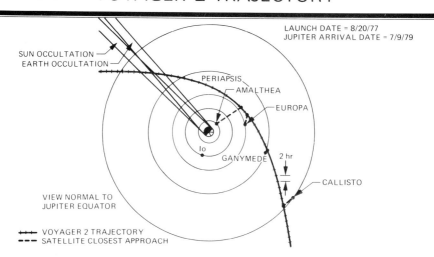

Fig. 6.21 — A trajectory-plane view of the path of Voyager 2 through the Jupiter system. Voyager 2's closest approach to Jupiter occurred on July 9, 1979. (260-533A)

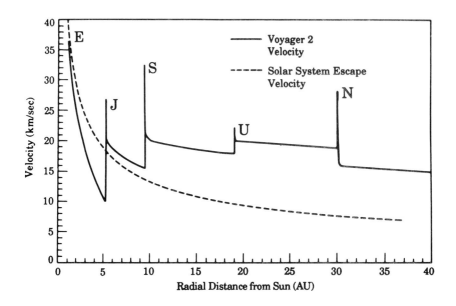

Fig. 6.22 — The Sun-centered velocity of Voyager 2 (solid line) is plotted against radial distance from the Sun in astronomical units, where 1 AU = 149 597 900 km. The dashed line represents the velocity needed to escape the Sun's gravity. The large effects of 'gravity assists' at Jupiter and Saturn are apparent. Jupiter increased Voyager's velocity beyond that necessary to escape the Sun's gravity. (JPL-12723)

(2) The Great Red Spot (Fig. 6.23), the white ovals (Fig. 6.24), and other major cloud disturbances rotate clockwise in the northern hemisphere and counter-clockwise in the southern hemisphere, indicating that they are regions of higher pressure within the atmosphere.

(3) There exists a stable zonal pattern of winds (Fig. 6.25) which is an indication that the wind patterns are a more fundamental characteristic of the atmosphere than the bright zones and dark belts.

(4) The zonal wind pattern extends into the polar regions, earlier thought to be a region where vertical motions would dominate over horizontal flows.

(5) The interactions of the Great Red Spot and the zonal wind flows are extremely complex, and major changes in those patterns occurred over the four-month period between the encounters of Voyager 1 and Voyager 2.

(6) Lightning superbolts (Fig. 6.26) near the cloudtops are common, and give rise to radiowave emissions called 'whistlers', so called because the generated radio frequencies change in much the same way as an audible whistle decreasing in pitch.

(7) The presence of a stratospheric temperature inversion was verified. Temperatures rise from a minimum of about 110 K near the 100-mbar level to about 160 K near 10 mbar.

(8) There is a concentration of ultraviolet-absorbing haze in the polar regions of the planet.

(9) Acetylene (C_2H_2) and ethane (C_2H_6) are present in ratios that vary both with time and with latitude on the planet.

(10) Both visible and ultraviolet emissions (Fig. 6.26) are seen in the nightside polar atmosphere. They are likely due to charged particles generated near Io's orbit and streaming down Jovian magnetic field lines.

(11) There is a strong ultraviolet emission from the entire sunlit disk of the planet. The large scale height of this emission is indicative of extreme upper atmospheric temperatures in excess of 1000 K.

Jupiter ring system

(12) A equatorial ring of finely divided material (Fig. 6.27) surrounds Jupiter. The main component of this ring extends from about 51 000 km to 57 000 km above the cloudtops.

(13) A gossamer extension of the main ring may extend to Amalthea's orbit, about 109 000 km above the cloudtops. A halo of diffuse ring material whose source may be the main ring extends downward to within about 20 000 km of the cloudtops.

(14) Typical ring particles are only a few micrometers across, and they scatter light predominantly in a forward direction. The ring would be extremely difficult to detect from Earth-based telescopes.

Jupiter satellites

(15) Two previously undetected satellites (Metis and Adrastea) orbit near the outer edge of the ring. Metis (Fig. 6.28) has a diameter of about 40 km; Adrastea (Fig. 6.29) is smaller and nonspherical, ranging from 15 to 25 km in diameter.

(16) Amalthea (Fig. 6.30) was shown to be an irregular, elongated, dark, and very

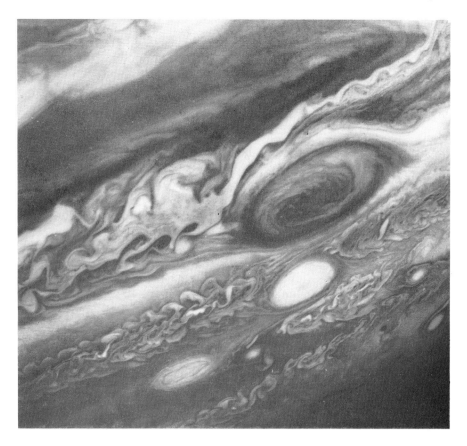

Fig. 6.23 — The Great Red Spot, an anti-cyclonic storm which has been observed in Jupiter's southern hemisphere for over 200 years. (P-21742)

      red object, ranging in size from 150 to 270 km. Its surface is likely contaminated by sulfur from Io.

(17) Thebe (Fig. 6.31), with a diameter near 100 km, was found between the orbits of Amalthea and Io.

(18) Nine nearly continuous volcanic eruptions (Fig. 6.32) were observed on Io, apparently generated by a combination of orbit perturbations by Europa and tidal distortions from Jupiter. At least seven of the nine were still erupting when Voyager 2 passed four months later. (One had stopped, and one could not be observed by Voyager 2.)

(19) The surface of Io (Fig. 6.33) is void of impact craters, extremely young, colored by sulfur and sulfur compounds, and able to undergo large-scale surface changes in a few months.

(20) At least two hot spots on Io's surface were detected, further confirming the presence of active volcanism.

(21) Io possesses a thin and probably transient atmosphere of sulfur dioxide (SO_2).

Fig. 6.24 — Although the white ovals are smaller and less spectacular than the Great Red Spot, they too are enormously large anti-cyclonic storms in Jupiter's southern hemisphere. (P-21754)

(22) The surface of Europa (Fig. 6.34) is very smooth and nearly void of impact craters. A deep water ocean may exist beneath a 10-km thick crust of water ice.

(23) A network of light and dark lines crisscross Europa. These suggest major stressing of the surface in the geologic past.

(24) The surface of Ganymede (Fig. 6.35) is highly variegated. Some areas are very dark, while others are relatively bright. Both heavily cratered and more lightly cratered areas exist.

(25) Extensive parallel ridge and valley systems (Fig. 6.36) were seen on Ganymede, perhaps indicating past periods of subsurface activity or motion of crustal plates.

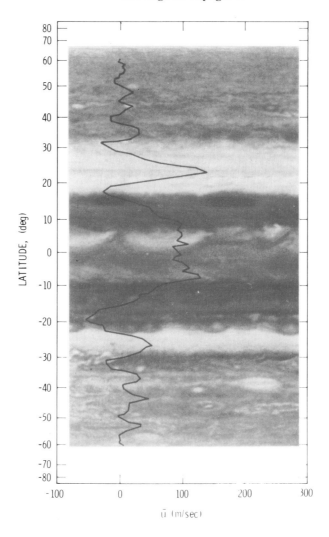

Fig. 6.25 — The zonal pattern of winds in Jupiter's atmosphere is relatively stable and of much longer duration than most cloud features. (260-1126A)

(26) The surface of Callisto (Fig. 6.37) is covered by impact craters, and no indications of significant recent geologic activity were seen.

(27) The sizes, masses, and densities of the four largest satellites have been determined (Table 6.3). Ganymede is the largest satellite in the Solar System. There is a downward progression of densities with distance from Jupiter, implying that heat from Jupiter may have helped to remove water from the inner satellites.

Jupiter magnetosphere

(28) An electric current of more than a million amperes flows along the magnetic flux tube linking Jupiter and Io.

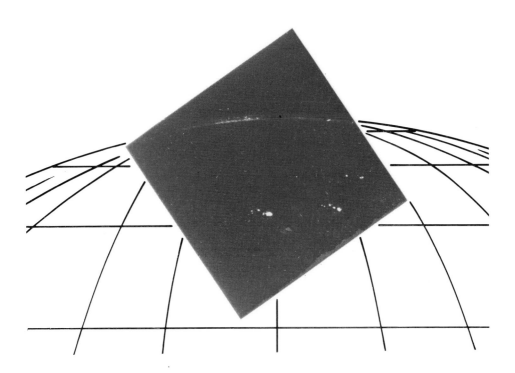

Fig. 6.26 — Lightning superbolts (bright clusters of spots) and visible auroral emissions (near the edge of the disk) are seen in this Voyager image of Jupiter's dark side. (260-929A)

(29) A doughnut-shaped torus (Fig. 6.38) containing sulfur and oxygen ions surrounds Jupiter at the orbit of Io. This torus emits ultraviolet light, has temperatures of up to 100 000 K, and is populated by more than 1000 electrons/cm^3.

(30) A 'cold' (i.e. forced to rotate with the magnetic field) plasma exists between Io's orbit and the planet (see Fig. 6.39). It has larger than expected amounts of sulfur, sulfur dioxide, and oxygen, all probably derived from Io's volcanic eruptions.

(31) The Sun-facing magnetopause (outer edge of the magnetosphere) responds rapidly to changing solar wind pressure, varying from less than 50 Jupiter radii to more than 100 Jupiter radii in distance from the planet's center.

(32) A region of 'hot' (i.e. not forced to rotate with the magnetosphere) plasma exists in the outer magnetosphere. It consists primarily of hydrogen, oxygen, and sulfur ions.

(33) Jupiter emits low-frequency radiowaves (wavelengths of one to several kilometers). The amount of radiation is strongly latitude-dependent.

(34) There exists a complex interaction between the magnetosphere and Ganymede. This results in deviations from a smooth magnetic field and charged particle distributions which extend up to 200 000 km from the satellite.

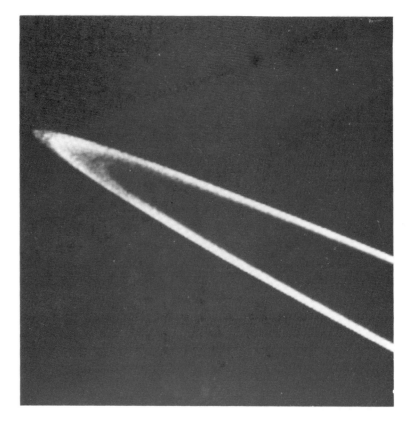

Fig. 6.27 — This equatorial ring of finely divided material around Jupiter was discovered by
Voyager. (260-674)

(35) About 25 Jupiter radii behind the planet, the character of the magnetosphere
      changes from the 'closed' magnetic field lines to an extended magnetotail
      without line closure. This occurs as a result of downstream interaction with the
      solar.wind.
(36) Jupiter's magnetotail (Fig. 6.40) probably extends to, and may extend beyond,
      the orbit of Saturn, more than 700 million km 'downwind' from Jupiter.

6.9 'SATURN THE GEM'

The second leg of the journeys of Voyagers 1 and 2 culminated with the Saturn
encounters on November 12, 1980, and August 26, 1981, respectively. The trajectories of the two spacecraft through the Saturn system are shown in Figs 6.41 and 6.42.
Two of the primary objectives of the Voyager 1 encounter with Saturn required a
close approach to Titan (Fig. 6.43) and a spacecraft passage behind the full radial
extent of the rings (Fig. 6.44). Because of the geometry of those two requirements,
Voyager 1 was pulled by Saturn's gravity into a path northward of the orbital planes
of the planets. Voyager 1's post-Saturn mission will be discussed later.

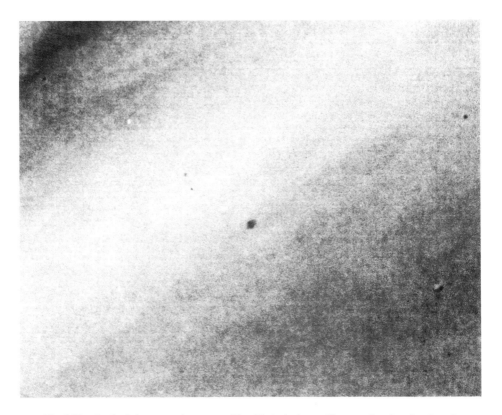

Fig. 6.28 — Jupiter's innermost known satellite, Metis, is shown silhouetted against the planet's clouds.

Following the successful acquisition of critical Titan and ring data by Voyager 1, project personnel were given permission by NASA Headquarters to target Voyager 2 for a Uranus trajectory. The appropriate gravity assist path required that Voyager 2 pass very close to Saturn's G ring. (Saturn's rings, from innermost to outermost, are called D, C, B, A, F, G, and E.) There were some concerns that the spatial density of ring particles might be sufficient to damage the spacecraft, but the potential gains were judged to outweigh the risks. Ring-particle and radiation environments during the Saturn encounters proved to be relatively benign. No damage was done to the spacecraft or any of its instrumentation.

One major problem did occur during the Saturn encounter. As the experience of the flight team increased, more efficient use of spacecraft resources resulted. The 48-h period surrounding Saturn closest approach was the most active observing period attempted. In some respects, the period was almost frenetic, as the scan platform was pointed in rapid succession at a dozen different satellites, at the rings, and at Saturn itself. To accomplish this, the platform was moved extensively at its most rapid rate, 1°/s. About 110 min after Saturn closest approach and an hour after passing by the G ring, the platform slowed and then stopped. Voyager 2

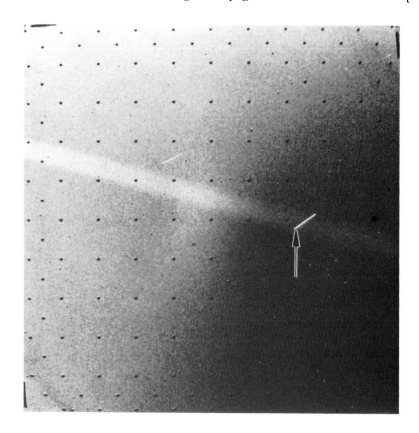

Fig. 6.29 — Adrastea is the bright streak in this smeared image of Jupiter's ring. (260-806)

was behind Saturn at the time, and news of the problem did not reach JPL until after the spacecraft had exited the planet's shadow and the additional 90-min one-way light time had elapsed. By that time, the spacecraft had automatically disabled further 'slewing' of the platform. Commands were sent to the spacecraft to try to free the platform by moving it at its low rate (5° per min), but the spacecraft refused to respond. Platform slewing commands were stripped from the computer sequences scheduled for transmission to Voyager 2. Special commands to alternately heat and cool the faulty scan platform azimuth actuator were generated and transmitted to the spacecraft. In the meantime, Voyager 2 was receding rapidly from Saturn, and the disabled platform was pointed uselessly at dark sky. At the urging of the Voyager scientists, commands were generated to try to turn the platform back in the direction of Saturn. Project management agreed to the transmission of the commands. A unanimous sigh of relief occurred as Saturn reappeared in the NA camera field of view about 2.8 days after closest approach (Fig. 6.45).

One other important objective was dependent on the ability to point the ISS cameras. About 10 days after Saturn closest approach, Voyager would fly within 2 million km of Phoebe, Saturn's outermost satellite. A special sequence of commands

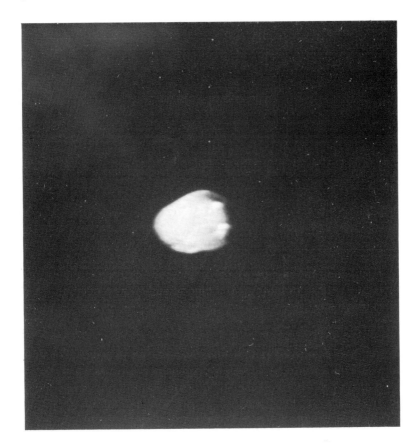

Fig. 6.30 — The innermost of the satellites discovered from Earth-based telescopes is Amalthea. It is seen here to be a dark, irregular body. (P-21223)

was generated to obtain images of Phoebe during the close passage, and that sequence also executed successfully. Following the Phoebe imaging, the only platform slewing allowed for more than a year was that required to analyze the problem and to deduce the failure mechanism. The prognosis was guarded, but encouraging. Extensive high-rate slewing had driven lubricant out of the critical parts of the platform gears, and galling of the gear shafts had caused them to seize. Temperature cycling had squeezed the shafts sufficiently to create some clearance and free the gears; migration of lubricant back into the interface between gear and shaft also helped. Further use of high-rate slewing is now prohibited, and the platform has worked flawlessly since early 1983.

When asked what percentage of its anticipated science Voyager 2 returned prior to its platform difficulties, Project Scientist Stone quickly replied, 'About 200%!' It is true that some pictures of Tethys and some darkside Saturn data were not obtained, but what was obtained far exceeded the expectations of all associated with the Voyager Project. Analysis of the data is likely to continue for decades. Formal

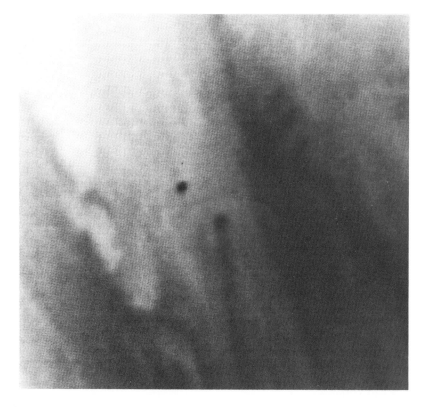

Fig. 6.31 — Thebe was discovered by Voyager. It orbits between Amalthea and Io. With its diameter of about 100 km, it is the largest of the three Jupiter satellites discovered by Voyager. (P-22580)

Voyager papers on the Saturn results are contained in the following dedicated issues or large groupings of scientific papers in *Science, Nature, Icarus, Journal of Geophysical Research* [20]. The University of Arizona books, *Saturn* [21], and *Planetary Rings* [22] also draw heavily on Voyager results at Saturn. Morrison [23] again was instrumental in documenting the highlights of the scientific findings. A summary of those findings is given below.

Saturn atmosphere
(1) As was the case with Jupiter, Saturn's atmosphere above the cloudtops is composed almost exclusively of hydrogen and helium. There are 6±5 g of helium for every 100 g of hydrogen.
(2) Excess thermal energy is radiated by Saturn at a rate 1.78±0.09 that received from the Sun, probably due to the gravitational separation of helium and hydrogen in the interior.
(3) Oval cloud features with clockwise rotation in the northern hemisphere and counterclockwise rotation in the southern hemisphere are limited to latitudes poleward of ±45°.

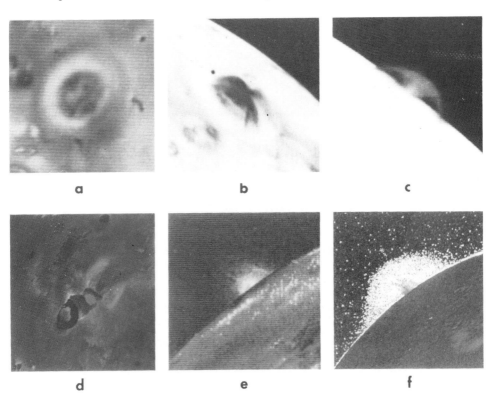

a

b

c

d

e

f

Fig. 6.32 — Active volcanic eruptions were observed on Jupiter's Io, evidence of the strong
tidal heating caused by Jupiter's gravity. (260-451)

(4) Maximum and minimum zonal wind speeds are located in the middle, rather
than at the edges, of dark belts and bright zones.

(5) A definite north–south symmetry of zonal winds was observed (Fig. 6.46). This
symmetry may imply deep circulation within the atmosphere and perhaps
differentially rotating concentric cylinders whose axes are parallel to the
rotation axis of Saturn.

(6) As at Jupiter, Saturn has a temperature inversion in its stratosphere. The
temperature rises from a minimum of about 80 K near the 100-mbar level to
140 K near 10 mbar.

(7) Ultraviolet light is emitted from the polar regions. These auroral emissions are
associated with charged particles from the solar wind spiraling down magnetic
field lines and striking atoms in the upper atmosphere.

(8) Ultraviolet emissions are also seen at lower latitudes. Since these are seen only
in sunlit portions of the atmosphere, they have been termed 'dayglow', and
their precise source is still a matter of debate.

(9) Large scale heights for hydrogen lead to the conclusion that temperatures in the
extreme upper atmosphere reach values of 600 to 800 K.

Fig. 6.33 — Io is the innermost of the four Galilean satellites of Jupiter. (P-34590)

Saturn rings

(10) Saturn possesses a ring system with enormously complex radial structure and few empty gaps, even at a scale of less than a km (Fig. 6.47).

(11) The D ring (Fig. 6.48), first proposed by Earth-based observers, but too faint to have been seen by them, was found to extend from the inner edge of the C ring to about 3200 km above the cloudtops.

(12) The first images of the G ring (Fig. 6.49), proposed on the basis of charged particle absorptions observed by Pioneer 12, were obtained. The ring lies between the orbits of Janus and Mimas.

(13) Structure which has the appearance of braiding was seen in the narrow F ring (Fig. 6.50), likely due to complex interactions between the ring particles and the shepherding satellites Prometheus and Pandora (Fig. 6.51).

(14) Both elliptical and discontinuous rings were found within gaps in the main ring system, generally at distances associated with strong satellite orbital resonances.

(15) The outer edge of the B ring is a centered ellipse (Fig. 6.52) which rotates at a rate that keeps its short axis pointed approximately at Mimas.

(16) Spiral density waves (Fig. 6.53) and spiral bending waves also attest to gravitational interactions of the ring particles with Saturn's satellites.

Fig. 6.34 — Streaks across Europa's surface may be fossil remnants of ancient fractures.
(P-21752)

(17) The outer edge of the A ring is less than 10 m thick; the main rings of Saturn, except for the 'corrugations' associated with bending waves, may be uniformly thin.

(18) Typical ring particle sizes range from micrometers to tens of meters or larger, and show radially varying size distribution.

(19) Radial 'spokes' of micrometer-sized particles were seen in the outer half of the B ring (Fig. 6.54). Their main periodicity is identical to the 10.657-h rotation period of Saturn's magnetic field, implying that the spokes are electrically charged and driven in part by the magnetic field.

Saturn satellites

(20) Three satellites (Atlas, Prometheus, and Pandora) were discovered in Voyager imaging.

Fig. 6.35 — Ganymede is the largest of the Galilean satellites, and the largest satellite in the Solar System. (P-21751)

(21) The sizes of all 17 known satellites of Saturn were first determined by Voyager imaging, leading to improved density estimates for those whose masses were known.

(22) More accurate determination of the masses of Tethys, Rhea, Titan and Iapetus were obtained, leading in addition to a revised estimate for the masses of Mimas and Enceladus.

(23) An enormous crater (Herschel), with a diameter of about 130 km, was discovered on Mimas (Fig. 6.55). Herschel, named for Sir William Herschel, is fully one-third the diameter of Mimas itself.

(24) The surface of Enceladus (Fig. 6.56) has undergone more recent geologic alterations than any other Saturn satellite surface seen by Voyager.

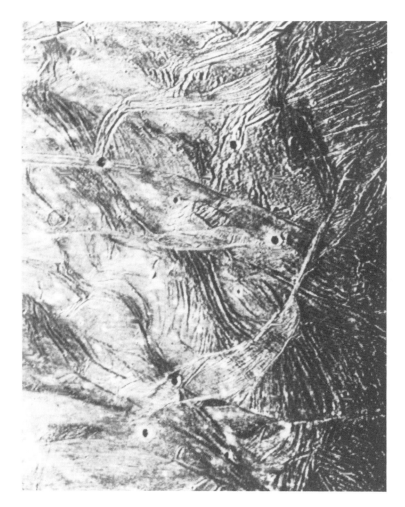

Fig. 6.36 — These parallel ridge and valley systems on Ganymede may be evidence for crustal movements on this large satellite. (P-21279)

(25) Tethys (Fig. 6.57) also has a large 400-km diameter crater, Odysseus, and a canyon, Ithaca Chasma, which girds two-thirds of the satellite's circumference.
(26) Wispy terrain and sharp albedo contrasts are characteristic of the surface of Dione (Fig. 6.58). Extensive surface fracturing and horizontally variable crater size distributions are also seen.
(27) Bright, wispy terrain is imbedded in an otherwise dark trailing hemisphere of Rhea (Fig. 6.59). Fractures and variations in crater size distribution are seen here as well.
(28) Titan's diameter is 5150±4 km, making it second in size to Ganymede among Solar System satellites.
(29) Titan has a near-surface atmospheric pressure and temperature of 1.6 bars and

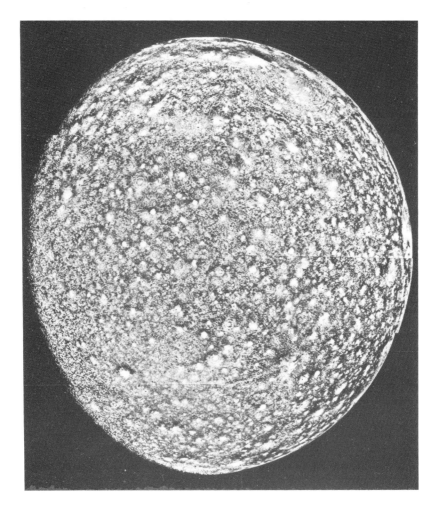

Fig. 6.37 — Callisto appears very little changed over its geologic history. Its surface bears the unobscured markings of an early bombardment by meteors and other cosmic debris. (P-21746)

95±1 K, respectively. Fig. 6.60 shows a comparison of the atmospheres of Titan and Earth.

(30) Titan's atmosphere is at least 90% nitrogen (N_2); methane (CH_4) and (probably) argon (Ar) are the other main constituents. Trace amounts of acetylene (C_2H_2), ethylene (C_2H_4), ethane (C_2H_6), methyl acetylene (C_3H_4), propane (C_3H_8), hydrogen cyanide (HCN), cyanoacetylene (HC_3N), cyanogen (C_2N_2), carbon monoxide (CO), and carbon dioxide (CO_2) were also detected.

(31) There are strong indications that Titan's surface may be covered with a liquid ethane (C_2H_6) ocean, combined with 25% liquid methane (CH_4) and 5% liquid nitrogen (N_2).

(32) A main haze layer (Fig. 6.61) extends to 200 km above Titan's surface.

Table 6.3 — Earth, Jupiter, Saturn satellite data

Satellite	Diameter (km)	Mass (10^{19} kg)	Density (g/cm^{-3})
Moon (E)	3476±1	7349±7	3.342±0.001
Io (J)	3630±10	8920±40	3.55±0.03
Europa (J)	3138±20	4870±50	3.01±0.07
Ganymede (J)	5262±20	14900±60	1.95±0.02
Callisto (J)	4800±20	10750±40	1.86±0.02
Mimas (S)	392±6	4.55±0.54	1.44±0.18
Enceladus (S)	500±20	7.4±3.6	1.13±0.57
Tethys (S)	1060±20	75.5±9.0	1.21±0.16
Dione (S)	1120±10	105.2±3.3	1.43±0.06
Rhea (S)	1530±10	249±15	1.33±0.08
Titan (S)	5150±4	13457±3	1.882±0.004
Iapetus (S)	1460±20	188±12	1.15±0.09

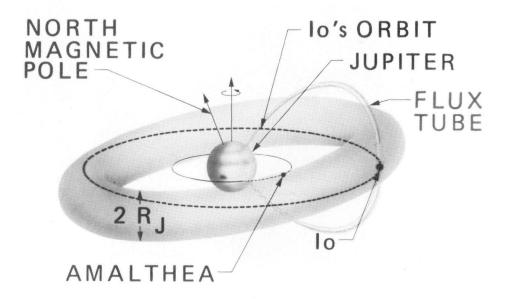

Fig. 6.38 — Encircling Jupiter is a doughnut-shaped torus of charged particles near the orbit of Io. The source of these particles may be Io's volcanic eruptions. (P-21218)

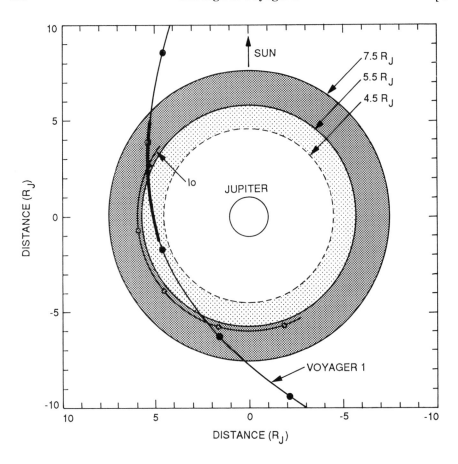

Fig. 6.39 — Between Io's orbit and the planet is a region of "cold" plasma.

Detached haze layers extend upward at least another 500 km. These haze layers made it impossible for Voyager to obtain images of Titan's surface.

(33) Particles in Titan's main haze layer have a mean diameter of 1 μm, but they are probably nonspherical or consist of layers of particles with different sizes.

(34) Hyperion (Fig. 6.62) possesses a highly irregular, ancient surface. Its surface is relatively dark, and gravitational interaction with Titan causes Hyperion to tumble chaotically.

(35) Iapetus (Fig. 6.63) has an extremely dark surface on its leading hemisphere. Its edges are sharply defined and no craters are visible in the interior of the dark region. The source of this dark material is still a matter of debate.

(36) Phoebe (Fig. 6.64) has a 9-h rotation period. (Its orbital period was previously known to be 550 days.) Its dark surface and inclined, retrograde orbit make it likely that Phoebe is a captured asteroid.

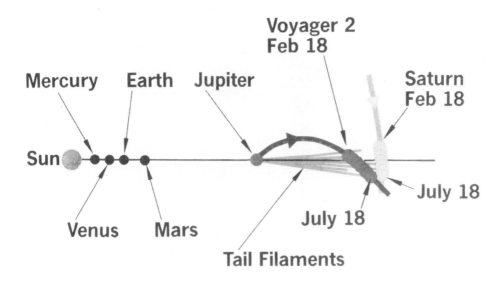

Fig. 6.40 — Jupiter's magnetotail may extend more than 700 million km 'downwind' of Jupiter to the orbit of Saturn. (260-1264A)

Saturn magnetosphere

(37) Kilometric radiowave radiation is pulsed from Saturn at intervals of 10.657 h, which is presumably the rotation period of Saturn's deep interior and its magnetic field.

(38) Saturn's magnetic field (Fig. 6.65) is basically dipolar, with a surface field strength of 0.21 Oe and a tilt of less than 1° from the rotation axis.

(39) The sunward magnetopause extends about 22 Saturn radii outward from the center of the planet. It undergoes variations from this value of about ±40% due to solar wind variations.

(40) An inner torus (inside the orbit of Rhea), with a population of singly charged hydrogen and oxygen ions, probably originates from the sputtering of water ice from the surfaces of Dione and Tethys.

(41) A hot ion region, with temperatures of 30 to 50 keV (which correspond to a temperature of about 500 000 000 K), exists near the outer edge of the inner torus.

(42) A thick plasma sheet of ions of hydrogen, helium, carbon, and oxygen extends outward nearly to the orbit of Titan.

(43) A torus of neutral hydrogen atoms extends from just outside the orbit of Mimas outward to the orbit of Hyperion. Its source is probably the atmosphere of Titan; it is replenished at an estimated rate of about 10^{27} atoms/s.

6.10   THE POST-SATURN MISSION OF VOYAGER 1

On November 23, 1980, shortly after the Saturn encounter period, the Voyager 1 PLS stopped transmitting useful data. A similar problem occurred prior to the

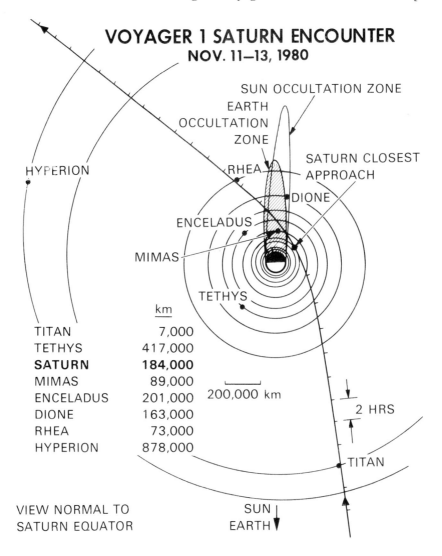

VOYAGER 1 SATURN ENCOUNTER
NOV. 11–13, 1980

SUN OCCULTATION ZONE

EARTH
OCCULTATION
ZONE

HYPERION

RHEA

SATURN CLOSEST
APPROACH

DIONE

ENCELADUS

MIMAS

TETHYS

	km
TITAN	7,000
TETHYS	417,000
SATURN	**184,000**
MIMAS	89,000
ENCELADUS	201,000
DIONE	163,000
RHEA	73,000
HYPERION	878,000

200,000 km

2 HRS

TITAN

VIEW NORMAL TO
SATURN EQUATOR

SUN
EARTH

Fig. 6.41 — A trajectory-plane view of the path of Voyager 1 through the Saturn system. Voyager 1's closest approach to Saturn occurred on November 12, 1980. (260-845A)

Jupiter encounter. On that occasion, temperature cycling of the instrument restored normal operation. Similar tactics in late 1980 and early 1981 were unprofitable, and the Voyager 1 PLS remained inoperative. By late 1988, some progress toward understanding the failure mechanisms had been made, and hope has not disappeared that the PLS might be restored to normal or nearly normal operation. This instrument could provide useful information for the long-term post-Saturn studies of the fields and particle environment of the outer Solar System.

Voyager 1 continues to transmit data 24 h a day. DSN tracking of the spacecraft

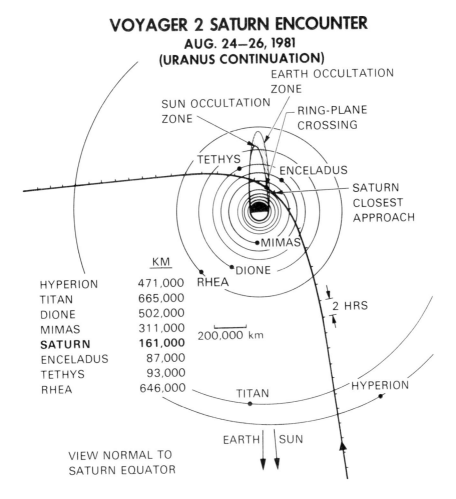

VOYAGER 2 SATURN ENCOUNTER
AUG. 24–26, 1981
(URANUS CONTINUATION)

EARTH OCCULTATION ZONE

SUN OCCULTATION ZONE

RING-PLANE CROSSING

TETHYS

ENCELADUS

SATURN CLOSEST APPROACH

MIMAS

KM

DIONE

HYPERION	471,000
TITAN	665,000
DIONE	502,000
MIMAS	311,000
SATURN	**161,000**
ENCELADUS	87,000
TETHYS	93,000
RHEA	646,000

RHEA

200,000 km

2 HRS

HYPERION

TITAN

EARTH   SUN

VIEW NORMAL TO SATURN EQUATOR

Fig. 6.42 — A trajectory-plane view of the path of Voyager 2 through the Saturn system. Voyager 2's closest approach to Saturn occurred on August 25, 1981. (P-23349)

has averaged between 10 and 20 h per day since Saturn. These data have served four useful purposes:

(1) they provided continued monitoring of the magnetic field and charged particle environments of the outer Solar System in support of a goal to search for the outer boundaries of the solar wind and measure the true interstellar environment;

(2) they provided ultraviolet astronomical data unobtainable from any other Earth-bound or spacecraft-borne instruments;

(3) they provided distant, disk-integrated information on the four gas giant planets in support of the higher-resolution encounter data; and

(4) they provided an engineering test bed for new techniques used by Voyager 2 during the Uranus and Neptune encounters.

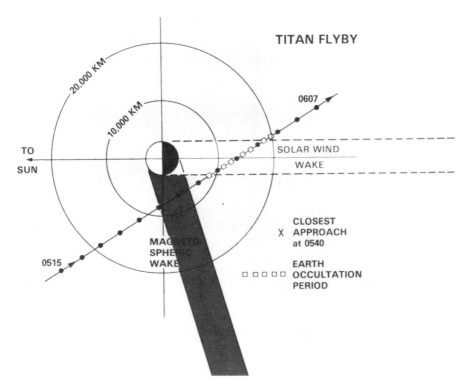

Fig. 6.43 — A trajectory-plane view of Voyager 1's passage of Titan. (260-1043A)

Each of these purposes was of value in ensuring the continued success of the Voyager missions as well as increasing our knowledge of the outer Solar System.

The primary Voyager Mission ended on September 30, 1981, and the Voyager Uranus/Interstellar Mission (VUIM) started the next day. VUIM extended through March 31, 1986, and had as its primary objectives [24]:

(1) To explore the Uranus system and to extend comparative studies of the outer planets to include Uranus exploration results on
   (a) the environment, atmosphere, surface, and body characteristics of the planet;
   (b) one or more of the satellites; and
   (c) the nature of the rings of Uranus.
(2) To investigate the interplanetary and interstellar media.

In addition, mission objectives to maintain the option for a continuing mission were:

(3) To preserve the capability to extend interplanetary and interstellar media investigations well beyond the orbit of Uranus, including a search for the heliopause boundary between the Solar System and interstellar space.
(4) To preserve the capability to extend these investigations to the Neptune system.

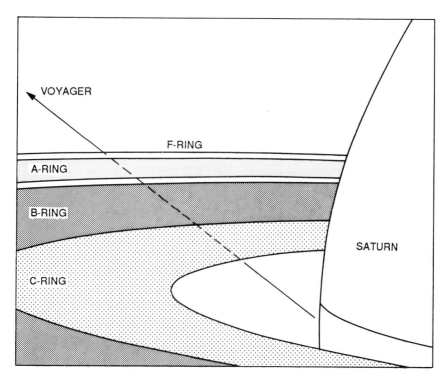

Fig. 6.44 — As viewed from Earth, Voyager 1 passed behind the full radial extent of Saturn's rings.

Voyager 1 played a supportive role in objectives (1) and (4), but was directly involved in objectives (2) and (3). The Voyager Neptune/Interstellar Mission (VNIM) started on April 1, 1986, and is scheduled to continue through September 30, 1990. The Voyager Interstellar Mission (VIM) is slated to follow VNIM. Each of these extended Voyager missions repeat, almost verbatim, objectives (2) and (3) above, replacing 'Uranus' with 'Neptune' in objective (3). It is now anticipated that Voyager 1 will continue its extended mission well into the twenty-first century, barring additional catastrophic electronic failures of critical science instruments or engineering subsystems. Once Voyager 2 has completed its Neptune-related observations, its primary mission will also be dedicated to these same two objectives, and should also continue to return useful data until beyond the year 2015.

NOTES AND REFERENCES

[1] Flandro, G. A. (1966) Fast reconnaissance missions to the outer Solar System utilizing energy derived from the gravitational field of Jupiter. *Astronautica Acta*, **12**, 329–337.

[2] Humes, D. H., Alvarez, J. M., O'Neal, R. L., Kinard, W. H. (1974) The

Fig. 6.45 — This dim image of Saturn about 2.8 days after closest approach served as verification that Voyager 2's scan platform was again responding to commands from Earth. (P-23969)

interplanetary and near-Jupiter meteoroid environments. *Journal of Geophysical Research*, **79**, 3677–3684.

[3] Voyager Project Plan, Voyager Document 618–5, Revision B, dated 1 April 1977, Jet Propulsion Laboratory, California Institute of Technology, Pasadena, p. 2–2.

[4] Sagan, C., Drake, F. D., Druyan, A., Ferris, T., Lomberg, J., Sagan, L. S. (1978) *Murmurs of Earth*, Random House, New York.

[5] A series of 12 papers on the Uranus encounter, including an overview by Stone and myself, and preliminary reports from each of the twelve science investigation teams, appears in *Science*, **233**, 39–109, dated 4 July 1986.

[6] Eshleman, V. R., Tyler, G. L., Anderson, J. D., Fjeldbo, G., Levy, G. S., Wood, G. E., Croft, T. A. (1977) Radio Science investigations with Voyager. *Space Science Reviews*, **21**, 207–232.

[7] Smith, B. A., Griggs, G. A., Danielson, G. E., Cook, A. F. II, Davies, M. E., Hunt, G. E., Masursky, H., Soderblom, L. A., Owen, T. C., Sagan, C., Suomi, V. E. (1977) Voyager Imaging experiment. *Space Science Reviews*, **21**, 103–127.

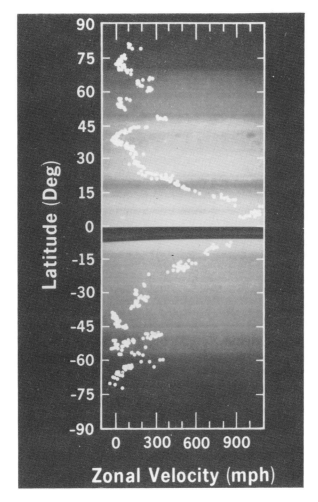

Fig. 6.46 — Zonal winds on Saturn showed a greater degree of north–south symmetry than Jupiter. (260-1529B)

[8] Hanel, R., Conrath, B., Gautier, D., Gierasch, P., Kumar, S., Kunde, V., Lowman, P., Maguire, W., Pearl, J., Pirraglia, J., Ponnamperuma, C., Samuelson, R. (1977) The Voyager Infrared Spectroscopy and Radiometry investigation. *Space Science Reviews*, **21**, 129–157.

[9] Broadfoot, A. L., Sandel, B. R., Shemansky, D. E., Atreya, S. K., Donahue, T. M., Moos, H. W., Bertaux, J. L., Blamont, J. E., Ajello, J. M., Strobel, D. F., McConnell, J. C., Dalgarno, A., Goody, R., McElroy, M. B., Yung, Y. L. (1977) Ultraviolet Spectrometer experiment for the Voyager mission. *Space Science Reviews*, **21**, 183–205.

[10] Lillie, C. F., Hord, C. W., Pang, K., Coffeen, D. L., Hansen, J. L. (1977) The Voyager mission Photopolarimeter experiment. *Space Science Reviews*, **21**, 159–181.

Fig. 6.47 — The extremely complex radial structure within Saturn's B ring was one of the great surprizes of the Saturn encounters of Voyager. (P-23946)

[11] Warwick, J. W., Pearce, J. B., Peltzer, R. G., Riddle, A. C. (1977) Planetary Radio Astronomy experiment for Voyager missions. *Space Science Reviews*, **21**, 309–327.

[12] Scarf, F. L., Gurnett, D. A. (1977) A Plasma Wave investigation for the Voyager mission. *Space Science Reviews*, **21**, 289–308.

[13] Behannon, K. W., Acuna, M. H., Burlaga, L. F., Lepping, R. P., Ness, N. F., Neubauer, F. M. (1977) Magnetic field experiment for Voyagers 1 and 2. *Space Science Reviews*, **21**, 235–257.

[14] Bridge, H. S., Belcher, J. W., Butler, R. J., Lazarus, A. J., Mavretic, A. M., Sullivan, J. D., Siscoe, G. L., Vasyluinas, V. M. (1977) The Plasma experiment on the 1977 Voyager mission. *Space Science Reviews*, **21**, 259–287.

[15] Krimigis, S. M., Armstrong, T. P., Axford, W. I., Bostrom, C. O., Fan, C. Y., Gloeckler, G., Lanzerotti, L. J. (1977) The Low Energy Charged Particle (LECP) experiment on the Voyager spacecraft. *Space Science Reviews*, **21**, 329–354.

[16] Stone, E. C., Vogt, R. E., McDonald, F. B., Teegarden, B. J., Trainor, J. H., Jokipii, J. R., Webber, W. R. (1977) Cosmic Ray investigation for the Voyager

Fig. 6.48 — This image of Saturn's faint D ring is the best ever obtained. (P-23967)

missions; energetic particle studies in the outer heliosphere and beyond. *Space Science Reviews*, **21**, 355–376.

[17] An isotope is an atom with the same number of protons in its nucleus but a different number of neutrons. Deuterium (also called heavy hydrogen) is an isotope of normal hydrogen. Atoms with the same number of protons in their nucleus (i.e., the same nuclear charge) also have the same 'atomic number', Z. Different isotopes of the same element will have the same atomic number, but different 'atomic mass'. Atomic mass units (amu) are scaled such that a carbon-12 atom, consisting of six protons, six neutrons, and six electrons, weighs precisely 12.0000 amu.

[18] Voyager 1 Jupiter results are outlined in (1979) *Science*, **204**, 945–1008; (1979) *Nature*, **280**, 725–806; and (1980) *Geophysical Research Letters*, **7**, 1–68.

Fig. 6.49 — Voyager 2 passed through Saturn's equatorial plane very close to the narrow G ring.
(P-23968)

Voyager 2 Jupiter results are outlined in (1979) *Science*, **206**, 925–996. Combined Voyager Jupiter results are contained in *Journal of Geophysical Research*, **85**, 8123–8841.

[19] Morrison, D., Samz, J. (1980) *Voyage to Jupiter*, NASA Special Publication #439, National Aeronautics and Space Administration, Scientific and Technical Information Branch, 199 pp.

[20] Voyager 1 Saturn results are outlined in (1981) *Science*, **212**, 159–243, and (1981) *Nature*, **292**, 675–755. Voyager 2 Saturn results are outlined in (1982) *Science*, **215**, 499–594. Combined Voyager Saturn results are contained in (1983) *Journal of Geophysical Research*, **88**, 8625–9018, in (1983) *Icarus*, **53**, 165–387, and in *Icarus*, **54**, 160–360.

[21] Gehrels, T, Matthews, M. S. (eds.) (1984) *Saturn*, The University of Arizona Press, Tucson, 968 pp.

Fig. 6.50 — A complex structure which has the appearance of braiding was seen in Saturn's narrow F ring. (P-23099)

Fig. 6.51 — Prometheus and Pandora exert a 'shepherding' influence on the particles of Saturn's F ring. (260-1135A)

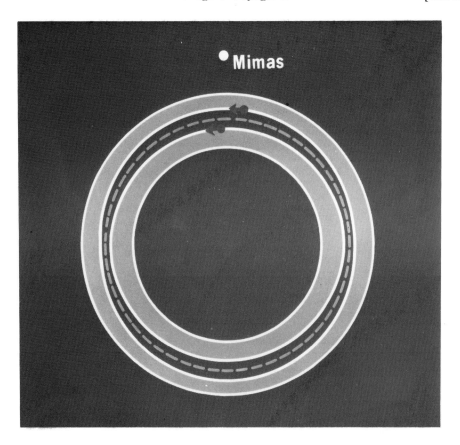

Fig. 6.52 — Owing to gravitational interaction with Mimas, particles near the outer edge of the
B ring form a 'centered' ellipse. (260-1450)

[22] Greenberg, R., Brahic, A. (eds.) (1984) *Planetary Rings*, The University of Arizona Press, Tucson, 784 pp.

[23] Morrison, D. (1982) *Voyages to Saturn*, NASA Special Publication #451, National Aeronautics and Space Administration, Scientific and Technical Information Branch, 227 pp.

[24] The science objectives are as stated in Section 2.0 of Voyager Document 618-5, Voyager Project Plan, Part II: Voyager Uranus/Interstellar Mission, dated July 22, 1981.

BIBLIOGRAPHY

Burns, J. A. (1986) Some background about satellites. In Burns, J. A., Matthews, M. S. (eds.) *Satellites*, The University of Arizona Press, Tucson, pp. 1–38.

Gehrels, T., Matthews, M. S. (eds.) (1984) *Saturn*, The University of Arizona Press, Tucson, 968 pp.

Fig. 6.53 — Spiral density waves within the A ring are also a result of interaction with some of
the satellites of Saturn. (260-1135B)

Morrison, D. (1982) *Voyages to Saturn*, NASA Special Publication #451, National
    Aeronautics and Space Administration, Scientific and Technical Information
    Branch, 227 pp.

Morrison, D., Samz, J. (1980) *Voyage to Jupiter*, NASA Special Publication #439,
    National Aeronautics and Space Administration, Scientific and Technical Infor-
    mation Branch, 199 pp.

Stone, E. C., *et al.* (1977) *Space Science Reviews*, **21**, 75–376 (Contains a series of 13
    papers, including a science overview, a mission description, and individual
    papers on each of the 11 scientific investigations).

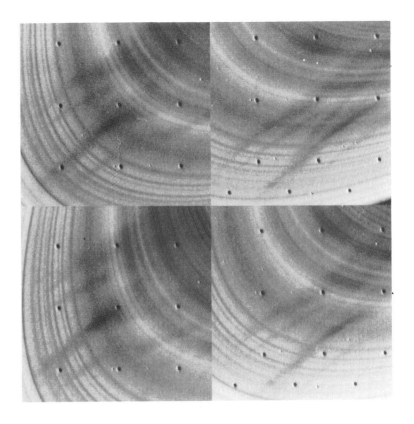

Fig. 6.54 — Radial 'spokes' were seen in the outer half of the B ring. Their source may be interaction of tiny ring particles whose electrostatic charges allow them to be influenced by Saturn's magnetic field. (260-1499)

Fig. 6.55 — Mimas may have been very nearly broken apart by the impact that caused Herschel
Crater. (P-23210)

Fig. 6.56 — Flow lines, fractures, and reduced numbers of craters on Enceladus testify to
geologically recent surface alteration on this small Saturnian satellite. (P-23956)

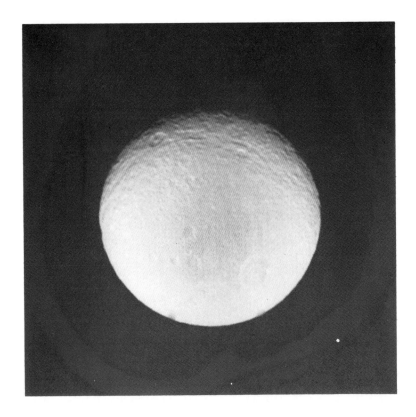

Fig. 6.57 — Ithaca Chasma very nearly encircles Tethys. It may be causally related to the 400-km diameter Odysseus Crater. (P-23948)

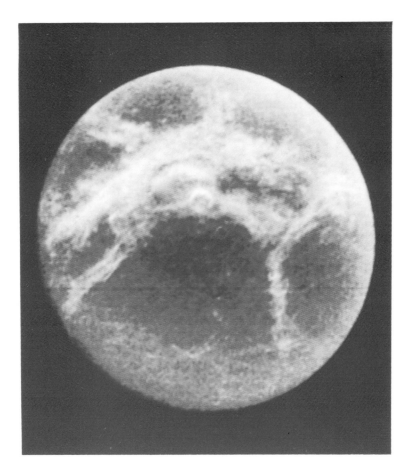

Fig. 6.58 — Dione, although close in size to Tethys, has wispy markings on its surface unlike anything seen on Tethys. (P-23269)

Fig. 6.59 — Voyager 1 passed relatively close to Rhea and obtained this mosaic of its north polar region. (P-23177)

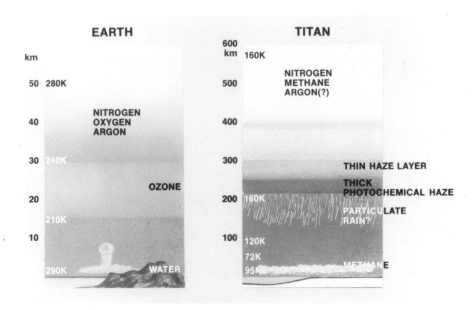

Fig. 6.60 — Many similarities exist between the atmospheres of Titan and Earth. Because of Titan's lower gravity, its atmosphere is much more extended than Earth's. (260-1526A)

Fig. 6.61 — A main haze layer obscures the surface of Titan. Detached haze layers above the main haze layer are also apparent. (260-1131)

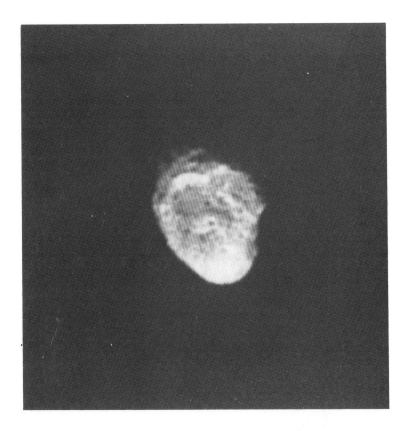

Fig. 6.62 — Hyperion is remarkably nonspherical for a satellite comparable in size to Mimas. Its elongated shape interacts gravitationally with Titan to cause Hyperion's spin rate and orientation to change dramatically and chaotically. (P-23936)

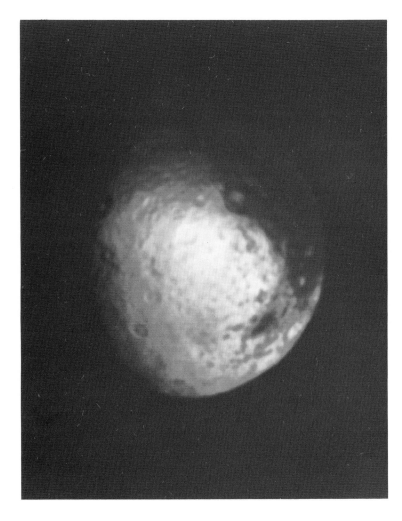

Fig. 6.63 — Earth-based observations of Iapetus showed that one side of this strange satellite is 6 times as bright as the other. (P-23961)

Fig. 6.64 — Voyager 2 passes within several million km of Phoebe, whose darkness and unusual orbit lead astronomers to believe that it is a captured asteroid rather than a primordial Saturn satellite. (P-24137)

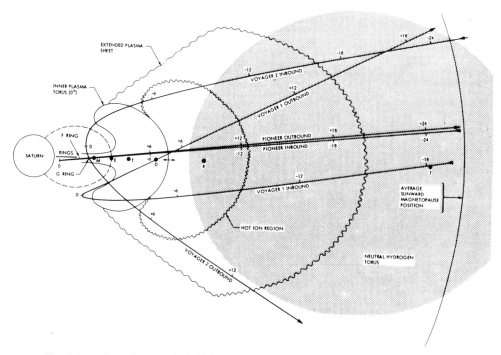

Fig. 6.65 — Saturn's magnetic field is closely aligned with its rotation axis. This schematic diagram shows the approximate sunward cross-sectional structure within Saturn's magnetosphere.

7

The Voyager 2 encounter with Uranus

7.1 PREPARING VOYAGER 2 FOR URANUS

Permission from NASA and funding from the United States Congress were the major political hurdles to be cleared for Voyager 2 to continue an extended mission to Uranus. By the start of the Voyager Uranus/Interstellar Mission (VUIM) on October 1, 1981, many of the engineering and scientific problems associated with such a mission had only begun to be addressed. Contrary to popular belief, Voyager 2 did not remain basically idle during the four years between the start of VUIM and the beginning of the Uranus encounter period. Like its twin, Voyager 1, the spacecraft continued to transmit useful scientific and engineering data 24 h a day. A relative dearth of other deep-space probes with competing requirements made it possible to track the two spacecraft with DSN antennas and collect their data for an average of 12 to 16 h a day. Some of the data was designed for post-Saturn calibrations, but the majority was either unique scientific information or the results of engineering tests being done in preparation for the Uranus encounter.

NASA-imposed funding constraints dictated that operating costs for the 'cruise' portion of VUIM be substantially reduced from those levels extant during the Saturn encounters. As a consequence, a large fraction of the Voyager Project staff at JPL was forced to find other flight projects to utilize talents and expertise gained during the Jupiter and Saturn encounters. Many joined the Galileo Project (Jupiter orbiter and atmospheric probe mission) or the Magellan Project (Venus radar mapping mission). Budgets for the Voyager investigation teams were also reduced, and several co-investigators whose primary scientific interests lay in the Jupiter and Saturn systems had their funding either severely cut or entirely eliminated.

The smaller remaining project staff at JPL was kept inordinately busy conducting the cruise science and engineering activities as well as doing the long-range planning for the upcoming Uranus encounter. The more detailed Uranus encounter science planning began in early 1984 after the Voyager science and engineering staff had been rebuilt to more robust proportions. At that time, many of those who had worked on Voyager during the Saturn encounter returned to the project.

7.1.1 Detecting Voyager's weak signals

Voyager's increasing age and distance from Earth combined to create some major challenges to the VUIM staff at JPL. Voyager electrical power, provided by the plutonium oxide RTGs, was dropping at a rate of about 7 W per year, owing both to the slow depletion of radioactive plutonium and to the slowly decreasing efficiency of the aging generators. The spacecraft transmitted the same 22-W telemetry signal over its high-power X-band radio signal that had been used at Jupiter and Saturn, but it became apparent that it would not be possible during the Uranus encounter to keep IRIS's health-preserving flash-off heater on while transmitting both at high-power X-band and low-power S-band. The low-power S-band signal served two useful purposes: (1) it carried its own back-up 40-bps (bits per second) stream of engineering data, and (2) it provided a second frequency for radio science experimentation, greatly enhancing the value of that investigation.

Even with the IRIS flash-off heater turned off, the desired dual-frequency atmospheric occultation experiment, utilizing S-band in its high-power mode and X-band in one of its two power modes, required more power than would be available. If this important experiment were to be available for use at Uranus, some major changes would have to be made in the power utilization of Voyager 2.

Spacecraft Team engineers studied the situation and proposed that a number of unneeded electronic components of the spacecraft be powered down. They also provided power usage rules to ensure that Voyager 2 did not enter 'PWRCHK', a programmed response to excessive usage of the available electrical power which automatically shut down certain parts of the spacecraft. These rules stated that the IRIS flash-off heater would have to be off during periods when both S- and X-band transmitters were needed, and that S- and X-band transmitters could not be in their high-power modes simultaneously. The former rule required special coordination to assure that IRIS had sufficient sensitivity for the Uranus measurements, while still permitting limited use of the dual-frequency radio transmitter capability of the spacecraft. The latter rule forbade doing the RSS occultation experiments in the manner that they had been done at Jupiter and Saturn. The compromise position was to use X-band at high power and S-band at low power during the Uranus ring occultation periods, and to reverse the power states to S-high and X-low during the atmospheric occultation period.

From Jupiter's distance, Voyager had been able to transmit its scientific and engineering data at a rate of 115 200 bps with high confidence that little or none of the data would be unintelligible to the sensitive 64-m tracking antennas of the DSN. However, each time Voyager's distance from the Earth doubles, the signal received at Earth by a given tracking antenna drops by a factor of four. Uranus was nearly 20 AU from Earth during the Uranus encounter, very nearly twice the distance of Saturn, which in turn was twice the distance of Jupiter. With its fixed 22-W high-power X-band telemetry stream, Voyager could not increase its radiated power and thereby compensate for the increased distances needed to continue sending data at a rate of 115.2 kbps (kilobits per second). Partial compensation for the lower received power at the DSN stations came from a lowering of the average data rates. At Saturn, the maximum data rate was 44.8 kbps; at Uranus, the maximum rates used were 21.6 kbps and 14.4 kbps.

The 44.8 kbps rate at Saturn was higher than the theoretical reduction of a factor of four below Jupiter's 115.2 kbps. Voyager's maximum rate of 115.2 kbps transmitted from Jupiter arrived at DSN stations at received power levels nearly a factor of two above the threshold limits, mainly as a result of continuing efforts since launch to improve the sensitivity of the 64-m stations. The Saturn rates were much closer to DSN limits.

Higher than predicted data rates at Uranus again were made possible through special techniques employed by the DSN. A new software system created a large number of unexpected problems during its testing and implementation stages, but eventually improved the overall efficiency of the tracking stations. The greatest improvements were a result of 'antenna arraying', so that data from Voyager 2 was received simultaneously from two or more tracking antennas, the signals from which were electronically synchronized and combined to achieve a single, stronger signal. In effect, the two or more antennas acted together as a single larger antenna. The 64-m antennas at Madrid, Spain, and at Goldstone, California, were arrayed with one or two adjacent 34-m antennas.

The premier tracking site for the Uranus encounter was the Canberra Deep Space Communications Complex (CDSCC), shown in Fig. 7.1. Because of the

Fig. 7.1 — Pictured is the Deep Space Network tracking complex near Canberra, Australia.
(355-4039)

southerly latitude of Uranus in Earth's skies in 1986, Voyager 2 passed almost directly over the Australian tracking stations, and tracking was possible for more than 12 h per day. The Madrid (MDSCC) and Goldstone (GDSCC) complexes could view Voyager 2 for only 8 h per day. At CDSCC, the 64-m antenna could be arrayed with both 34-m antennas and with the Parkes Radio Telescope (Fig. 7.2). In return

Fig. 7.2 — The Parkes Radio Telescope near Parkes, Australia, assisted in the collection of Voyager 2 Uranus data. The signal was electronically combined with that at the Canberra complex by means of a 400-km, high-quality microwave link. (355-4675A)

for the Voyager 2 tracking time offered at the Parkes Radio Telescope, NASA provided a low-noise X-band receiver and compensatory radio astronomy observing time on the Canberra 64-m DSN antenna. With some NASA-provided funding assistance, the Australian government built a high-quality, 400-km-long microwave link between Canberra and Parkes to facilitate arraying of these two 64-m antennas. When combined with the two 34-m stations at Canberra, the result was a collecting area equivalent to a single antenna with a diameter of nearly 100 m. In practice, the two 64-m antennas were arrayed with a single 34-m station, reserving the remaining 34-m dish for Voyager 1 or another deep space probe. The tracking capability afforded by this long-distance 'array' of receiving stations made it possible to return approximately 50 additional images (or equivalent quantities of non-imaging science data) in any given 24-h period during the Uranus encounter. That represented an increase of 20 to 25% over the capability of the unassisted worldwide Deep Space Network.

7.1.2 Reprogramming Voyager's computers

Each of the Voyager spacecraft were built with three pairs of computers, as discussed in Chapter 6. These are the two Computer Command Subsystem (CCS) computers, the two Flight Data Subsystem (FDS) computers, and the two Attitude and Articulation Control Subsystem (AACS) computers.

The two CCS computers issue to other spacecraft subsystems the time-sequential commands that constitute the Voyager sequence of events. At Jupiter, both CCS computers contained the same commands, one acting as prime and the other as backup. At Saturn and Uranus, except for a few critical commands, the CCS computers were used primarily in a non-redundant fashion. This offered the advantage of a nearly-doubled memory space, which could be used either to carry out a more complex sequence or to operate the spacecraft for a longer period of time before reprogramming was needed. CCS 'loads' operated the spacecraft for periods of from two days to six months before new sets of instructions had to be transmitted.

The two FDS computers issue timing pulses to the individual science instruments to control their routine operation modes. Changes were made prior to Uranus to the operation mode of the PLS to improve its sensitivity to the lower plasma levels experienced in the outer Solar System. The timing of PLS modes was also changed to make PLS less susceptible to interference caused by the stepper motor within the LECP. The FDS computers also control the formatting of data to be sent back to Earth. A number of different data formats were chosen for the Uranus encounter; data format is selected by CCS command. These new modes utilized the two FDS computers in parallel to accomplish some on-board processing of imaging data. This on-board processing reduced the number of bits without any loss of imaging information by a technique to be described later. Appropriately, the set of data formats was called the 'dual-processor program'. In addition to image data compression (IDC), the dual-processor program also utilized for the first time an item of hardware on the spacecraft known as the Reed–Solomon encoder to provide a more efficient data encoding scheme than had been used at Jupiter or Saturn. Another first for Voyager at Uranus was a data scheme employing simultaneous real-time imaging and tape-recorder playback. Prior to Uranus, only real-time non-imaging data could be returned with playback data. The dual-processor program permitted complete image readout to the tape recorder in either 48 s or 4.0 min. Complete real-time image readouts could be programmed to take either 4.0 or 8.0 min. One dual-processor mode permitted reading only the top 60% of an image in 2.4 min.

The two AACS computers may not be used simultaneously, since each controls a separate system of attitude control thrusters and associated plumbing. Only the primary AACS computer on Voyager 2 has been used in flight. It has been reprogrammed on a number of occasions to correct systematic errors detected in the scan-platform pointing, to improve spacecraft stability for the purpose of reducing image smear, to conserve attitude control gas, and to permit more flexibility in turning the spacecraft around one or more of its three principal axes.

7.1.2.1 Image data compression

Image data compression (IDC) in the FDS computer permitted the same amount of real information in the 14.4-kbps Uranus data stream as was contained in a 29.9-kbps data stream at Saturn. The Uranus 14.4-kbps rate, like the Saturn 29.9-kbps rate, permitted transmission of real-time non-imaging science and engineering data at its maximum rate (3.6 kbps), plus transmission of one real-time image every 4 min. At Saturn, the full 5.12 million bits per image had to be transmitted. The dual-processor program for Uranus processed imaging data to intercompare adjacent picture elements and transmit only the differences, cutting by a factor of almost three the total number of bits per image.

Long-term planning accounts for the majority of the successes of Voyager. Quick responses to unexpected problems were also necessary on rare occasions. As an example, barely a week before Voyager 2 reached its closest approach to Uranus, both WA and NA images began to show intermittent bright and dark streaks. Image processing personnel and ISS team members initially thought the problem might be in the computer processing done after the images were received at JPL. When no problem could be found in the ground software, the spacecraft FDS computer memory was checked. It was discovered that one memory location in the processor that performed IDC was the culprit. In less than three days, FDS experts reprogrammed the IDC routines to use an alternative memory location, and the streaks disappeared just six days before the January 26, 1986, closest approach date.

7.1.2.2 Reducing image smear

A major complication added by the remoteness of Uranus from the Sun was the large reduction in light levels available for imaging. The brightness of sunlight at Uranus was a factor of 4 lower than that at Saturn, a factor of nearly 14 below that at Jupiter, and a factor of 369 below that at Earth! Furthermore, the rings and inner satellites of Uranus are among the darkest objects in the Solar System. The source of this darkness may be the result of the bombardment of methane (CH_4) ice by high-energy protons. Proton bombardment of methane ice results in the release of the hydrogen, leaving a residue of black carbon.

Voyager's orientation in space is maintained by keeping the Sun and a reference star in its celestial sensors. Any deviation of more than 0.05° from centered viewing of the Sun in the Sun Sensor or of the star from centered viewing in the Canopus Star Tracker will trigger the firing of the appropriate attitude control thruster jet. Motion within the spacecraft attitude 'deadband' is generally relatively slow. At Saturn, for example, this gentle rocking motion averaged only about 1/10 the angular speed of the hour hand on a clock. Even that was too rapid for the combination of high resolution and long exposures required at Uranus. A 15-s exposure required angular rates a factor of 10 slower to avoid smearing over more than two pixels of the NA camera. SCT engineers again came up with a solution: shorten from 10 to 5 ms the duration of the jet pulses used to stabilize the spacecraft, effectively giving a gentler correction to detected orientation errors. SIS and SEQ personnel also assisted by building into the CCS sequence as much time as possible for the spacecraft to 'settle' after maneuvers, scan platform slewing, or tape recorder motion. The result was a high percentage of long-exposure images whose clarity was exceptional.

Even at Jupiter and Saturn, tape recorder motion caused image smear. To reduce that smear, Voyagers 1 and 2 were programmed to fire the yaw [1] thrusters to counteract the torque caused by starting or stopping the recorder. The routine was called DSSCAN, for *D*igital *S*torage *S*ubsystem motion *CAN*cellation. Although DSSCAN stopped the major effects of tape recorder starts and stops, yaw-thruster firing results in small motions in the pitch axis as well. For the short exposures used at Jupiter and Saturn, this added motion was negligible, but the longer Uranus system exposures made even second-order effects important. DSSCAN was reprogrammed to include pitch-axis compensation at Uranus.

At Saturn, a new technique called image motion compensation (IMC) was first tried. The spacecraft had been built to reference its orientation to internal motor-driven gyroscopes at times when celestial references could not be used (during solar occultation periods or at times when roll, pitch, or yaw manoeuvers were being executed). Voyager software makes it possible to compensate for slow drifts in the orientation of the gyroscopes. SCT engineers determined that the spacecraft could be turned slowly and accurately (rates up to 110° per hour in each axis) by offsetting the gyroscope drift rate factors in the software from their actual drift rates. In essence this tricked the software into believing that the gyroscopes were drifting rapidly. Such Gyro Drift Turns (GDTs) could be used to turn the entire spacecraft to pan the cameras to compensate for apparent angular motion of a nearby satellite. During this process, images and other data would be recorded on the digital tape recorder for later playback. IMC was used for one Voyager 1 observation of Saturn's satellite Rhea. At Uranus, IMC was used for eight separate observations, one of which (Miranda closest approach imaging) utilized nine separate GDT rates. All GDTs executed exactly as planned.

7.1.2.3 *Encoding the data*
Although the DSN is generally able to collect almost flawless Voyager data, the quality of the received data can be adversely affected by poor weather conditions at the tracking stations, changes caused by Earth's ionosphere or solar wind conditions, or data transmission which must pass close to the Sun in transit to Earth. Encoding of the data prior to its transmission from the spacecraft can help to overcome this problem and permit errors in the data to be detected and corrected. This encoding adds extra bits to the data stream, often in amounts comparable to the intrinsic data prior to encoding. At Jupiter and Saturn, for example, the encoding scheme essentially doubled the required bit rate for non-imaging data.

To conserve bits during the Uranus encounter, the Reed–Solomon data encoder hardware was used for the first time. Prior to the Uranus encounter, one of two redundant 'Golay' encoders had been used. As long as it functioned properly, the single Reed–Solomon encoder could cut by more than 50% the number of code bits needed for data streams of 14.4 kbps or less. In its initial flight testing, the Reed–Solomon hardware functioned flawlessly.

Before the Uranus dual-processor FDS program was used, imaging data had always been transmitted without either Golay or Reed–Solomon coding. Infrequent bit errors would result in lighter or darker pixels, comparable to 'snow' in a television picture of substandard quality. Bit errors in the IDC images at Uranus could potentially affect entire lines of imaging and were therefore more serious than corresponding errors at Jupiter or Saturn. For this reason, whether transmitted in real time or played off the tape recorder at a later time, IDC images required the imposition of Reed–Solomon coding. The Reed–Solomon encoder could not handle data rates higher than about 14.4 kbps. The combined playback/imaging data format required a data rate of 21.6 kbps, so only a portion of the data could be encoded. Since the recorded non-imaging data was Golay-encoded prior to recording, the logical choice was to Reed–Solomon-encode only the real-time data. Recorded IDC imaging data could then be saved for playback at lower data rates, which would permit Reed–Solomon encoding of the entire data stream. This choice resulted in added complications for tape recorder data management, but experienced SEQ personnel handled the problem without breaking stride. The 14.4-kbps limitation also explains why the Reed–Solomon encoder was not used at Jupiter and Saturn, since the prime data rates at those planets were considerably higher than 14.4 kbps.

7.1.2.4 Sensing and handling problems automatically
The large distances and correspondingly long communications times make it prudent to program the two Voyager spacecraft to perform periodic health checks and to take appropriate actions if faults are detected. For Voyager 2 it is also important to consider what should be done if the remaining radio receiver should fail. Since no further commanding of the spacecraft would be possible, all future actions of the spacecraft would have to be preprogrammed into the spacecraft computers prior to the receiver failure. For this reason, Voyager science and engineering personnel periodically load a part of the CCS computer with a skeleton sequence known as a Back-up Mission Load (BML). Only during the busiest portions of an encounter, when CCS memory space is at a premium, is BML removed. If Voyager 2's radio receiver had failed more than two weeks before closest approach to Uranus, the spacecraft would have obtained a series of pictures of Uranus and its rings, collected fields and particles data, and performed a simple radio science occultation experiment as Voyager 2 passed through the shadow of Uranus. Concurrent IRIS, PPS, and UVS data would also have been obtained, though the optimum viewing geometries actually used during the encounter would not have been accommodated. No high-resolution satellite data would have been obtained.

A large number of other 'Failure (or Fault) Protection Algorithms' (FPAs) were programmed into Voyager 2. In most of these FPAs, several preliminary steps are tried before drastic steps are taken to correct the problem. All of the FPAs have been thoroughly tested on the ground to be certain they work properly, but few have occurred in flight as a result of actual problems on the spacecraft. A relatively detailed description of these FPAs has been given by Jones [2].

If excessive electrical current is used, a drop in power-supply voltage would be detected. In such a circumstance, the spacecraft would have shut down all non-essential systems, rechecked the voltage, and then continued along one of a number of different courses of action, depending on the conditions detected. In one of these scenarios, for example, the imaging cameras (largest power users among the science instruments) would be shut off, the other three scan platform instruments would be pointed in a near-optimum direction in space, the scan platform motors would have been turned off, and Voyager would have continued to transmit non-imaging data at a rate of 4800 bps.

Leaks in the plumbing system associated with the attitude control jets (not correctable by lesser actions) would have been countered by shutting down the primary plumbing and activating an entirely separate backup plumbing system.

If a non-catastrophic collision or a passing bright particle were to cause the Sun Sensor (SS) to lose the Sun, the spacecraft would perform a series of yaw and pitch turns to systematically search the sky to relocate the Sun, switching to a backup SS if necessary. Once the Sun was located in the SS field of view, a slow roll turn would be executed until the predesignated star is sighted in the Canopus Star Tracker (CST). A backup CST is also available for usage if no star is sensed by the primary CST. Seizure of the scan platform can also result in turning off power to the platform motors to prevent damage to these critical pieces of hardware.

Should one of the spacecraft transmitters fail, backup transmitters for both the S-band and X-band transmitters are available. The precise actions taken by the spacecraft depend on whether or not Voyager is in a critical part of the encounter operations.

Some automatic responses to science instrument problems are also available. Since IRIS needs to be at a precisely controlled temperature to operate successfully, it has a backup heater in the event the CCS computer senses that the primary heater is not operating properly. Occasionally the PRA erroneously senses an overload and inserts signal attenuators. This condition is called PRA 'Power-On Reset' (POR). When Voyager senses that a PRA POR has occurred, it increments a counter and sends the appropriate commands to the PRA to restore normal operation. Since the PPS could be damaged or destroyed by high light levels, certain types of responses to other problems on the spacecraft will often include switching the PPS to its smallest field of view and its lowest gain state, as well as pointing the scan platform to a predetermined 'safe' position.

The CCS computer itself performs frequent self-tests. If a command is transmitted from Earth that the CCS is not programmed to receive, or if partial memory failure occurs, the response could result in termination of the executing sequence of events until corrective commands are received from Earth.

These are just a few samples of the Voyager's pre-programmed 'thought' processes. It is comforting to know that these silent sentinels stand ready to guard against such a wide variety of potentially life-threatening conditions on Voyager. Though few have actually been called to service, such safeguards have undoubtedly contributed to Voyager's lifetime extending far beyond early expectations.

7.1.3 Planning a sequence of events

Those who have driven a vintage automobile for a lengthy period of time are well aware of the idiosyncrasies and strong points of an 'ancient' machine. Each such machine develops its own personality, and as its owner and operator becomes intimately familiar with that personality, loving and tender care can often elicit responses far beyond those available to 'strangers'. So it has been with Voyager. Years of experience have taught the Voyager science and engineering personnel how to utilize the capabilities of the spacecraft more and more efficiently to collect unique science data about these distant giant worlds. For those closely involved with the long-term deliberations and planning, Voyager has offered an intellectually stimulating challenge to find new and innovative methods and techniques for investigation of the outer Solar System.

7.1.3.1   Assessing what was already known

Because of the great distance of Uranus and the fact that no Pioneer spacecraft had blazed the trail for Voyager 2, much less was known about this giant planet than was known about Jupiter and Saturn prior to their encounters. A group of over 100 scientists gathered in Pasadena, California, on February 4 through February 6, 1984, to discuss the state of knowledge of the Uranus system (and the Neptune system). This conference resulted in the book *Uranus and Neptune* [3], which was referenced extensively in Chapter 5.

The rotation periods of Uranus and Neptune were virtually unknown. There was no direct information on the nature of their magnetic and associated radiation fields. The Uranian rings were known to be very dark and could not be resolved in Earth-based images; at that time it was supposed than Neptune had no ring system. Only the barest fragments of information existed on the two satellite systems.

The high tilt of Uranus's rotation axis offered special challenges. Voyager was approaching Uranus at a time when only its southern hemisphere (and the southern hemispheres of its five known satellites) would be illuminated and visible during the four-month encounter period. Since all the satellites and the rings orbit Uranus in or near its equatorial plane, prime data-collection periods for all the bodies would occur within a short 6-h period. Corresponding events were spread over many days at Jupiter and Saturn. Fitting the most important observations into such a short time period would require careful planning and judicious use of the available time and resources.

7.1.3.2   Selecting major scientific goals

Armed with the latest available Uranus Earth-based data and a meager understanding of Voyager's capabilities, the Uranus Science Working Groups (USWGs) held a series of meetings during the months of April, May, and June of 1984. Their primary purpose was to recommend to the Voyager Science Steering Group a set of time-sequential observations to address as many of the scientific unknowns at Uranus as possible. Their final report was presented to the SSG in July, 1984, and published a few months later [4].

The three USWGs were the Atmospheres, the Rings, and the Satellites and Magnetospheres Working Groups. Quite naturally, their recommendations were divided into four areas: atmosphere, rings, satellites, and magnetosphere.

The atmospheric goals, from highest to lowest priority, included the following:

(1) Bulk atmospheric composition: hydrogen (H_2), helium (He), and methane (CH_4) abundances in Uranus's upper atmosphere.
(2) Global energy budget: thermal emission and bolometric Bond albedo of Uranus to determine the amount of heat coming from the interior of the planet, if any.
(3) Vertical structure: atmospheric temperatures and pressures, both vertically from the troposphere to the exosphere as well as horizontally with latitude and longitude.
(4) Horizontal cloud and temperature structure, composition, physical processes, and rotation rate: meteorology.
(5) Composition and variable constituents: methane (CH_4), ethylene (C_2H_4), acetylene (C_2H_2), atomic hydrogen (H), *ortho/para* [5] hydrogen (H_2) ratio, etc.

(6) Auroral regions: structure and appearance at both poles.
(7) Clouds and hazes: vertical structure and particle properties.

Goals for the Voyager observations of the Uranian rings, again in order of priority, included the following:

(1) Ring structure: both radial and azimuthal profiles and their changes with time.
(2) Ring satellite search: moonlets in or near the rings.
(3) Search for additional ring material.
(4) Orbital kinematics: improve orbit models, compare ring motions to satellite motions.
(5) Ring particle motions: importance of self-gravity, vertical thickness of rings, collisions between particles.
(6) Ring particle properties: size, shape, reflectivity, temperature, composition.
(7) Ring environment: interaction with magnetospheric plasma or with an extended neutral atmosphere.

Prioritized observational objectives for Uranus satellites Voyager studies of the satellites included:

(1) Global properties: sizes, shapes, masses, densities, variations with longitude and time.
(2) Surface characteristics: types of surface structures, characterization of surface processes, variations with solar phase angle.
(3) Satellite system: searches for satellites both interior and exterior to the orbit of Miranda.

Magnetospheric objectives were stated in a slightly different format, since the fields (MAG), particles (CRS, LECP, PLS), and wave (PRA, PWS) investigations operate continuously and generally do not need to be pointed at their targets. The primary long-range remote-sensing observations of interest in magnetospheric studies of Uranus were ultraviolet observations of possible auroral activity and searches for more extensive distributions of hydrogen molecules, atoms, and ions within the Uranus system. Other objectives, in time order, included observations of: the magnetopause (inbound), the outer magnetosphere, magnetosphere–satellite interactions, a possible sheet of plasma above Uranus's equator, the magnetic tail structure, and the magnetopause (outbound).

7.1.3.3 *Matching scientific goals with spacecraft capabilities*
With the assistance of Voyager FSO personnel at JPL, the USWG members then proceeded to describe the instruments, instrumental configurations, and times when Voyager 2 observations contributing to the selected science objectives could be accomplished. This task was initially done independently within the three separate working groups. Each working group was encouraged to select observation designs which would simultaneously contribute to more than a single science objective. A particular design for an observation was known as a 'link', (representative of a link in a chain of events). An individual link might be used several times, but always with

either the same design or the same science objective. Details of the link designs were not included in this early stage of planning, but were developed later.

Once the sets of links and execution times were defined, then individual link execution times were incorporated in a computerized time-ordered listing of events and laid out on graphical timelines. These timelines also included the estimated capabilities of the communications link between Voyager and the DSN tracking stations. Assuming full station arraying, the maximum data rate available while Voyager was over the Madrid tracking stations was 8.4 kbps; the Goldstone tracking stations could receive data at rates up to 14.4 kbps; Canberra had both the longest tracking passes (12 h) and the highest data rate capability (21.6 kbps). Available rates were lower near the beginning and end of each pass. Early guidelines suggested that for all but a two-week period including Uranus closest approach, only a single 64-m station should track Voyager 2. For 64-m coverage, maximum data rates were 7.2 kbps for MDSCC, 8.4 kbps for GDSCC, and 14.4 kbps for CDSCC. Real-time imaging was only possible for rates of 8.4 kbps or greater. Where the data rates required for a particular link exceeded the stated DSN capabilities, decisions had to be made about whether to shift the link execution times, delete particular link executions, delete the link entirely, or recommend augmentation of the DSN station coverage.

In an iterative fashion, each of the three working groups eventually put together a recommended sequence of atmospheric (or ring, or satellite, or magnetospheric) observations. Each also categorized their recommended links as first priority, second priority, or third priority science. Although these relative priorities existed only within a given discipline, they nevertheless served a very useful purpose in the later resolution of intergroup recommendation conflicts. Although most USWG members were investigators on one of the science teams, it is fortunate indeed that all were objective enough to consider the relative merits of scientific observations from all disciplines. The processes required to optimize the overall science return would otherwise have been enormously more difficult.

7.1.3.4   *Working Out conflicts*
Conflicts between the requested observations of the three working groups became immediately apparent once their requests were merged into a single data base. For all but a three-week period from 14 days before to seven days after closest approach to Uranus, these conflicts were mainly resolved by relatively minor time shifts of one or more of the observations. Occasionally, routine periodic observations would have to be stopped prior to a unique, many-hour-long observation and then resumed afterwards.

The highest-quality observations were concentrated within a few days of close approach to Uranus, especially on the approaching leg where the planet, rings, and satellites were nearly fully illuminated. Here conflict resolution was considerably more painful, because many good observations had to be discarded to retain the highest-priority science and to provide a relatively uniform mix of atmospheric, ring, and satellite science. Some observations proved to be impossible, due to the combination of geometrical factors and other spacecraft and sequence constraints.

Among the more difficult compromises was the one associated with two high-priority observations of Miranda. Determination of Miranda's mass required real-time tracking of the spacecraft during the period surrounding closest approach to this small satellite. High-resolution imaging requirements dictated that image motion compensation (IMC) be provided, and this could only be accomplished by turning the entire spacecraft to keep the narrow-angle camera pointed at the satellite near the time of closest approach to Miranda. Doing IMC unfortunately also turns the spacecraft antenna off Earth-pointing, making real-time radio tracking of the spacecraft impossible. It seemed at first that the two observations were incompatible. Upon closer examination, the imaging objectives were slightly weighted toward the pre-closest-approach time period, while the radio science Miranda mass determination experiment was most sensitive to data at and following closest approach. If both experiments were somehow to be accommodated, it was apparent that the imaging should precede the mass determination. The question then became one of determining the time at which the spacecraft should be back on Earth-line.

One innovative idea provided about 5 min of time and the key to a workable solution. Up to then, IMC designs had assumed that the observation started with an Earth-pointed antenna and ended with the antenna several degrees off Earth-line. The amount of time required to return the spacecraft antenna to Earth-line was approximately 30 s for every degree. Since the Miranda IMC maneuver took the spacecraft nearly 10° off earth-line, nearly 5 min would be required to return. Suppose the software could be fooled into designing the IMC maneuver backwards in time: start at the end time of the imaging with the spacecraft antenna pointed at Earth and end at the beginning of the imaging with the 10° offset. A pitch/yaw prewind before the observation could then replace the normal pitch/yaw unwind after the observation. The ploy worked. The best Miranda imaging, including critical IMC maneuvering, continued until about four minutes before closest approach to Miranda. Thanks to the backwards design of its maneuvers, Voyager 2's antenna was close enough to Earth-line about five minutes before Miranda closest approach to begin the mass determination experiment, and both experiments successfully returned high-quality data.

A multitude of other similar (though generally less dramatic) compromises characterized the design of the Uranus encounter sequence of events. The sense of pride in a job done well was felt by most of those intimately involved in the preparation of Voyager 2 for the encounter. The feelings bordered on euphoric during those dramatic few days when the major discoveries about Uranus were being received, analyzed, and disseminated to the waiting press and public. For many it was difficult to go home to get some much-needed rest each day or night for fear that the receipt of some new tidbit of information might be missed.

The working groups were successful in resolving all but two or three conflicts they identified. Resolution of the two or three unresolved items (including the Miranda mass versus imaging mentioned above) and of conflicts identified after completion of the working group meetings was left for Voyager Project personnel with the assistance of the Science Steering Group.

7.1.3.5 *Translating commands to Voyager-readable language*
The end product of the working group activities was a 'strawman' time-ordered listing of events for the entire encounter period. Starting from this strawman, Voyager Flight Science Office personnel, assisted by Flight Engineering Office and

Mission Planning Office personnel, began the development of a 'Scoping Product' which provided the first quantitative estimate of spacecraft resources needed to accomplish the recommended observations. The first result of scoping was the selection of boundaries between encounter phases and between the CCS computer 'loads' within those phases (see Table 7.1). As shown in the table, the encounter

Table 7.1 — CCS computer load boundaries for the Uranus encounter

Phase/load	Phase/load start time [date], [h:min (GMT)], [day/h:min (from C/A)]	Phase/load duration
Observatory	[1985 Nov 04], [12:42.4], [−81/05:17.6]	66d 20h
B701	[1985 Nov 04], [12:42.4], [−81/05:17.6]	28d 22h
B702	[1985 Dec 03], [10:42.4], [−52/07:17.6]	26d 23h
B703	[1985 Dec 30], [09:42.4], [−25/08:17.6]	10d 23h
Far Encounter	[1986 Jan 10], [08:42.4], [−14/09:17.6]	11d 18h
B721	[1986 Jan 10], [08:42.4], [−14/09:17.6]	7d 20h
B723	[1986 Jan 18], [04:42.4], [−06/13:17.6]	3d 22h
Near Encounter	[1986 Jan 22], [02:42.4[, [−02/15:17.6]	4d 09h
B751	[1986 Jan 22], [02:42.4], [−02/15:17.6]	2d 01h
B752	[1986 Jan 24], [03:42.4], [−00/14:17.6]	2d 08h
Post Encounter	[1986 Jan 26], [11:42.4], [+01/17:42.4]	30d 02h
B771	[1986 Jan 26], [11:42.4], [+01/17:42.4]	8d 11h
B772	[1986 Feb 03], [22:42.4], [+10/04:42.4]	21d 15h
end	[1986 Feb 25], [13:42.4], [+31/19:42.4]	

phases were specified as Observatory Phase, Far Encounter Phase, Near Encounter Phase, and Post Encounter Phase. The nature of the science observations in each phase is discussed below. During scoping of the encounter, three CCS loads were specified for the Far Encounter Phase. It was later found that the activities of this time period would fit within the constraints for only two loads, so the middle load (B722) was absorbed into loads B721 and B723. Scoping of the encounter was completed during the period from June 11, 1984, to January 11, 1985.

Each of the successive levels of sequence preparation contained several opportunities for new inputs, incorporation of those inputs, generation of revised sequence products, and critical review of those products. It was this iterative process which enabled Voyager Project personnel to ferret out both major and minor problems, in addition to improving the scientific and engineering efficiency of the final sequence of events.

Following scoping, the next level of complexity was introduced. During this 'Integrated Timeline (IT) Product' development period, supporting engineering, spacecraft configuration, and science calibration and configuration commands were

added. DSN station coverage needed to support the observations was also nego-
tiated. For the first time, link designs were specified in full-blown detail. The
software used to store and manipulate the computer files was known as ASSET (for
Automated Science/Sequence Encounter Timeline), and utilized a specially programmed version of dBASE II. Each IT Product was critically reviewed by all
elements of the Voyager Project. Once the IT Product was formally approved, often
with a short list of matters still needing resolution, the sequence development baton
was passed from the Science Investigations Support Team (SIS) to the Sequence
Team (SEQ), with SIS, the Spacecraft Team, the Navigation Team, and the Mission
Planning Office in major supporting roles. The IT Product development for the
Voyager 2 Uranus encounter took place in the time interval from October 15, 1984,
through May 10, 1985.

From IT Product approval forward, sequence development was dependent on
several Voyager computer programs which have existed since before Voyager launch
(ASSET was initially developed in 1984 especially for the Uranus and Neptune
encounters). Some of these programs have been inherited from other planetary
exploration projects, but all were adapted to Voyager, and all have since been
updated on numerous occasions to correct problems, improve their efficiency, or add
new capabilities. Each change required approval by a Voyager Change Board (VCB)
consisting of the Mission Director and managers of each of the Voyager Offices. The
set of approved software is known as the 'Mission Build'.

The first in this series of programs is POINTER, which translates sequence inputs
into detailed science instrument commands, determines in scan platform coordinates
the appropriate pointing for each target, and performs simple verification of the
sequence validity. The output of POINTER includes pictorial representations of the
scenes to be viewed by ISS, IRIS, PPS, or UVS (output by the TARGET module of
POINTER); it also includes a detailed time-ordered printout of the sequence of
commands, generated by a module of POINTER known as OPSGEN (Observation
Pointing Sequence GENerator). Both OPSGEN printout and the TARGET plots
are reviewed thoroughly, and errors in implementation are corrected. If the
TARGET plots or OPSGEN printout show the need for a change of strategy or other
modifications, written sequence change requests must first be approved by the VCB.
Approved changes are incorporated into the SERF (Science Events Request File), a
computer-readable POINTER output file which is used as input for the next stage of
sequence development. The SERF is therefore representative of the 'Final Timeline'
(FT) Product. Another module of POINTER (known as VERIFY) is used near the
conclusion of sequence preparation to provide verification that the intent of each
science observation has been preserved through the end-to-end series of program
runs. Observatory Phase FT Product development started on February 11, 1985; the
Post-Encounter FT Product was completed on July 19, 1985.

The SERF and several other files (the moment-by-moment orientation of the
spacecraft, initial conditions, and other information needed to properly configure
the spacecraft systems) are processed by a program called SEQGEN (SEQuence
GENerator). In addition to the sequence integration task performed by SEQGEN,
one of its main purposes is to verify that none of the constraints on sequence structure
imposed by spacecraft limitations or accepted usage rules are violated. The primary
outputs of SEQGEN are the ESF (Event Sequence File) and the SRF (Sequence

Request File). Computer printouts of each of these files are again thoroughly reviewed to ascertain their correctness. An additional output of SEQGEN is a file specifying scan platform pointing as an input to the POINTER VERIFY process mentioned above.

The corrected SRF serves as input to the program SEQTRAN (SEQuence TRANslator), which puts the instructions into language the spacecraft can store, understand, and execute. The output of SEQTRAN is the DMWF (Desired Memory Word File), which is then processed through the COMSIM (COMmand SIMulation) program to do a final verification that all commands are correctly formulated and in accord with the understood operation modes of the spacecraft. The GCMD (Ground CoMmanD) file which is generated by COMSIM is the file which can be transmitted to the spacecraft. COMSIM also generates an EVTSDR (EVenT System Data Record). Together with the corrected ESF from SEQGEN, the EVTSDR permits the generation of an SOE (Sequence Of Events) listing , which details the spacecraft events that will occur and is a reminder of ground event timing in support of Voyager. The SEQGEN, SEQTRAN, and COMSIM programs together generate what is called the 'Uplink' (UP) Product. UP Product development for Uranus spanned the period from April 19 through October 9, 1985.

The POINTER VERIFY module is used near the end of UP product development as a verification that instrument pointing is as desired. VERIFY uses inputs from SEQGEN and COMSIM to specify scan platform pointing, spacecraft attitude, and ISS operation timing. Since IRIS, PPS, and UVS essentially operate continuously, times for individual instrument pointing verification are critical and must be specified by the respective Experiment Representatives.

To accommodate new information on Uranus received during the sequence generation, or to correct or improve the observations, a limited update of each of the UP products was scheduled. This update process occupied most of the final six weeks prior to transmission of each CCS computer load to Voyager 2. The update included automatic pointing adjustments to account for any detected changes in the spacecraft trajectory.

The critical Near Encounter sequence (B752) required special treatment. Allowances were made for adjusting the pointing specifications and timing of several observations within the last five days of the update process. Even this 'Late Ephemeris Update' (LEU) was insufficient for some of the most critical observations; many required pointing and timing adjustments even after the B752 sequence was in Voyager 2's CCS computers and had started its execution. The 'Late Stored Update' (LSU) process actually replaced several commands stored in the CCS computers with more precise commands transmitted to Voyager 2 at the latest possible moment.

The entire process, as complex as it was, was completed on schedule. Voyager 2 performed all its encounter observations exactly as expected. A single exception was the problem mentioned in section 7.1.2.1, which involved streaks in some of the images taken about a week before Uranus closest approach. The streaks were due to failure of one memory location in Voyager 2's secondary FDS computer memory, and the problem was corrected within a few days.

7.2 CHARACTERISTICS OF THE URANUS ENCOUNTER

The actual Uranus encounter spanned 113 days (see Table 7.1). A number of encounter-related science activities occurred during periods immediately preceding and following the encounter. Closest approach to Uranus occurred at 1759 Greenwich Mean Time (9:59 AM Pacific Standard Time) on January 24, 1986. The one-way light time at Uranus closest approach was 2 h 45 min. Radio signals transmitted from Voyager 2 when it was closest to Uranus actually arrived at the Canberra Deep Space Communications Complex at 2044 GMT (12:44 pm PST) and were relayed within fractions of a second to JPL in Pasadena, California. The Voyager science activities associated with each of the the encounter phases, the geometry of the Uranus encounter, and a brief outline of the handling of the data from Voyager are described below.

7.2.1 Encounter science activities

Generally, all Voyager spacecraft activities between the beginning of October 1985 and the end of March 1986 were science activities, since all were aimed at providing the best possible science return from Uranus, its rings, and its satellites. An extensive network of supporting science activities and observations from Earth was also an important factor in the success of the encounter. Some of those observations were discussed in Chapter 5. Here we will concentrate on the encounter activities of the Voyager spacecraft itself, including engineering activities in support of science observations, science calibrations, and the science observations themselves. Even these activities were so extensive that only the barest of details can be given here. For convenience in discussing the activities, they have been grouped by mission phase, and the phases are discussed in chronological order.

7.2.1.1 *Saturn–Uranus Cruise*

A series of five trajectory correction maneuvers (TCMs) was planned for Voyager 2 between the close approaches of Saturn and Uranus. The first of these (TCMB10) occurred just 34 days after Saturn closest approach. Its accuracy was sufficient that the scheduled TCMB11 was cancelled. More than three years passed before another TCM was needed. On 13 November 1984 TCMB12 adjusted the speed and direction of Voyager 2 such that it would pass within 29 000 km of Miranda 55 min before its 1800 GMT closest approach to Uranus. A final Uranus encounter TCM (B13) was performed during the encounter Observatory Phase, again with such precision that TCMB14, scheduled during the Far Encounter Phase, was also unneeded.

Occasional imaging of the planet was performed during the time period between the end of the Saturn encounter and the beginning of the Uranus encounter. The frequency of such imaging increased as Voyager approached Uranus. By July 1985 Voyager's pictures were of higher resolution than the best Earth-based telescopic views. On July 15, 1985, at a distance from Uranus of 247 million km, Voyager 2 obtained the image shown in Fig. 7.3. The smallest features that could be seen at this distance would be 4600 km in diameter. Uranus is featureless in this view from above its south pole other than the general darkening toward the limb (the edge of the planet's disk) and toward the day/night terminator at the right. The four largest satellites could also be seen from this distance, and were useful for optical determination of Voyager's path relative to Uranus.

Fig. 7.3 — Of July 15, 1985, at a distance from Uranus of 247 million km, Voyager 2 obtained
this image. The smallest features that could be seen at this distance are 4600 km in diameter.
Uranus is featureless in this view from above its south pole, other than the general darkening
toward the limb and toward the day/night terminator at the right. Ariel may be seen above and
to the left of the planet. Umbriel is below and slightly to the right of the planet. Titania is in the
lower right-hand corner. Oberon is near the left edge. (P-28944)

One of the more important activities just prior to the start of the Observatory
Phase was the 'Near Encounter Test' (NET). Because of the complexity of the
sequence of activities surrounding Uranus closest approach, the most critical of these
activities were tested on Voyager 2 during the NET. Maneuvers to provide image
motion compensation and to track the limb of the planet during the radio occultation
were tested. The power system was also checked to better determine the power
margins under circumstances comparable to those near closest approach to Uranus.
There was also an attempt to determine how the spacecraft radio receiver would drift
in frequency during the series of commands needed to adequately perform the radio
science occultation experiment. The simulated time period covered by the NET was
from 2 h before Uranus closest approach ('U−2:00') to U+6:00. Wide-angle camera
images of the star background served to check the accuracy of the various maneuvers
during their execution. Since the spacecraft scan platform had not been moved at its
medium rate ($\frac{1}{3}$°/s) since shortly after the Saturn encounter, the four medium-rate

slews to be used during Near Encounter were also tested. Only minor flaws were found in the NET, and appropriate corrections were made for the actual Near Encounter. In addition to testing the spacecraft Near Encounter sequence of events, the NET also served as a good test of the readiness of both the DSN tracking stations and personnel and of the Ground Data System at JPL.

One other major type of activity performed during the latter portions of Saturn–Uranus Cruise was the series of instrument calibrations performed in preparation for the encounter observations. Measurements were made to more precisely determine the sensitivity, pointing direction, and relative response across the fields of view of the PPS. A target plate attached to the spacecraft was illuminated by canting the spacecraft to one side. Because the target plate had precisely known reflective properties, and because the distance and brightness of the Sun were also well-known, ISS and IRIS radiometer responses were calibrated by viewing the target plate. The MAG was calibrated several times during the Saturn–Uranus Cruise period by rotating the spacecraft through several complete turns, first in the roll direction, and then in the yaw direction. The response of the magnetometers themselves was not expected to change appreciably, but differing power loads and distribution on Voyager give rise to spacecraft magnetic fields that must be measured and subtracted from the total field in order to determine the pre-existing ambient field. Most of the instruments have calibration or conditioning modes which permit less dramatic periodic checks of their responses and health during cruise and the encounter.

The primary data type during cruise was the 12 to 16 h per day of low-rate (160 bps) science data. The UVS was used to observe stars; CRS, LECP, MAG, and PLS were used to study magnetic fields and charged particles between the planets; PRA and PWS were assisting in the study of the interplanetary environment as well as searching for radio signals which might indicate the presence of a Uranian magnetic field.

7.2.1.2 Observatory phase

The actual encounter period began with the Observatory Phase (OB), extending from 4 November 1985 to 10 January 1986. OB was divided into three CCS computer loads: B701, B702, and B703. Although Voyager imaging of Uranus exceeded Earth-based resolution as early as March 1985, OB offered the first opportunity for nearly continuous observations of the planet and its system. The imaging strategy for Uranus differed from that at Jupiter or Saturn. At the two prior planets the bulk of OB imaging consisted of five-color narrow-angle camera (NA) imaging of the planet every 72° of longitude (i.e., five times each rotation period). Since the south pole of Uranus was pointed very nearly at the Sun in 1985 and 1986, the viewable area changed very little as the planet rotated. Instead, the bulk of Uranus OB imaging was done in a series of continuous 38-h 'movies' in which NA images were shuttered every 4.8 min over the Goldstone and Canberra tracking stations, and every 9.6 min over the Madrid tracking stations. Generally, these imaging sequences started over GDSCC and then progressed sequentially across CDSCC, MDSCC, and GDSCC, finishing with the CDSCC pass. In this fashion, the first 16 h and the last 16 h could be done   at   the   4.8-min   rate   with   only   the   central   6   h restricted to the slower 9.6-min rate. Five such movies done in OB were spaced in such a way that resolution had improved by a factor of 1.4 in each successive series. These movies were designed to look for cloud features which might be followed as the planet rotated to determine prevailing wind speeds at various latitudes.

PRA continued to search for evidence that radio signals were being generated by solar wind interaction with Uranus's magnetic field. UVS searched for ultraviolet emissions from Uranus's atmosphere and from the space between Uranus and its satellites. These were designed to detect hydrogen and other gases escaping from the planet or its satellites. The chemical composition of those gases provided clues about the chemical composition of Uranus's atmosphere or the surfaces of its satellites.

Near the midpoint of OB, Voyager 2 passed behind the Sun as viewed from Earth. This provided an opportunity for a special RSS experimental test of Albert Einstein's General Theory of Relativity. The experiment was done by transmitting both X-band and S-band signals through the Sun's extended atmosphere (the solar corona). Einstein's General Theory predicts that the gravity of the Sun will delay the signal slightly. While the delay of the radio signal can be measured with great precision, free electrons in the solar corona also delay the radio signal. The experiment thus involves determination of the free-electron effect, subtraction from the total observed delay, and comparison of the difference with Einstein's predictions. Results of the first step of this process were discussed by Anderson et al. [6]; final results had not been published by early 1989, but Anderson reports [7] that the data are consistent with the predictions of Einstein to a high degree of accuracy.

The Sun and other stars to be used in Near Encounter observations were measured with the PPS and UVS during OB as a calibration of their brightnesses. Similar calibrations would later be done in Post Encounter, so that the critical observations would occur midway between two high-quality calibration points. The IRIS Flash-Off Heater (FOH) was turned off in OB to allow time for cool-down of the instrument for critical Far Encounter and Near Encounter observations. Periodic calibrations of the fields and particles instruments continued, including an abbreviated version of the roll and yaw maneuvering to calibrate MAG. In connection with its Solar Conjunction measurements, RSS performed a number of 'occultation-like' tests to exercise both the spacecraft radio system and the ground systems and personnel in preparation for the Uranus occultation experiment.

Engineering activities in OB included two measurements called 'Torque Margin Tests' (TMTs). These TMTs were devised by spacecraft engineers to measure how freely the scan platform was turning. Normally, the stepping motors on the scan platform are driven by electronic pulses approximately 0.04 s in duration. By reprogramming the AACS computer the motor-drive pulses can be made as short as 0.001 s. Post-Saturn testing showed that a healthy motor will still drive at full rate with 0.006-s pulses, but will slow perceptibly with 0.005-s pulses. The TMTs in OB utilized 0.006-s pulsing to ascertain whether the spacecraft scan platform motors would still operate at full speed. No slowing was noted. Other OB engineering activities included TCMB13, conditioning of the spacecraft gyroscopes, calibration of the gyroscope drift rates, and a check to ascertain whether the FDS and CCS computer clocks were synchronized.

7.2.1.3 Far Encounter Phase

By January 10, 1986, the apparent size of Uranus was increasing rapidly and soon would no longer fit reliably within a single NA field of view. Four-image (or larger) NA imaging mosaics were needed during most of the Far Encounter Phase (FE) to reliably capture the planet. FE activities ended on January 22. The activities of FE were split into two CCS computer loads: B721 and B723.

Two additional 38-h movies of the type done in OB were inserted early in FE when the planet could still be captured in a single NA frame. Most of the atmospheric imaging emphasis centered on WA coverage of the entire disk plus scattered higher-resolution NA mosaics. Detailed imaging of the rings began in the FE period. Imaging coverage of the five known satellites also began in FE; although all of the additional ten satellites had been discovered by early FE, there had not been time to alter the sequence to specifically point the cameras at any of the newly discovered moonlets.

The FE time period was a critical one for the Voyager Navigation Team. The position of the spacecraft relative to the planet is best determined by imaging the satellite disks against a background of known stars. The highest quality optical navigation frames were obtained in FE. The gravitational pull of Uranus on Voyager 2 was beginning to grow to the point where measures of Voyager's acceleration toward the planet now permitted refinement of the planet's mass; prior to FE, the value of Uranus's mass had been estimated from Earth-based measurements of the satellite orbits. The accuracy of those estimates and the precision of TCMB13 during OB made it unnecessary to execute the Trajectory Correction Maneuver scheduled for FE.

The temperature of IRIS, which had been turned on in OB, stabilized sufficiently to begin its planetary observations by the middle of FE. Its most important measurements in FE were the temperature measurements of the disk of Uranus about nine days before planetary closest approach. At that time the apparent angular size of Uranus matched the $\frac{1}{4}$°-diameter IRIS field of view. UVS observations similar to those done in OB continued during FE, but at a reduced frequency. Although the PPS had to be used sparingly to preserve its useful lifetime, it also began definitive measurements of the planet, its satellites, and its rings during FE. Daily high-rate samples of PRA and PWS data supplemented the lower-rate data from these two investigations and from the four fields and particles investigations.

PRA and PWS first began to sense the Uranian magnetic field during the FE period. Although actual penetration of the field would not occur until the Near Encounter Phase, radio waves generated by interaction of the solar wind with the planetary magnetic field were seen during the last few days of FE.

The final RSS Operational Readiness Test (ORT) was done in early FE. Both the spacecraft and the tracking stations at Canberra and Parkes, Australia, performed well during this ORT and during the actual planetary occultation measurements in Near Encounter. Near the end of FE a final check of the scan platform was done using the TMT procedure described for OB, and the platform successfully passed that final hurdle. The stage was now set for the critically important Near Encounter Phase observations.

7.2.1.4 *Near Encounter Phase*

The Near Encounter Phase (NE) included more than 90 of the highest priority Uranus science observations. NE extended from January 22 through January 26, 1986, with Uranus closest approach at 1800 GMT on January 24. CCS computer load B751 lasted 49 h; the critical B752 load lasted 56 h. Early in B751, the roll orientation of the spacecraft was changed to accommodate PLS and LECP desires. Prior to that time, the star Alkaid, which forms the end of the handle of the 'Big Dipper' (the star constellation Ursa Major), was used as a roll reference. From U−2d6h until U+0d6h, Canopus (the brightest star in the southern constellation Carina) would serve as the primary stellar reference. Thereafter, Fomalhaut (the brightest star in the constellation Piscus Austrinus) and Achernar (the brightest star in the constellation Eridanus) would be used to control the roll orientation of Voyager 2.

Continuous data collection was typical of all the Voyager investigations during NE. Comparison of the investigation findings yielded far more information about the Uranus system than would otherwise have been possible. Each observation link had a designated leading investigation; other investigations were 'riders' whose observation objectives also had to be considered. In the discussion below, the observations will be grouped by target (atmosphere, rings, satellites, or magnetosphere) and by lead investigation (RSS, IRIS, UVS, ISS, etc.), so the interested reader needs to remember that useful data were received by more than just the leading investigation.

The most important (and most complex) of the atmospheric observations was the RSS radio occultation experiment. When telemetry data were being transmitted to Earth, the data consisted of 'dots and dashes' in a binary code, which were superimposed on the basic X-band radio frequency. Just before the spacecraft passed behind Uranus as viewed from Earth (Fig. 7.4), the telemetry stream was turned off

VIEW FROM EARTH

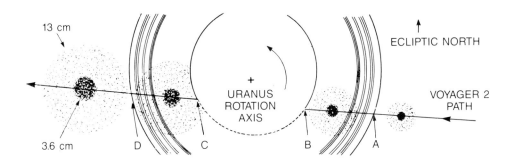

Fig. 7.4 — The path of Voyager 2 behind Uranus as viewed from Earth is depicted. The RSS atmospheric and ring occultation experiments occurred during this period of time. (260-1824)

so that more power could be concentrated in the main X-band and S-band 'carrier' frequencies used by RSS. During passage behind the planet, the spacecraft was continuously maneuvered to keep the antenna pointed at the closest point along the planetary limb to the spacecraft–Earth line. The atmosphere of the planet refracted (i.e., deflected) the radio beam toward Earth, and radio signals were successfully received at the Australian tracking stations during the entire occultation period. The experiment provided pressure, temperature, and composition information about the deep atmosphere of Uranus.

Atmospheric temperature, pressure, and composition information was also provided by a series of IRIS observations. IRIS measurements near the latitude and longitude of the RSS occultation exit point enabled the two experiments to be coupled to provide more accurate atmospheric composition data (especially, the helium abundance) than could be provided by either RSS or IRIS alone. Composition of the south polar region (during the inbound leg of Voyager's journey) and of the north polar region (during the outbound leg) were obtained during long periods of staring at the planet. Horizontal temperature variations were measured at closer range by scanning the IRIS field of view in latitude or in raster patterns across the illuminated and dark hemispheres.

PPS and ISS were used to obtain a series of six combined measurements of the atmospheric brightness under a variety of solar illumination and viewing conditions. These would later be used in conjunction with the IRIS whole-disk temperature measurements near $U \pm 9$ days to determine the planet's heat budget: what fraction of the observed thermal energy comes from solar heating, and what fraction is from internal heat sources. ISS also had a series of seven inbound mosaics of WA images designed to provide relatively high-resolution views of the illuminated atmosphere and its cloud structure.

UVS observations of Uranus's atmosphere in NE included a series of 12 different auroral emission studies, two limb profiles, and observations of occultations of the Sun and of the stars Algenib (gamma Pegasi) and Nucatai (nu Geminorum). The auroral emission studies helped to verify the location of the magnetic pole and to characterize interactions of the atmosphere and the magnetic field. The limb profiles and occultation measurements provided information on the vertical distribution of hydrogen and other gases in the extreme upper atmosphere.

Ring observations were primarily the purview of ISS. Seven three-frame mosaics were shuttered during approach to the planet. Full ring system mosaics were obtained both on approach and during the outbound portion of the flyby. Four WA frames were shuttered near the time Voyager passed through the equatorial plane of the planet to provide a relatively high-resolution view of the entire radial extent of the rings. One WA image of the rings was obtained while the unusually dark rings were silhouetted against the illuminated hemisphere of the planet. A series of one WA and three NA frames was shuttered while the spacecraft was in the shadow of Uranus. This permitted viewing the backlighted rings without the normal concurrent problem of glare associated with sunlight on the camera optics. It was also an ideal viewing geometry to look for finely divided material in the rings. Since these images were shuttered during spacecraft maneuvering in support of the RSS atmospheric occultation experiment, the timing and pointing of these images had to be carefully chosen to minimize image smear due to the combined effects of maneuvering, spacecraft motion, and ring motion.

Other ring observations were primarily of the occultation variety. PPS and UVS combined to watch the stars Nunki (sigma Sagittarii) and Algol (beta Persii) as the motion of the spacecraft caused the rings to appear to move in front of them. The apparent path of Nunki (Fig. 7.5) cut only the epsilon and delta rings of Uranus;

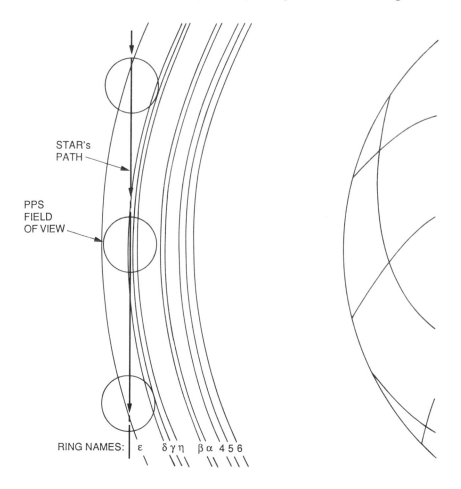

STAR's
PATH

PPS
FIELD
OF VIEW

RING NAMES: | ε   δ γ η   β α 4 5 6

Fig. 7.5 — The apparent path of the star Nunki (sigma Sagittarii) due to the motion of Voyager 2 provided information on the high-resolution radial structure of the epsilon and delta rings of Uranus.

Algol's apparent radial speed was much faster and traversed the entire radial extent of the rings twice (Fig. 7.6). The spacecraft radio signal as viewed from Earth passed behind the rings both before and after the planetary occultation, enabling RSS to obtain two high-resolution radial profiles of the rings. The RSS, PPS, and UVS data were used to disclose radial variations in the optical thickness of the rings at three

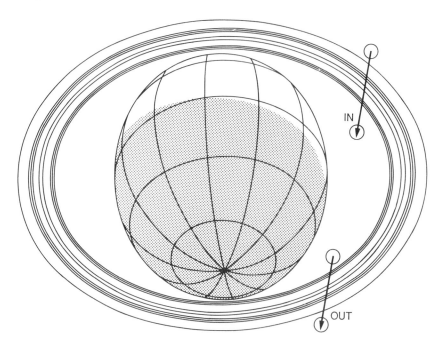

IN

OUT

Fig. 7.6 — The star Algol (beta Persii) passed behind all the known rings of Uranus twice, but because it was dimmer than Nunki and its apparent radial speed was much higher, the resulting data was of somewhat lower quality.

different wavelengths. As Voyager 2 crossed the plane of the rings about 44 min prior to Uranus closest approach, PWS obtained high-rate samples of its data. Tiny particles striking the spacecraft at high relative velocities will be instantly vaporized and ionized (electrically charged) by the impact. The ionization process creates radiowaves in the frequency range sampled by PWS; the bursty radio signals detected by PWS during ring-plane crossing are thus a record of particle impacts on the spacecraft.

Three investigations contributed most heavily to satellite studies. RSS used variations in the velocity of the spacecraft during the Miranda flyby to determine that tiny satellite's mass. RSS used a similar technique to deduce the mass of Uranus itself and of the entire system. With these improved mass determinations, ground-based satellite orbital period measurements could be translated into more precise orbit sizes. PPS studied the variations in brightness of the five major satellites under a variety of illumination and viewing conditions to determine the surface light-scattering properties.

ISS obtained periodic imaging of the satellites during approach. Their gravitational effects on each other could be detected in their motions, providing a means for calculating the ratio of the masses of Ariel, Umbriel, Titania, and Oberon. Combined with the RSS Uranus and Uranus system mass determinations, this led to improved values for the masses of all the major satellites of Uranus.

ISS also mapped the illuminated surfaces of the five major satellites at high resolution in color and at higher resolution through the clear filter. NE also provided an opportunity to obtain moderate-resolution imaging of Puck, the largest of the ten satellites discovered by Voyager 2. This was made possible by diverting to Puck a single image planned for Miranda near U−8.5 hours. It is interesting to note that the image was recorded for later playback and that the first attempt to return the image failed due to antenna-pointing problems at the Canberra 64-m tracking station. The problem was traced to errors in the antenna pointing files generated by a new and incompletely validated software system. Fortunately, only a few hours were required to generate commands and transmit them to Voyager 2 to replay the data before it was overwritten on the tape recorder. The second playback was successful.

Next to the RSS Uranus occultation experiment, the satellite imaging sequences in NE were the most complex science observations of the Uranus encounter. High-resolution images were obtained in full color for all five of the previously known satellites. Monochromatic images with better resolution were obtained for all but Oberon. Of these nine imaging sequences, all but one (Umbriel full color) utilized Image Motion Compensation (IMC), and the highest resolution mosaic of Miranda tailored the IMC rates individually to each of its eight constituent NA images. IMC, as discussed in Chapter 6, utilizes slow simultaneous rotations of the spacecraft about its pitch, yaw, and roll axes to pan the cameras during relatively close passage of the target. All IMC's, including the complex Miranda sequence, executed as planned. Because the satellites were in widely divergent directions in the sky, efficient camera pointing dictated the use of medium rate (0.33°/s) slewing on four occasions during NE. Without these medium rate slews, some of the important satellite imaging would not have been possible.

NE was also the prime period for collection of CRS, LECP, MAG, PLS, PRA, and PWS data on the magnetosphere of Uranus. Frequent high-rate samples of PRA and PWS data were recorded for later playback. The LECP telescope systems were held fixed during 10.4 min of each 12-min period from U−13.5 hours until U+20 hours. During the remaining 96 s of each 12-min period the telescope system turned 45° each 6 s, completing one full forward rotation and one reverse rotation in that time. Canopus (alpha Carinae) was used as a spacecraft roll reference on approach to the planet; Fomalhaut (alpha Piscus Austrini) was used from U+6 hours until U+19 hours. These roll orientations provided the best alignment for the PLS and LECP sensors to measure charged particles rotating synchronously with the planet and its magnetic field.

Deviations from expected Voyager and DSN performance did occur during NE, but the deviations were few in number and minor in their consequences. The Voyager Team and its aging brainchild had once again performed nearly miraculous feats.

7.2.1.5 Post Encounter Phase

The Post Encounter Phase (PE) was in some ways a little like the cleaning crew that does its work the morning after an all-night party. While the peak of exciting discoveries had passed, there was still much in the way of important observations and calibrations that remained to be done. PE extended from January 26 through February 25, 1986. It was divided into CCS computer loads B771 and B772.

Among the more important activities of PE was the completion of Digital Tape

Recorder (DTR) playback of the important recorded NE events. The highest priority events recorded on the DTR were transmitted to Earth two complete times to improve both the probability of successful return and to decrease data noise and data gaps generally inherent in single playbacks. No recurrences of the antenna-pointing mishap associated with the initial attempt to return an image of Puck were experienced during these playback periods.

Voyager 2 exited Uranus's magnetosphere during the first day of PE. It crossed the oscillating bowshock wave seven times in the following three days, finally leaving it on January 29. PRA continued to monitor the rotation of Uranus's magnetic field for an additional few days until the pulsed radio waves were too weak to be detected. A combination of PRA data and MAG data was used to determine the rotation period of Uranus's magnetic field (and presumably its interior) to an accuracy of less than a minute.

During PE, IRIS continued its observations of the dark hemisphere of Uranus, basically repetitions in reverse time order of the IRIS observations done in FE. After completion of the disk temperature measurements about nine days after closest approach, IRIS was turned off, and its Flash-Off Heater was turned on to prevent further degradation of IRIS sensitivity.

ISS collected a series of atmospheric images over a period of 36 h to look for evidences of cloud features in the viewable crescent of Uranus. A ring 'movie' was also attempted, but the ring proved to be too dark, and these images yielded no useful data.

UVS repeated its scans of the system and its auroral emission searches using observation designs similar to those done in OB and early FE. The Sun and stars used for the NE occultation experiments by UVS and PPS were recalibrated. Periodic calibrations of the fields and particles instrumentation also resumed, including the series of roll and yaw turns used to calibrate the MAG experiment. Just 21 days after closest approach, TCMB15 corrected the course of Voyager 2 to ensure that it was headed for an August 1989 encounter with Neptune. AACS, FDS, and CCS computer programs were revamped for the long Uranus–Neptune cruise period and its associated reduction in level of activity. The Uranus encounter period officially ended.

7.2.1.6 *Uranus–Neptune cruise*

Many of the science and engineering calibrations in support of an encounter are performed during the early cruise period following an encounter. Following the Uranus encounter only a minimal amount of such activity occurred because of another Solar System object: Comet Halley. The encounter period of the European Space Agency's Giotto spacecraft with Comet Halley began less than a week after the end of Uranus PE for Voyager. Since both spacecraft were in the same general direction in the sky there was competition for 64-m tracking stations, and in such circumstances a spacecraft in its encounter period is given precedence. To complicate matters, the Japanese Suisei and Sakigake spacecraft and the Russian Vega 1 and Vega 2 spacecraft were also nearing their encounters with Comet Halley. Voyager 1 and Pioneers 10, 11, and 12 were not in critical mission phases and had been practically ignored during the four-month Voyager 2 encounter, and needed DSN tracking coverage as well. It was a realization of these factors that led to the attempt to complete as many of the calibrations as possible prior to the end of PE, unless their nature would permit a delay of several months.

One lesson learned from the Uranus planning experiences was that it was unwise to cut the staffing levels so low that extensive retraining would be needed when the staffing buildup occurred. Neptune preparations were planned as a more relaxed, long-term effort which would enable a relatively large fraction of the experienced personnel to be retained through the long cruise period between Uranus and Neptune encounters. The Neptune planning started very shortly after the end of the Uranus encounter.

7.2.2 Encounter geometry

Let us now briefly examine the physical and geometric characteristics of the encounter. A listing of the geometric event timings and distances is given in Table 7.2. At the beginning of OB, Voyager 2 was approaching Uranus latitude $-73°$ at a

Table 7.2 — Uranus encounter characteristics

Voyager 2 event	Time from Uranus C/A ($\pm$h:min)	Spacecraft event time (GMT h:min)	Distance (km)
Titania closest approach	$-2:49$	Jan 24, 1986 15:10	365 200
Oberon closest approach	$-1:47$	Jan 24, 1986 16:12	470 600
Ariel closest approach	$-1:38$	Jan 24, 1986 16:21	127 000
Miranda closest approach	$-0:55$	Jan 24, 1986 17:04	29 000
Ring-plane crossing	$-0:43$	Jan 24, 1986 17:16	115 300
Uranus closest approach	$0:00$	Jan 24, 1986 17:59	107 100
Epsilon ring RSS occ.	$+1:44$	Jan 24, 1986 19:43	142 000
6 ring RSS occ.	$+2:03$	Jan 24, 1986 20:02	161 200
Enter Uranus Sun occ.	$+2:25$	Jan 24, 1986 20:24	177 600
Enter Uranus RSS occ.	$+2:36$	Jan 24, 1986 20:35	186 500
Umbriel closest approach	$+2:53$	Jan 24, 1986 20:52	325 000
Exit Uranus Sun occ.	$+3:45$	Jan 24, 1986 21:44	242 800
Exit Uranus RSS occ.	$+4:02$	Jan 24, 1986 22:01	257 600
6 ring RSS occ.	$+4:35$	Jan 24, 1986 22:34	286 200
Epsilon ring RSS occ.	$+4:54$	Jan 24, 1986 22:53	302 700

Distances are from spacecraft to center of target body. Radiowave travel time from Voyager 2 to Earth was 2 h 45 min.

speed of 14.74 km/s. The range from Voyager 2 to the center of Uranus was 104 000 000 km. By the beginning of FE, the range had dropped to 18 500 000, but the latitude and speed were essentially unchanged. The dawning of NE found an increase in speed to 14.85 km/s over $-71°$ latitude. The range was now a mere

3 440 000 km. At Uranus closest approach, the speed had reached its maximum value of 18.03 km/s. The latitude was +23° and the range was 107 100 km from the center of Uranus, whose equatorial diameter (at the 1-bar pressure level) is 25 559 km. PE started at a range of 2 270 000 km, a speed of 14.90 km/s, and a latitude of +56°. The encounter ended with the spacecraft at a range of 40 600 000. The latitude was +53°, and the speed was back down to its OB value of 14.74 km/s. Fig. 7.7 presents a view of Voyager's path through the closest portions of the Uranus encounter.

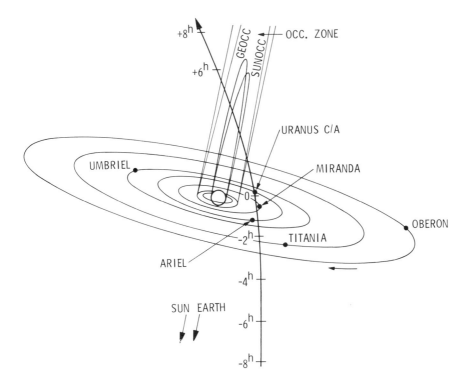

Fig. 7.7 — A schematic of the path of Voyager 2 through the closest portions of the Uranus encounter. The projected orbits of the five major satellites are shown, together with the positions of these satellites at the time of Voyager's closest approach to Uranus. Also shown are the limits (in the plane of Voyager's trajectory) of the Sun and Earth 'shadows,' both for the planet and for the outermost (epsilon) ring. (260-1798)

It was not always possible to obtain images of the five largest satellites at their closest approaches. Table 7.3 lists the ranges in thousands of km, theoretically achievable resolutions in kilometers per imaging line pair (km/lp), and the solar phase angles for the best color imaging obtained, the best resolution (generally monochromatic) imaging obtained, and the corresponding values at satellite closest approach. The imaging team sought to obtain the best achievable resolutions at phase angles of between 60° and 70°. The table shows that within the limitations of the selected trajectory, the actual satellite imaging came close to achieving that goal.

Table 7.3 — Imaging range (10^3 km)/resolution (km/line pair)/phase angle (°)

Satellite	Best color	Best resolution	Closest approach
Miranda	143/ 2.8/16°	33/ 0.6/40°	29/0.6/ 67°
Ariel	165/ 3.2/32°	127/ 2.5/69°	127/2.5/ 73°
Umbriel	1040/20.3/31°	560/10.9/56°	325/6.3/124°
Titania	499/ 9.7/34°	368/ 7.2/70°	365/7.1/ 78°
Oberon	660/12.9/39°	660/12.9/39°	471/9.2/ 79°

7.2.3   Encounter operations

The difficulties associated with the allocation of DSN tracking antennas for the many interplanetary spacecraft has already been mentioned. Early in the sequence development period Voyager Flight Operations Office (FOO) personnel arranged equitable agreements with other flight projects for Voyager's use of the DSN. Because of the uniqueness of the Uranus data, Voyager 2 was given top priority for most of its four-month encounter period. FOO personnel were also responsible for transmitting CCS computer loads and individual commands to Voyager. For transmissions from GDSCC in California, a microwave link connecting JPL and Goldstone was used. For transmissions from Australia or Spain, a NASA communications (NASCOM) satellite link had to be used.

Data transmitted from the spacecraft underwent a similar routing. Because of the possibility that data might be garbled or lost in the process of transfer to JPL, each tracking complex recorded the raw data from the spacecraft while they were being relayed to JPL's waiting computers. Replays of the raw data recordings were obviously much simpler than requiring Voyager to reacquire original data or replay recorded data. Again, relays from GDSCC to JPL utilized the microwave link from Goldstone. Data from CDSCC or MDSCC were relayed via the NASCOM satellite to Goddard Space Flight Center in Greenbelt, Maryland, and the data were then forwarded to JPL by means of telephone lines and microwave links.

As the data were received at JPL, they were timetagged and any coding imposed by Voyager was removed. The received data were split into two basic streams: (1) imaging data (including high-rate PWS and PRA data), and (2) general science and engineering data.

The imaging data were transferred to the Multi-mission Image Processing Laboratory (MIPL) to be converted from digital picture elements into images. If the imaging had undergone compression, MIPL reversed the compression process. Imaging processing included, among a variety of other things, blemish removal, image enhancement, and intercomparisons of images. The comparisons included image differencing to bring out changes or image combinations to produce photographs either in true or exaggerated color. Actual production of hardcopy pictures or slides was done by JPL Photographic Services personnel.

General science and engineering data were processed to provide graphical displays and printouts of the data for 'quick-look' analyses and to produce the detailed Experiment Data Records (EDRs) for the individual science investigation teams, generally as magnetic computer tapes. In addition to the EDRs, some portions of the data are transmitted to remote investigator facilities for more timely processing. Supplementary Experiment Data Records (SEDRs) were produced on the basis of received engineering data and predicted trajectory constants to provide the geometry data (coordinates, distances, angles, configurations, etc.) corresponding to the science data.

Beginning on January 22, 1986, daily press conferences were held at JPL to announce the latest findings of each of the investigations. Hundreds of press members and a number of television networks covered the press conferences and conducted private interviews with the scientists and others associated with Voyager. The science teams met jointly twice a day to discuss and intercompare their findings, to report on the status of each investigation and its instrumentation, to prepare for the press conferences, and to plan the future course of their data analyses and publication. Spirits were high, and the excitement of each new discovery created a collective state of mind bordering on euphoria. The first of a planned series of NASA space achievements for the year 1986 was progressing well. What happened then would shake the world in general and NASA in particular, and it would abruptly turn the happy cheers into tears and quiet head-shaking disbelief.

7.2.4   The Challenger disaster

The final press conference during the Voyager Uranus encounter was to take place Tuesday, January 28, 1986. The morning science meeting had convened on the fourth floor of Building 264 at JPL at 8 am PST. There were three television monitors in the meeting room. Two displayed the latest images received from Voyager 2, and the third was a public information channel. The meeting was completed in 35 min, because many of those involved in this most successful unmanned planetary exploration mission ever launched were also avid followers of NASA's manned space efforts, and the launch of the Space Shuttle Challenger was set for shortly after 8:35 am PST. Challenger lifted off the launch pad at Cape Kennedy, Florida Cape Kennedy, Florida , at 8:38 am PST, and those sitting in the Voyager conference room cheered along with millions of others around the world. Then, just 72 s later, the Challenger's main engine erupted in an enormous ball of fire which instantly destroyed the Challenger and its seven astronauts. The successes of the space program had not prepared us for such an event. We sat in silent disbelief that the events on the television screen had actually occurred. Not even the NASA Shuttle launch commentator spoke for over a minute, until the gravity of the situation had fully entered the consciousness of all who watched. Many in that JPL conference room had close friends in the manned space program; some knew one or more of Challenger's astronauts personally.

Voyager's wrap-up press conference was postponed. After one or two attempts at press interviews, all such interviews were cancelled. There was a spontaneous day of silence for fallen friends and blasted hopes. It was a much more somber group of scientists and press that gathered on January 29 to pay their respects to the seven Challenger astronauts and say a few words about the Voyager successes.

The Challenger disaster was more than a personal disaster for the astronauts and their families, friends, and co-workers. Those close to NASA knew of the inherent dangers of manned space flight, and no conceivable Shuttle design would ever guarantee the absolute safety of its occupants. The Challenger disaster was also a temporary death knell for the continued unmanned exploration of the planets. No new planetary probes had been launched for nearly a decade (Pioneer 12's launch to Venus in August, 1978, was the last). NASA had abandoned its unmanned launch capability after Pioneer 12 in order to concentrate its limited resources on Shuttle development. The Galileo mission to Jupiter was scheduled to be the first interplanetary probe to be launched from the Shuttle (in May, 1986). More than 30 months were to elapse before another Shuttle would be launched. The Galileo mission suffered a launch delay of nearly $3\frac{1}{2}$ years, and a Jupiter arrival delay of more than five years. An entire line of interplanetary missions (including Ulysses to Jupiter and the Sun, Mars Observer, Magellan to Venus, Comet Rendezvous/Asteroid Flyby, and Cassini to Saturn) would suffer corresponding delays. In essence, Voyager would be the sole planetary explorer from the time of the Jupiter encounters in 1979 through the Neptune encounter in 1989. Among other responses, the situation called for a positive future for Voyager. Both NASA and the Voyager Project were determined to do everything possible to ensure the continued success of this 'Grand Tour' to Neptune and beyond, and the Voyager Team was equal to that challenge.

NOTES AND REFERENCES

[1] Rotational motion of a spacecraft can be described in terms of rotations around one or more of three axes: roll, pitch, and yaw. Roll motion turns the spacecraft around the axis of the 3.7-m antenna, very nearly preserving antenna-pointing in space. Pitch and yaw motions are perpendicular to roll and to each other and change the antenna-pointing direction.

[2] Jones, C. P. (1979) Automatic fault protection in the Voyager spacecraft. *American Institute of Aeronautics and Astronautics, Paper No. 79-1919*, 11 pp.

[3] Bergstralh, J. T. (1984) *Uranus and Neptune*, NASA Conference Publication 2330, National Aeronautics and Space Administration, Scientific and Technical Information Branch, 636 pp.

[4] Miner, E. D., Ingersoll, A., Esposito, L., Johnson, T., Wessen, R. (1985) Science objectives and preliminary sequence designs for the Voyager Uranus and Neptune encounters. Voyager Project Document #1618-57, Jet Propulsion Laboratory, Pasadena, California.

[5] The hydrogen molecule, consisting of two hydrogen atoms, has two different structures. In one of these (*ortho*-hydrogen), the nuclear spins of the two atoms are parallel. In the other (*para*-hydrogen), the nuclear spins are anti-parallel. Ortho-hydrogen preferentially occupies odd-numbered molecular energy levels; para-hydrogen occupies even-numbered levels. The fraction of para-hydrogen in hydrogen gas is dependent on the temperature of the gas; for a temperature of 64 K, the equilibrium fraction of para-hydrogen is about 62%.

[6] Anderson, J. D., Krisher, T. P., Borutzki, S. E., Connally, M. J., Eshe, P. M., Hotz, H. B., Kinslow, S., Kursinski, E. R., Light, L. B., Matousek, S. E., Moyd, K. I., Roth, D. C., Sweetnam, D. N., Taylor, A. H., Tyler, G. L., Gresh, D. L., Rosen, P. A. (1987) Radio range measurements of coronal electron densities at 13 and 3.6 centimeter wavelengths during the 1985 solar conjunction of Voyager 2. *The Astrophysical Journal*, **323**, L141–L143.

[7] Anderson, J. D. (1989) private communication.

BIBLIOGRAPHY

Finnerty, D. F., Martin, J., Doms, P. E. (1987) Asset: an application in mission automation for science planning. *Journal of the British Interplanetary Society*, **40**, 461–470.

Kohlhase, C. (1985) The Voyager Uranus Travel Guide. Voyager Project Document #618-150. National Aeronautics and Space Administration, Jet Propulsion Laboratory, California Institute of Technology, Pasadena, 171 pp.

McLaughlin, W. I. (ed.) (1985) Mission systems. Special issue of *Journal of the British Interplanetary Society*, **38**, 433–480. Contains the following articles: Haynes, N. R. 'Planetary mission operations: an overview.' 435–438; Miner, E. D., Stembridge, C. H., Doms, P. E. 'Selecting and implementing scientific objectives.' 439–443; Jordan, J. F. 'Navigation systems.' 444–449; Linick, T. D. 'Spacecraft commanding for unmanned planetary missions: the uplink process.' 450–457; Smith, J. G. 'Communicating through deep space.' 458–464; Jones, C. P. 'Engineering challenges of in-flight spacecraft — Voyager: a case history.' 465–471; and Ebersole, M. M. 'The Space Flight Operations Center development project.' 472–480.

Miner, E. D., Stone, E. C. (1988) Voyager at Uranus. *Journal of the British Interplanetary Society*, **41**, 49–62.

8

The interior of Uranus

8.1 CONSTRAINTS IMPOSED BY VOYAGER RESULTS

Modeling of the interior of a planet is predominantly a mathematical process. The same physical laws that govern processes in the laboratory are assumed to apply within Uranus. Those laws can be expressed as mathematical relationships which describe the temperatures, pressures, and other chemical and physical characteristics within the planet. Unfortunately, not all the relevant laws are well-understood. It is often difficult or impossible to reproduce in the laboratory the conditions (high pressures and temperatures and chemical makeups) which characterize the interior of a giant planet. In such circumstances planetary scientists must rely on calculations and extrapolations from more familiar conditions.

With so many unknowns one might be tempted to say that modeling of the interiors of the planets is little more than educated guesswork. Fortunately, Uranus provides many outward clues which scientists in true Sherlock Holmes fashion can piece together to deduce the nature of things they cannot directly observe. Voyager 2 has helped in this process by providing several additional clues and improving the precision of previously known clues.

For example, as a result of the Voyager 2 encounter the total mass and the average density of Uranus are known to greater precision than ever before. One constraint imposed upon models is that the hypothetical internal mass distribution must be consistent with the observed mass. Mass values derived from Voyager data [1] are given in Table 8.1.

The mass of Uranus is not strictly spherical in its distribution within the planet. As a result, the gravitational forces exerted by the planet on its rings, its satellites, and the Voyager 2 spacecraft include gravity harmonic terms. The most important of these are J_2 and J_4. The values for these harmonic coefficients have been determined with high precision [2] from a combination of Voyager and ground-based observations; they are given in Table 8.2. Their constraining effect on interior models is on the distribution of density, especially in the outer half of the distance from the center to the cloud tops.

Table 8.1 — Uranus mass and density values from Voyager 2

Quantity	Units	Value
GM of the Uranian system	km^3/s^2	5 794 560$\pm$10
Sun's mass/Uranian system mass	—	22 902.94$\pm$0.04
GM of Uranus	km^3/s^2	5 793 947$\pm$23
Mass of Uranus	kg	8.687$\times$10^{25}
1-bar equator radius of Uranus (=*A*)	km	25 559$\pm$5
1-bar polar radius of Uranus (=*C*)	km	24 973$\pm$25
Ellipticity of Uranus disk (1—*C*/*A*)	—	0.0229$\pm$0.0001
Mean density of Uranus	g/cm^3	1.27$\pm$0.01

Table 8.2 — Uranus gravity harmonics

$$J_2=(3.34343\pm0.00032)\times10^{-3}$$
$$J_4=(-2.885\pm0.045)\times10^{-5}$$

Prior to the Voyager encounter, the rotation period of Uranus was essentially unknown. Because of interactions between the magnetic field of the planet and the influx of solar wind particles, both MAG and PRA were able to detect the periodic changes associated with those interactions. From that data, the body of Uranus is known to complete a rotation in 17.24 h [3]. This rotation period is longer than was expected on the basis of the J_2 'distortion' of the gravitational field [4], and therefore implies more mass in the outer one-third of the planet than previously had been supposed.

The chemical composition and pressure/temperature structure of the detectable atmosphere of Uranus were determined by the RSS occultation experiment during passage behind the planet as viewed from Earth [5]. Higher levels in the atmosphere were sensed by the two Voyager spectrometers (IRIS [6] and UVS [7]). The derived atmospheric structure and chemical composition Uranus chemical composition will be discussed in Chapter 9. The variations of temperature, pressure, and composition within the planet must coincide near the sensible atmosphere with the observed atmospheric characteristics, which therefore provide additional constraints which must be met by useful interior models.

Earth-based measurements of the microwave brightness of Uranus at many wavelengths also serve to constrain models. Microwaves generally originate in deep layers of the atmosphere, and the microwave brightness provides a measure of the temperature at those levels, modified by absorption of the radiation in the overlying atmosphere. Detailed interior models should also be consistent with the observed absence of heat escaping from the interior of the planet as measured by Voyager's

IRIS [8] and should allow for the presence of the unusual magnetic field of Uranus detected by Voyager's MAG [9] (see Chapter 10).

The Voyager measurements have resulted in the abandonment of earlier interior models. On the one hand, the measurements are insufficient to constrain possible models to a single unique set of internal characteristics; on the other, no model has yet been proposed which meets all of the constraints implied by the Voyager observations. As is often the case, Voyager results have answered many questions about the interior of the planet, but they have also resulted in many new questions for which planetary scientists still have no satisfactory answers.

8.2 MODELS CONSISTENT WITH VOYAGER DATA

The mathematical details associated with modeling the interior of Uranus are presented elsewhere [10]. They include specification of the relationships between mass, density, pressure, temperature, composition, and physical state at various depths within the planet. It is not the purpose of this chapter to duplicate those mathematical derivations, but rather to summarize the salient features of the resulting models.

Detailed chemical composition of the interior of Uranus is not specified in current models. Rather, chemical components are grouped into three broad classes: gas, ice, and rock. These general names do not refer to the phase (gaseous, liquid, or solid) of the material, but only to the types of material included in each class. Gas as defined by planetary scientists is taken to mean hydrogen and helium. It is generally assumed that the ratio of helium to hydrogen is the same as that observed for the Sun. Ice is defined to be a solar mix of water (H_2O), methane (CH_4), and ammonia (NH_3). Water is undoubtedly the most abundant of the ices. The rock component would be a mix of silicon dioxide (SiO_2), magnesium oxide (MgO), and either sulfur and oxygen compounds of iron (FeS and FeO) or metallic iron (Fe) and nickel (Ni). Silicon dioxide is generally assumed to be the most abundant of the rock components, comprising 35% to 40% of the rocky materials. Because of the high temperatures and pressures in the interior of the planet the gas, ice, and rock components are all expected to behave as liquids.

Post-Voyager models of the interior of Uranus by different authors [11] all possess some characteristics in common. In each of the models the rock core is relatively small (or nonexistent), constituting less than 3% of the total mass of the planet. The density of the rock core is near $9 \, g/cm^3$.

An intermediate region with density between that of rock and that of gas overlies the core region. The composition of this region varies between models, but ice is the primary component in each. About 85% of the total mass of the planet is contained in this intermediate region, which varies in density from about $1 \, g/cm^3$ near its outer boundary to nearly $5 \, g/cm^3$ near its inner boundary. Some authors propose distinct compositional boundaries within this region, others suggest a composition which varies continuously with depth.

The outermost region of the planet is predominantly hydrogen and helium, but with concentrations of water, methane, and ammonia enhanced by some factor, A, over solar abundances. The RSS occultation experiment showed that A is about 30 for methane at pressures greater than about 1 bar. This predominantly gaseous region has a density of about 0.3 g/cm^3 or less and occupies the outer 20% to 25% of the planet's radius. The mass contributed by this outer layer is less than 15% of the total mass of the planet.

Although the details of the models presented by various authors differ, their predictions of density variation with depth are remarkably similar (see Fig. 8.1). This

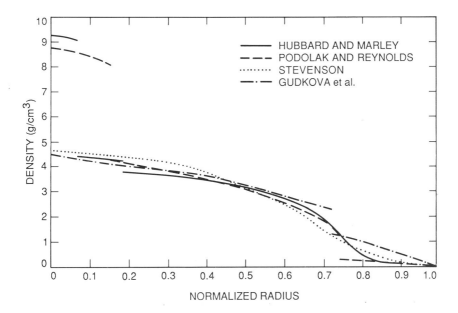

Fig. 8.1 — Variation of density with depth in the interior of Uranus for four different models. (Adapted from Podolak, Hubbard, and Stevenson [10]).

similarity in density profiles occurs because the distribution of density is reasonably well constrained by the mass and J_2 of Uranus. However, a large variety of mixtures of gas, ice, and rock can lead to the densities between 1 and 5 g/cm^3 that are characteristic of the bulk of the planet. The resulting models of Uranus's interior consequently are not unique: each author has a certain amount of freedom to insert his own personal biases regarding the internal composition of Uranus as long as his model results in the proper density profile.

The variation of temperature (and to a lesser degree, pressure) with depth is likewise poorly constrained. The pressure at the center of Uranus may be close to 10 000 000 bars. Central temperatures are generally calculated assuming a temperature increase with depth (lapse rate) which is adiabatic (see notes at end of Chapter 5 for a definition of adiabatic). With such lapse rates, the central temperature would be about 7000 K. Theoretical considerations lead to the conclusion that the internal lapse rates would not depart substantially from adiabatic, so it is quite likely that the temperature at Uranus's center lies between 6000 K and 10 000 K.

Heat from the interior of Uranus has been determined [8] to be less than 13% of that absorbed by the Sun, and the measurements are consistent with a complete absence of internal heat. Three possible reasons have been suggested [10] for the low amount of heat flowing from the interior: (1) Uranus might have been formed in some undetermined fashion which allowed for buildup of the planet without substantial heating; (2) the internal temperature lapse rate over much of the interior might be less than the adiabatic lapse rate so that the interior is not cooling; or (3) a very large number (millions) of immiscible stratified layers, or a composition stratification which varies smoothly with distance from the planet, prevent the heat from escaping from the hot interior of the planet. The consensus of planetologists' opinion is that the first two suggestions are unlikely; by a process of elimination, the third suggestion is favored. Thus the interior of Uranus is likely to be hot, and the overlying 'insulating' layers will keep it hot for an indefinite period.

In summary, our present understanding of the interior of Uranus is one in which there are three main regions (see Fig. 8.2). The intermediate region constitutes the majority of both the mass and the volume of the planet. This and the outer region are subdivided into a large number of stratified layers which act as insulators, so that little or no heat escapes from the deep interior. The outer region is composed of both gas and ice materials. The intermediate region contains mixtures of gas, ice, and rock materials. If a central core of molten rock exists, its size is much less than the size of the Earth.

NOTES AND REFERENCES

[1] Anderson, J. D., Campbell, J. K., Jacobson, R. A., Sweetnam, D. N., Taylor, A. H., Prentice, A. J. R., Tyler, G. L. (1987) Radio science with Voyager 2 at Uranus: results on masses and densities of the planet and five principal satellites. *Journal of Geophysical Research*, **92**, 14 877–14 883.

[2] French, R. G., Nicholson, P. D. Porco, C. C., Marouf, E. A. (1989) Dynamics and structure of the Uranian Rings. In Bergstralh, J. T., Miner, E. D. (eds.) *Uranus*, University of Arizona Press, Tucson (in preparation).

[3] Desch, M. D., Connerney, J. E. P., (1986) The rotation period of Uranus. *Nature*, **322**, 42.

[4] French, R. G. (1984) Oblateness of Uranus and Neptune. In Bergstralh, J. T. (ed.) *Uranus and Neptune*, NASA Conference Publication 2330, National Aeronautics and Space Administration, Scientific and Technical Information Branch, pp. 349–355.

[5] Lindal, G. F., Lyons, J. R., Sweetnam, D. N., Eshleman, V. R., Hinson, D. P., Tyler, G. L. (1987) The atmosphere of Uranus: results of radio occultation measurements with Voyager 2. *Journal of Geophysical Research*, **92**, 14 987–15 001.

[6] Flasar, F. M., Conrath, B. J., Gierasch, P. J., Pirraglia, J. A. (1987) Voyager infrared observations of Uranus' atmosphere: thermal structure and dynamics. *Journal of Geophysical Research*, **92**, 15 011–15 018.

[7] Herbert, F., Sandel, B. R., Yelle, R. V., Holberg, J. B., Broadfoot, A. L., Shemansky, D. E., Atreya, S. K., Romani, P. N. (1987) The upper atmosphere of Uranus: EUV occultations observed by Voyager 2. *Journal of Geophysical Research*, **92**, 15 093–15 109.

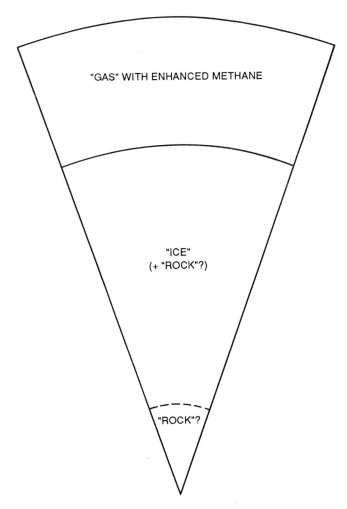

Fig. 8.2 — Representation of the interior structure of Uranus, depicting the extent of the three main regions. It is possible that no central core of molten rock exists.

[8] Conrath, B. J., Pearl, J. C., Appleby, J. F., Lindal, G. F., Orton, G. S., Bezard, B. (1989) Thermal structure and energy balance of Uranus. In Bergstralh, J. T., Miner, E. D. (eds.) *Uranus*, University of Arizona Press, Tucson (in preparation).

[9] Ness, N. F., Connerney, J. E. P., Lepping, R. P., Schulz, M., Voigt, H. (1989) Magnetic field and magnetosphere configurations. In Bergstralh, J. T., Miner, E. D. (eds.) *Uranus*, University of Arizona Press, Tucson (in preparation).

[10] Podolak, M., Hubbard, W. B., Stevenson, D. J. (1989) Models of Uranus' interior and magnetic field. In Bergstralh, J. T., Miner, E. D. (eds.) *Uranus*, University of Arizona Press, Tucson (in preparation).

[11] Interior models for Uranus have been described in papers by Hubbard and

Marley (1989), by Podolak and Reynolds (1989), by Stevenson (1988), and by Gudkova, Zharkov, and Leontyev (1988). References for these papers are given in Podolak, M., Hubbard, W. B., Stevenson, D. J. (1989) Models of Uranus' interior and magnetic field. In Bergstralh, J. T., Miner, E. D. (eds.) *Uranus*, University of Arizona Press, Tucson (in preparation).

BIBLIOGRAPHY

Bergstralh, J. T., Miner, E. D. (1989) *Uranus*, University of Arizona Press, in preparation (see chapter by Podolak, Hubbard, and Stevenson entitled, 'Models of Uranus' interior and magnetic field' and the references cited therein).

9

The atmosphere of Uranus

9.1 CHEMICAL COMPOSITION

Uranus is a gas giant planet. The main component of the atmosphere detectable by remote sensing techniques is hydrogen, the simplest of the elements. Most of this hydrogen exists in its molecular form, H_2, in which hydrogen atoms are bound together in pairs. Because of its predominance in the atmospheres of the Sun and the gas giant planets, hydrogen is often used as the measuring stick for the relative amounts of other atmospheric gases in those bodies.

The second most abundant gas is helium (He). Helium gas exists as atoms and is an inert, nonflammable gas that forms few chemical bonds with other elements. Both hydrogen and helium are colorless. They readily transmit visible light. Most of their selective absorption of light occurs at ultraviolet wavelengths.

The characteristic blue–green color of Uranus is due to methane (CH_4), the third most abundant component of Uranus's outer atmosphere. Methane ice crystals and gas readily absorb red light. The subtraction of red light from the incident sunlight changes its color from the yellows, reds, and browns of Saturn and Jupiter to the characteristic blues and greens of Uranus and Neptune.

Water (H_2O) and ammonia (NH_3) are also major constituents of the deep atmosphere of Uranus. Temperatures in the detectable atmosphere at pressure levels between 1 mbar and 1 bar are less than 90 K. Ammonia freezes at about 145 K and water freezes at about 275 K. Temperatures in the deep atmosphere of Uranus reach 145 K at pressure levels near 8 bars, far below the regions probed by Voyager instruments. Temperatures near 275 K are not reached until the pressure is near 120 bars! Upward motions in the atmosphere can carry small amounts of both ammonia and water to much higher levels in the atmosphere, but in quantities that are undetectable in the case of water and barely detectable in the case of ammonia. Vertical convection within the atmosphere of Uranus is much less than in the other gas giant planets because of the absence or near absence of escaping internal heat. At those higher levels in the atmosphere where pressures are less than about 1 μbar the temperatures increase with altitude until they again exceed the melting points of ammonia and water. Because of the rarified nature of the upper atmosphere and the water and ammonia 'cold traps' in the deeper atmosphere, these two constituents cannot be detected at these extreme altitudes.

9.1.1   Helium/hydrogen ratio

One of Voyager 2's main scientific objectives for atmospheric studies of the gas giant planets was determination of the helium abundance in each planet. This abundance is generally expressed as the ratio of the mass of helium to the mass of hydrogen in the atmosphere. There are (or were) sound reasons for the importance attached to the helium abundance measurements. The gas giants were expected to be relatively unaltered reservoirs of hydrogen and helium from the primitive solar nebula [1]. It was assumed that all the gas giant planets would have the same helium-to-hydrogen mass ratio, and that the measured value would be representative of the helium-to-hydrogen ratio in the material from which the Solar System was formed.

Measurements of the helium abundance in the outer atmosphere of the Sun [2] yield helium-to-hydrogen mass ratios near 0.28. Thermonuclear reactions within the Sun's core are constantly creating helium out of hydrogen, so the solar helium abundance was expected to represent an upper limit for the helium abundance in the primordial Solar System. Voyager determined the outer atmospheric helium abundance of Jupiter [3] to be 0.18 ± 0.04 and of Saturn [4] to be 0.06 ± 0.05. Each is considerably less than the solar value, and Saturn's helium abundance is much less than Jupiter's. Obviously, it was erroneous to assume that the outer atmospheres of the gas giant planets would be representative of the unaltered primordial solar nebula hydrogen and helium abundances.

The high (greater than 3 000 000 bars) pressures in the gas interiors of Jupiter and Saturn transform hydrogen into a metallic state [5]. Temperatures in the interiors of both Jupiter and Saturn are low enough that the helium separates from the hydrogen and forms liquid droplets. Because of their higher density, these droplets sink slowly through the metallic hydrogen, resulting in a depletion of helium at higher levels, which eventually manifests itself in lower helium abundance in the outer atmosphere. The high pressures needed for transformation of hydrogen to its metallic form occur at lower temperatures inside Saturn than inside Jupiter. Helium depletion should therefore be more pronounced in the outer atmosphere of Saturn than in the outer atmosphere of Jupiter helium depletion Jupiter, consistent with the Voyager measurements.

Pressures in the outer regions of Uranus are never high enough to create metallic hydrogen. As a consequence, helium depletion by the same mechanism as at Jupiter and Saturn is ineffective, and the helium abundance of Uranus may be representative of the primordial solar nebula. Only two mechanisms have been suggested which might change the picture somewhat. If Uranus was formed with an abundance of elemental carbon (C) or nitrogen (N), combination with hydrogen to form methane (CH_4) or ammonia (NH_3) would reduce the amount of hydrogen and consequently raise the mass ratio of helium to hydrogen [6]. Other authors [7] suggested that at high temperatures (above 2000 K) and high pressures (above 200 000 bars) methane has a tendency to break apart into its components carbon and hydrogen. This would lead to an increase of hydrogen in the outer atmosphere, effectively reducing the helium-to-hydrogen mass ratio.

The Uranus helium abundance measured by Voyager [8] is 0.26±0.05. This is close enough to other estimates of the helium abundance in the primordial solar nebula that neither of the mechanisms discussed above that effectively enrich or deplete the helium abundance is likely to be important for Uranus. The helium mass fractions for the outer atmospheres of Jupiter, Saturn, Uranus, and the Sun are shown with their uncertainties in Fig. 9.1.

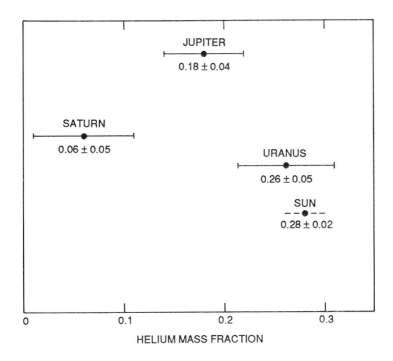

Fig. 9.1 — The helium mass fractions for the outer atmospheres of Jupiter, Saturn, Uranus, and the Sun. Uncertainties are shown by the superimposed error bars.

9.1.2 Methane abundance

Methane is the only 'ice' component of Uranus whose abundance can be determined directly from Voyager data. Water and ammonia are believed to be major components of the interior of Uranus, but they freeze at much higher temperatures than methane. The pressure levels within Uranus corresponding to those higher temperatures are about 8 bars for ammonia and 120 bars for water. Such pressures exist at levels deeper in the atmosphere than could be sensed by Voyager instruments (see Fig. 9.2).

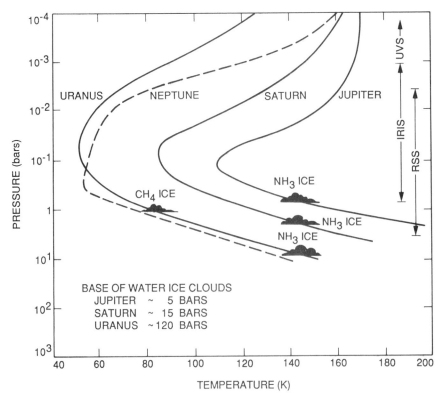

Fig. 9.2 — Temperature–pressure curves for the four gas giant planets. The approximate pressure levels probed by ultraviolet, infrared, and radio science investigations are shown at the right of the figure. The approximate condensation temperatures and corresponding pressure levels for methane, ammonia, and water clouds are indicated.

The abundance of methane was determined from Radio Science (RSS) experimentation during the period when Voyager flew behind Uranus as viewed from Earth. Fig. 9.3 depicts the path of Voyager and the raw RSS data for the same period. The intensity of the signal received at Earth dropped precipitously as the spacecraft dipped behind Uranus, but it did not disappear entirely. Signals were detected continuously for the entire 86-min period of the occultation. On two occasions during the occultation the signal strength momentarily decreased an additional amount and then returned to prior levels. The two events correspond to those times when the radio signals reached depths corresponding to a methane cloud deck. Between the times of the two events a part of the radio beam penetrated the Uranus atmosphere to depths below the methane cloud deck and part of the beam was scattered by the cloud itself.

Above the methane cloud deck the relative 'humidity' of methane gas was about 30%, meaning that methane constitutes about 0.0002% of the mass of the atmosphere above the cloud deck. Immediately below the methane cloud deck, about 2.3% of the molecules are methane, the methane has a relative humidity of about 78%, and the mass fraction of the heavier methane is about 14% [9].

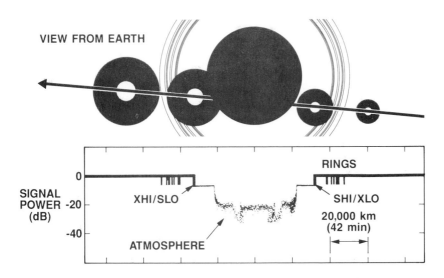

Fig. 9.3 — Path of Voyager 2 behind Uranus as viewed from Earth tracking stations. Radio science data for the occultation period are displayed for the same period. Abrupt dips in the x-band signal intensity occur as the radio signal passes through the methane cloud layer near the 1.3-bar pressure level.

9.1.3   Minor constituents

No chemical constituents other than methane (CH_4), helium (He), and molecular and atomic hydrogen (H_2 and H) were detected by Voyager in the lower atmosphere of Uranus. Acetylene (C_2H_2) was positively detected in the upper atmosphere, and there is evidence for the upper atmospheric presence of ethane (C_2H_6) as well [10]. Both of these gases are byproducts of the action of sunlight on methane gas: an example of photochemistry. Other expected byproducts of methane photochemistry at Uranus include the polyacetylenes ($C_{2n}H_2$, with $n=2$, 3, etc.), the most important of which is diacetylene (C_4H_2) [11]. Although these minor constituents of the Uranus atmosphere contribute little to the gaseous makeup of the atmosphere, they condense in the upper atmosphere to form hazes which affect the reflective characteristics of the atmosphere, raise the temperature of the upper atmosphere and lower the temperature of the underlying atmosphere.

9.2   GLOBAL ENERGY DISTRIBUTION

Uranus appears to be unique among the gas giant planets. Jupiter, Saturn, and Neptune are much warmer than would be expected on the basis of solar heating alone. The excess heat comes from the interiors of the planets and in each case contributes nearly as much energy to the outer atmosphere as is contributed by the Sun. The heat escaping from the interior of these three planets causes turbulence in their atmospheres; it may also be responsible for the atmospheric banding (the light belts and dark zones) which gives these planets their layered appearance (see Fig. 9.4). Uranus is no warmer than Neptune, even though it receives $2\frac{1}{2}$ times as much

Fig. 9.4 — Heat escaping from the interior of Jupiter may be responsible for the atmospheric banding (the light belts and dark zones) shown in this Voyager image of Jupiter. Similar effects cause atmospheric banding in the atmospheres of Saturn and Neptune. (P-20993)

sunlight. Uranus is slightly more reflective than Neptune, so that it absorbs less of the incident sunlight, but the difference is far too small to explain the similar temperatures of the two planets. On the basis of these observed characteristics, astronomers concluded that Uranus radiated much less internal heat than Neptune.

An important task for Voyager 2 was to carefully measure the temperature and reflective characteristics of Uranus under a variety of illumination conditions. These data, combined with the knowledge that Uranus's distance from the Sun diminishes the intensity of sunlight illuminating the planet by a factor of 368 relative to Earth, would enable scientists to calculate what fraction of the radiated heat was generated in the planet's interior. Precise determination of the internal heat helps differentiate between model predictions for Uranus.

9.2.1 Amount of sunlight absorbed

An accurate estimate of the amount of sunlight absorbed by Uranus demands (1) knowledge of the amount of sunlight incident on the planet and (2) determination of the amount of sunlight reflected back into space. The difference between these two represents the solar energy absorbed by the planet.

The total flux of solar radiation per unit area received outside the Earth's atmosphere at the average Sun–Earth distance (1.000 AU) is known as the solar constant. Its value is slightly variable [12], but has a value of about 1367 ± 2 W/m^2 for the years 1985–8 [13]. At greater distances from the Sun, the flux of solar energy per unit area drops by a factor of r^{-2}, where r is the distance from the Sun in AU. The mean Sun–Uranus distance as determined by Voyager is 19.26234 AU; the total flux of solar radiation at the distance of Uranus is therefore 3.684 ± 0.005 W/m^2. For reference, the Sun–Uranus distance at the time of the Voyager 2 closest approach to Uranus on January 24, 1986, was 19.12227 AU. The corresponding incident solar flux was 3.738 ± 0.005 W/m^2.

The brightness of sunlight scattered from the planet was carefully measured by Voyager for a variety of illumination and viewing conditions. From these measurements and earlier Earth-based measurements it was concluded [14] that, when averaged over the 84-year orbital period of Uranus, the ratio of solar energy scattered away to space relative to that incident upon Uranus was 0.300 ± 0.049. In other words, 0.700 ± 0.049 of the solar radiation incident upon Uranus during its passage around the Sun is being absorbed by the planet.

Sunlight illuminates the planet from one direction only. The heat escapes primarily as infrared radiation, which is emitted in all directions from both the illuminated and dark parts of the planet. One must therefore effectively reduce the solar flux per unit area by an additional factor of four (the ratio of the area of a sphere to that of a circle of the same radius) before comparing it to the heat energy being radiated by the planet. The effective planetwide solar flux input, averaged over a Uranus year, is thus $(3.684)(0.700)/4=0.645\pm0.046$ W/m^2. From the Stephan–Boltzmann equation [15], the solar heating alone would result in a planetwide average effective temperature [16] of 58.2 ± 1.0 K.

9.2.2 Amount of reradiated energy

On the basis of theoretical considerations the poles of Uranus were expected to be a few degrees warmer than the equator of Uranus. As with the other gas giant planets, the deep atmosphere of Uranus has an enormous heat capacity. It heats and cools over time periods many times longer than the 84-year orbital period. The temperatures encountered are the result of solar energy input and internal processes averaged over many orbits of the Sun. Because of the high tilt of Uranus's rotation axis the poles of the planet are very nearly vertically illuminated with sunlight for major fractions of each 84-year orbit. The same is also true for the equatorial regions, but planetary rotation turns the equatorial regions away from the Sun for half of each 17.24-h Uranus day. Rotation affects the polar regions very little. In the absence of other processes, the solar energy input alone should result in equilibrium temperatures about 6 K warmer in the polar regions than in the equatorial regions of the planet.

It was with some surprise that Voyager scientists measured equatorial temperatures which were nearly the same as those measured in both the dark north polar region and the illuminated south polar region. The inescapable conclusion is that transport of energy toward the equator must occur within the atmosphere. Fig. 9.5

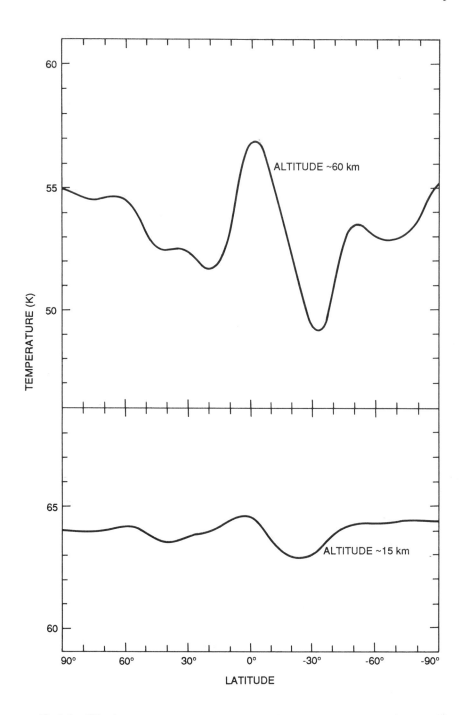

Fig. 9.5 — Effective temperatures of Uranus for latitudes extending from the north pole (+90°) to the south pole (−90°), as measured by the infrared investigation on Voyager 2. Temperatures are shown for altitudes close to 60 km (top) and 15 km (bottom).

depicts the measured effective temperatures of Uranus for latitudes extending from the north pole ($+90°$) to the south pole ($-90°$). Temperature variation with planetary longitude was neither expected nor detected.

If the atmosphere did not selectively absorb some of the emitted radiation, the brightness of Uranus at any thermal infrared wavelength would provide a measure of the temperature of the planet. The actual situation requires modeling of the atmosphere and the infrared spectrum to provide a measure of the total thermal flux escaping the planet. The errors introduced by such modeling are not large. For example, two different modeling techniques discussed by Conrath *et al.* [17] result in global effective temperatures which differ by only about 0.1 K. The Uranus effective temperature as derived from Voyager measurements is 59.1 ± 0.3 K. This number is derived from an estimated average thermal flux emitted from the planet of 0.693 ± 0.013 W/m^2.

9.2.3 Does Uranus have an internal heat source?

A simple comparison of the estimated emitted thermal flux to that provided by sunlight alone shows that very little of the emitted energy can be coming from the interior of the planet. The ratio of these fluxes is sometimes known as the 'energy balance' and is 1.06 ± 0.08. This is compared to values for the other giant planets in Table 9.1.

Table 9.1 — Energy balance for Jupiter, Saturn, Uranus, and Neptune

	Jupiter	Saturn	Uranus	Neptune
Mean solar distance (AU):	5.20256	9.55475	19.26234	30.1096
Incident flux (W/m^2):	50.50	14.97	3.68	1.51
Reflected flux (W/m^2):	17.41	5.15	1.10	0.47
Total reflectivity:	0.345	0.344	0.300	0.31
Absorbed flux/4 (W/m^2):	8.27	2.46	0.65	0.26
Equilibrium temperature (K):	109.9	81.1	58.2	46.2
Emitted flux (W/m^2):	13.56	4.61	0.69	0.69
Effective temperature (K):	124.4	95.0	59.1	59.1
Energy balance:	1.64	1.87	1.06	2.7
Internal flux (W/m^2):	5.29	2.15	0.04	0.43

It is apparent from Table 9.1 that the amount of internal heat escaping from Uranus is much less than for any of the other gas giant planets. The reasons for this deficiency are not obvious, but the consequences may be extensive. It is likely that the relative blandness of the Uranian atmosphere is related to a lack of vertical motions within the outer atmosphere. Such vertical motions would normally be caused by interaction between the escaping heat and the atmosphere.

As implied in Chapter 8, the near absence of escaping internal heat does not necessarily mean that internal heat sources do not exist. It is possible that the heat is trapped in Uranus's interior and effectively prevented from escaping by the structure of the overlying layers. Additional theoretical work and perhaps additional observational data will be needed to better understand the reasons for the low levels of escaping internal heat from Uranus.

9.3   VERTICAL STRUCTURE WITHIN THE ATMOSPHERE

A 'blackbody' has uniform temperature, absorbs all the sunlight incident upon it, and reradiates the same amount of energy in a mathematically precise fashion. The amount of thermal radiation at any given wavelength (color) of light is defined by a relationship known as the Planck function , after German physicist Max Planck. It shows that a very hot blackbody (several tens of thousands of degrees Celsius) emits most of its energy at blue, violet, and ultraviolet wavelengths. A cool blackbody (comparable in temperature to Uranus's effective temperature of 59.1 K) radiates most of its energy at red and infrared wavelengths. The Planck function also predicts the T^4 dependence (where T is the temperature) of total emitted energy. Typical blackbody curves for temperatures of 10 000 K, 1000 K, and 100 K are shown in Fig. 9.6.

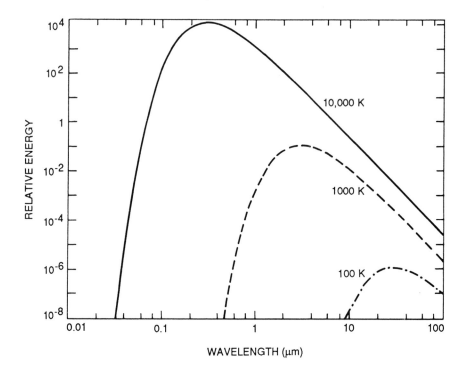

Fig. 9.6 — Typical blackbody curves for temperatures of 10 000 K, 1000 K, and 100 K. Note the decrease in energy and increase in the wavelength of maximum energy as temperature decreases.

The Uranian atmosphere, like most real bodies, departs substantially from blackbody behavior. The presence of hydrogen gas, helium gas, and methane gas and ices causes selective absorption of light at predictable wavelengths (colors). The measured thermal emission escaping the planet is affected by and in turn affects the relative chemical composition and temperature in the overlying atmosphere. As a result, the net thermal energy measured at any given wavelength is the sum of contributions (both emission and absorption) from many levels in the atmosphere. The relative amounts of energy from each level (at a given wavelength) are often referred to as the 'weighting function' for that wavelength. At wavelengths where the atmosphere is strongly absorbing, the weighting function peaks at high altitudes in the atmosphere; where little or no absorption occurs the weighting function has its maximum much deeper in the atmosphere. The nature of the weighting function at each wavelength is determined by a combination of theoretical and empirical methods and must include instrument response. Once determined, these weighting functions can be used to help derive the temperature and pressure of the atmosphere at levels near each weighting function peak. By selecting the appropriate wavelengths, one can in principle reconstruct vertical profiles of temperatures, pressures, and composition within the atmosphere. This was one of the primary goals of the infrared and ultraviolet investigations aboard Voyager.

The above procedures provide atmospheric vertical profiles for altitudes where the pressures are less than 0.5 bar. (A bar is very nearly the pressure of Earth's atmosphere at sea level.) The radio science occultation experiment probed the Uranus atmosphere to 2.3 bars and therefore extended to greater depths the vertical structure information obtained by the ultraviolet and infrared investigations.

9.3.1 Temperatures and pressures at different altitudes

For a planet with a solid surface the atmospheric altitude scale is usually referenced to the level of the average surface, or in the case of Earth to mean sea level. A useful reference point for a gas giant planet is the level in the (equatorial) atmosphere where the pressure is 1.0 bar. For Uranus this falls at a distance of 25 559 km from the center of the planet; the temperature at this level is 76.4 K. Table 9.2 presents the pressure, temperature, and methane relative abundance (in percentage of the total numbers of molecules) at intervals of 10 km above and below the 1-bar reference level [18]. The ratio of helium to hydrogen molecules is assumed to be constant at 15/85. The data were derived on the basis of two radio occultation measurements. The ingress occurred over latitudes from 2° to 6° south; the egress latitudes were between 6° and 7° south.

9.3.2 Variations with latitude

The radio science occultation profiles were limited to near-equatorial latitudes by the geometry of the flyby. Atmospheric observations by the infrared and ultraviolet investigations have no such limitation. Fig. 9.5, introduced earlier in discussions about Uranus's energy balance, depicts the latitudinal variation of temperature at altitudes close to 60 km (top part of figure) and 15 km (bottom part of figure). At altitudes near 15 km the temperatures are remarkably uniform, varying by less than 1 K from the mean temperature of 64 K. Much larger excursions are seen near 60 km

Table 9.2 — Atmospheric data for Uranus

Alt. (km)	Temp. (K)	Press. (bar)	Total (10^{17} molecules/cm^3)	CH$_4$ (% of molecules)
200.0	104.7	0.00052	0.36	0.00
190.0	88.4	0.00067	0.55	0.00
180.0	76.1	0.00088	0.84	0.00
170.0	70.3	0.00124	1.27	0.00
160.0	68.0	0.00174	1.84	0.00
150.0	66.8	0.00248	2.69	0.00
140.0	68.8	0.00354	3.72	0.00
130.0	67.9	0.00501	5.35	0.00
120.0	63.3	0.00722	8.25	0.00
110.0	63.9	0.0105	11.9	0.00
100.0	62.3	0.0152	17.7	0.00
90.0	57.2	0.0228	28.9	0.00
80.0	55.2	0.0342	44.9	0.00
70.0	53.8	0.0544	73.1	0.00
60.0	53.4	0.0852	115.6	0.00
50.0	53.2	0.1337	182.0	0.00
40.0	54.5	0.2089	277.9	0.00
30.0	56.5	0.3232	414.7	0.00
20.0	60.9	0.4887	580.9	0.00
10.0	67.7	0.7119	760.9	0.06
0.0	76.4	1.0000	948.3	0.31
−10.0	85.1	1.376	1171.0	2.26
−20.0	94.0	1.868	1439.0	2.26
−27.5	100.9	2.309	1657.0	2.26

altitude, where temperatures drop from 57 K near the equator to 50±2 K near ±30° latitude, returning to about 55 K at the poles.

The rotation of the planet also causes some variations with latitude. Oblateness (polar flattening) of the body of Uranus is one of the primary results. Whereas the 1-bar equatorial pressure level occurs at a distance of 25 559 km from the center of the planet, a pressure of 1 bar occurs at a radius of only 24 973±20 km at the poles. Other pressure levels in the polar regions occur at correspondingly reduced distances from the center of Uranus. The difference between polar and equatorial radii of Uranus is larger than would have been predicted for the measured rotation rate of 17.24 h. It is this fact that led to revisions in the interior model of Uranus.

9.3.3 Discrete cloud and haze layers

The pre-Voyager view of Uranus suggested the possibility of a thin layer of haze at altitudes near 200 km above the 1-bar pressure level. A more substantial but still relatively transparent cloud deck of methane ice crystals was expected near the 2-bar level. An opaque cloud of unknown composition near the 3-bar level was also suggested.

Voyager 2 confirmed the existence of the haze layer, but found that it was somewhat lower in the atmosphere, at altitudes below 180 km. The haze probably extends to altitudes below 50 km but is difficult to detect below about 70 km altitude. Detailed analysis of the data indicates that the haze is probably composed of tiny ice particles of ethane (C_2H_6), acetylene (C_2H_2), and diacetylene (C_4H_2) [19]. Particle sizes near 75 km altitude are about 10^{-7} m. The average particle size increases as the number density increases at lower altitudes. Absorption of about 1% of the incident sunlight by the upper haze layer causes the atmosphere to be heated by several K at those altitudes.

The methane (CH_4) cloud was found by Voyager to be 2 or 3 km thick with its bottom near 8.5 km below the 1-bar pressure level (i.e., near a pressure of 1.3 bar). The cloud absorbs about 20% of the incident sunlight near 25° south latitude, but may be more absorbent in the polar regions. Since methane is primarily absorbent at infrared wavelengths, it is likely that some additional impurities make the methane cloud absorbent at visible and near-infrared wavelengths. One possibility is that more complex hydrocarbons are produced in Uranus's stratosphere and settle into the methane cloud layer.

Fig. 9.7 shows a triplet of images of Uranus. The images were taken through the

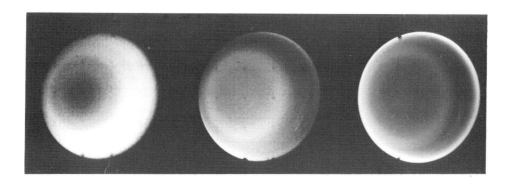

Fig. 9.7 — Violet, orange, and methane (red) images of Uranus. Image contrast of each has been enhanced to bring out details. The differences, although exaggerated by the contrast stretch, are nevertheless indicative of latitudinal variations in the cloud and haze structure.
(P-29517)

violet, orange, and methane (red) filters, respectively; the contrast of each has been enhanced to bring out details. The differences, although exaggerated by the contrast stretch, are nevertheless indicative of latitudinal variations in the cloud and haze structure. Fig. 9.8 shows a possible interpretation of circulation patterns within the atmosphere which might give rise to these variations [20].

The deepest cloud was not sensed by Voyager. A combination of Earth-based and Voyager observations lead to the conclusion that ammonia (NH_3) in the deep atmosphere constitutes less than 1% of the relative amount predicted by models of

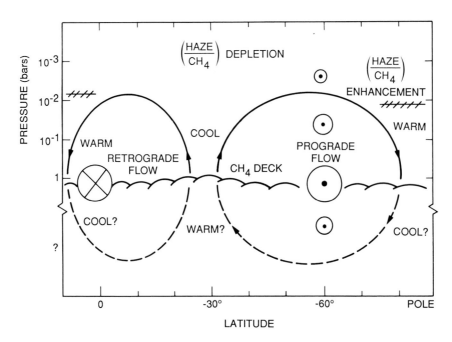

Fig. 9.8 — A schematic interpretation of circulation patterns within the atmosphere which attempts to explain many of the observed characteristics of the Uranian atmosphere. (Adapted from Allison *et al.* [20].)

the solar nebula. This low abundance of ammonia may permit hydrogen sulfide (H_2S) to be present in large enough quantities to form a cloud with its top near the 2.7-bar level [19]. This cloud absorbs at least 60% of the light incident upon it, and it may be virtually opaque. The model of Fig. 9.8 would predict a somewhat higher abundance of ammonia near 25° south latitude; if a hydrogen sulfide cloud exists, it would be at much deeper levels at this latitude.

9.3.4 The upper atmosphere
The structure of the extreme upper atmosphere is much less certain than that of the lower atmosphere, but some estimates are possible on the basis of ultraviolet spectrometer data. Table 9.3 lists temperatures, pressures, and number densities to altitudes of 7000 km. Above 500 km the relative abundance of helium decreases rapidly. At 1000 km, helium constitutes about 1% of the atmosphere; it has dropped to less than 0.1% at 2000 km altitude. Hydrogen, mainly in its diatomic molecular form, is the sole detectable gas at higher altitudes. The fraction of hydrogen in its atomic form continues to increase with altitude. At 7000 km altitude about one-quarter of the hydrogen is atomic (H) and three-quarters is molecular (H_2).

At even higher altitudes in the 'exosphere' atomic hydrogen becomes the dominant constituent. The 800 K temperature above 5000 km is representative of a very slow decrease in the number of atoms of hydrogen per volume as the altitude increases. Extrapolation to the altitudes of the Uranian rings (12 000 to 25 000 km

Table 9.3 — Upper atmospheric data for Uranus

Alt. (km)	Temp. (K)	Press. (bar)	Total (molecules cm^{-3})
7000	800	3×10^{-13}	3×10^{6}
6000	800	2×10^{-12}	2×10^{7}
5000	800	1×10^{-11}	1×10^{8}
4000	750	5×10^{-11}	5×10^{8}
3000	680	5×10^{-10}	5×10^{9}
2000	600	5×10^{-9}	1×10^{11}
1000	500	1×10^{-7}	2×10^{12}
500	150	5×10^{-6}	1×10^{14}

altitude) shows that enough atomic hydrogen must be present to cause the ring particles to experience orbital drag. This force and its implications for the lifetimes of the rings will be discussed further in Chapter 11.

As the atomic hydrogen is exposed to ultraviolet light from the Sun, much of it is ionized (outer electrons absorb enough energy to escape from the hydrogen atoms), resulting in a large population of free electrons. The resulting 'ionosphere' was detected during the radio occultation experiment. Differences of up to a factor of 100 in the number of electrons per volume were noted between the ingress and egress portions of the occultation at certain altitudes. Both showed number densities still well in excess of 100 electrons/cm^{3} at all altitudes up to 10000 km above the 1-bar pressure level. These high number densities are consistent with the presence of a very extended upper atmosphere of atomic hydrogen.

Atomic hydrogen emits ultraviolet light, predominantly at a wavelength of 1216 Å ($1 \text{Å} = 10^{-10}$ m). It is caused by the release of light energy as the electron in atomic hydrogen drops from its lowest excited energy level into its normal energy level. Ultraviolet light from the Sun is often responsible for providing the initial energy to 'excite' the hydrogen atom. Thereafter the atom spontaneously relaxes by re-emitting 1216-Å ultraviolet light. This process is known as solar fluorescence . The sunlit portion of Uranus's atmosphere appears to emit far more ultraviolet radiation than can be accounted for by solar fluorescence alone, and some atmospheric scientists have proposed that all or most of the radiation comes from 'electron excitation'. In this latter process, electrons in the extended ionosphere are energized by processes not yet fully understood. They collide with the hydrogen atoms, giving up a part of their energy to the atoms. These atoms then relax by emitting the observed ultraviolet radiation. The electron energizing is hypothesized to occur only on the sunlit side of Uranus, and the resulting ultraviolet emissions were called 'electroglow'.

There is still lively debate about the relative contributions to the ultraviolet emissions from Uranus of solar fluorescence and electron excitation. Until this debate is settled it is more descriptive and less judgmental to call the emissions 'dayglow' whether they are caused by solar fluorescence, electron excitation, a combination of the two, or some other as yet undetermined process. There are large uncertainties associated with the problem. The energy output of the Sun at ultraviolet wavelengths is highly variable and poorly determined. There is disagreement about whether solar fluorescence is sufficient to cause the observed emissions. Similarly, the process which energized the electrons sufficiently to cause electron excitation is not well understood. At present it is clear that some process is occurring to cause the daylight side of Uranus to glow brightly in the ultraviolet; explanation of the effect must wait for additional analysis.

9.4   HORIZONTAL CLOUD AND TEMPERATURE STRUCTURE

The dominant features of the Uranian atmosphere are axially symmetric, showing almost no variation with longitude on the planet. A space traveller approaching the planet would have great difficulty detecting the rotation of the planet by visual means. Contrast in latitude was also minimal, and many described the visual appearance of Uranus as that of a featureless blue ball.

This characteristic of Uranus is in sharp contrast with the atmospheres of Jupiter and Saturn. Tracking of discrete cloud features in the atmospheres of these two larger planets revealed much about the development of weather patterns and about the nature of east–west wind speeds. Prevailing winds remained relatively constant in speed and direction at each latitude. In the case of Saturn, there was additionally a definite north–south latitudinal symmetry in the wind speeds (see Fig. 9.9).

The 98° axial tilt of Uranus, its apparent absence of escaping internal heat, and its large distance from the Sun combine to create a planetary meteorology that is unique among the giant planets of the Solar System.

9.4.1   Cloud features
Only a few localized cloud features were detected in Voyager imaging. Contrast in both color and brightness of these features was low, and only in images processed to provide extreme contrast enhancements could they be seen at all. A total of eight discrete cloud features could be tracked with sufficient accuracy to provide measures of wind speeds. Fig. 9.10 shows an example of one such cloud feature. Latitudinal variations of cloud brightness and color were shown earlier in Fig. 9.7. If storms of the sort seen in the atmospheres of Jupiter and Saturn exist in the atmosphere of Uranus, they are either too deep or generate too little contrast to render them observable in Voyager images.

9.4.2   Zonal wind profiles
No longitudinal variation in temperature was detected by the infrared or ultraviolet investigations. Latitudinal variations were discussed earlier and were depicted schematically in Fig. 9.5. These temperature differences give rise to zonal winds , as implied by the model of Fig. 9.8.

In addition to the temperature data, several types of information provide clues about motions within the atmosphere. The rotation of the planet gives rise to forces which help to shape the winds. The unique orientation of Uranus during the Voyager flyby led to some speculation that the near-polar solar illumination and heating might generate a different type of wind regime than had been encountered in the

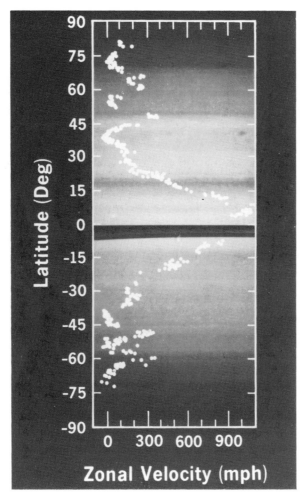

Fig. 9.9 — Zonal wind speeds for the planet Saturn, showing the strong north-south latitudinal symmetry. (260-1529B)

atmospheres of Jupiter and Saturn. One major conclusion of the Voyager encounter data at Uranus was that planetary rotation plays a much stronger role in determining wind directions than does the solar illumination.

Fig. 9.11 provides an estimate of the variation of wind speed with latitude near the 1-bar pressure level in the atmosphere of Uranus. The imaging data covers latitudes from 25° to 41° south latitude, with a single additional point near 71° south latitude. All of the wind speeds derived from imaging are prograde, that is the winds blow in the same direction as the planet rotates. Because 90° latitude is a single point at each pole, the wind speeds there must fall to zero. The radio science occultation experiment provided an additional wind speed estimate of about 100 m/s retrograde for 6° south latitude. No direct wind speed information was obtained for the

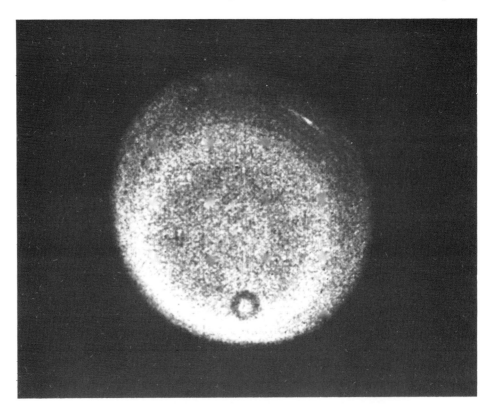

Fig. 9.10 — One of eight detected discrete cloud features used to provide measures of wind speeds is shown near the upper limb in this Voyager 2 image of Uranus. Contrast has been enhanced to bring out low-contrast details. Dark circular marks are due to dust specks in the camera optics. (P-29468)

unilluminated northern hemisphere. The wind speed model derived from infrared temperature data is also shown in Fig. 9.11 and applies to both hemispheres, but at an altitude about 50 km above the 1-bar pressure level. The difference in wind speeds with altitude is probably real. The model does not strongly constrain wind speeds near the equator of the planet.

Extrapolation of the 1-bar wind profile from low latitudes to the south pole is obviously very dependent on the single data point at 71° south latitude. Wind speeds represented by the dashed line are from the equation,

$$U(\text{m/s}) = 200 \ (0.4 \cos b - \cos 3b) \ ,$$

where U is the wind speed and b is the latitude. The form of this equation is purely empirical; it is solely a convenient method of interpolating between the wind speed data points. The actual wind profile may depart significantly from this simple model

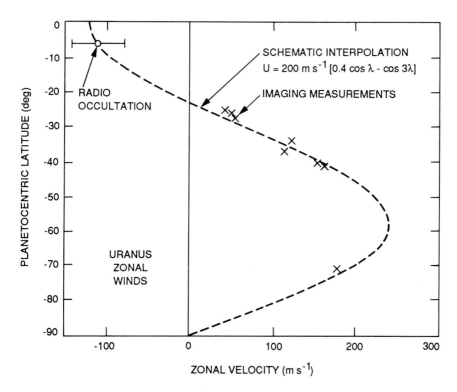

Fig. 9.11 — Variation of wind speed with latitude near the 1-bar pressure level in the atmosphere of Uranus. The imaging data covers latitudes from 25° to 41° south latitude, with a single additional point near 71° south latitude. The wind speed model derived from infrared temperature data is also shown in Fig. 9.11 and applies to both hemispheres, but at an altitude about 50 km above the 1-bar pressure level. The single retrograde point near 6° south latitude was derived from radio science occultation results.

for moderate to high latitude. The number of data points is insufficient to disclose small-scale structure in the profile at any latitude.

The sources which provide the power to drive the Uranian winds are not understood. In some respects, the wind fields on Earth are similar to those measured for Uranus: near Earth's equator winds are retrograde, at mid-latitudes the winds are prograde. On Earth, sunlight falls predominantly on low latitudes. This causes poleward expansion of the near-equatorial atmosphere. The Coriolis force [21] due to the rotation of the Earth turns this poleward expansion into clockwise atmospheric currents in the northern hemisphere and counterclockwise currents in the southern hemisphere. The net effect is a dominant westerly (prograde) wind at middle latitudes in each hemisphere and a dominant easterly (retrograde) wind near the equator.

Most of the solar heating in Uranus's atmosphere (both averaged over a year and near the time of the Voyager encounter) is in the polar regions. Expansion toward the equator would be expected in such conditions, and the Coriolis forces might then result in retrograde winds at middle latitudes. The observed winds are nevertheless prograde, and atmospheric scientists are left without an adequate explanation of the phenomenon. Additional data are badly needed.

9.4.3 The color of Uranus

Uranus is deficient in its reflectivity at red wavelengths. As mentioned in Chapter 5 this characteristic has long been attributed to methane in the atmosphere of the planet. Hydrogen and helium contribute almost nothing to the planet's coloration, primarily because very little visible light is absorbed by these gases. Because of the methane absorption and low contrast Uranus appears to the human eye to be a featureless blue–green globe.

Although of low contrast, color differences were seen in the Voyager images of Uranus. Fig. 9.7 was used earlier to display atmospheric banding; the subtle coloration of that banding provides additional clues about the distribution of clouds and hazes on a global scale.

The violet image is uniformly bright over most latitudes. Methane absorbs little of the violet light. Sunlight probably reflects uniformly from ice particles (haze) suspended in the atmosphere. Near the south pole, a slightly reddish and darkened region of the atmosphere can probably be explained by an enhancement in its content of more complex hydrocarbons, formed by the action of sunlight on methane gas.

In orange light the suspended ice particles are somewhat subdued in brightness because absorption by methane gas is beginning to be apparent. A darker region between about 20° and 40° south latitude corresponds to those latitudes where temperatures are depressed by a few degrees. Here the reflective hazes may be reduced sufficiently to permit viewing of some plumes (similar to the tops of thunderheads on Earth) emerging from the deeper methane clouds. Seven of the eight discrete cloud features used to determine atmospheric wind speeds were found in orange images of this latitude range. The hazes reappear between 40° and 60° south latitude but are obscured by the overlying hydrocarbons at more southerly latitudes. The haze reduction near 30° south latitude is less apparent in violet light because the reflectivity of the haze particles is higher and absorption by methane gas is less at violet wavelengths.

Methane gas absorption dominates the view of the planet at red (6190-Å methane filter) wavelengths. The hazes poleward of the 'gap' at 30° south latitude are seen dimly before they disappear behind the polar hydrocarbon layer at higher latitudes. The dominant feature in this view is the brightened edge ('limb') of the planet. Ice hazes in the atmosphere apparently have a larger scale height (i.e., their concentrations diminish more slowly with altitude) than does the methane gas. Absorption of the red light by methane is consequently lowest near the limb, where the haze reflectivity is unobstructed.

The above discussions are accurate in their descriptions of the subtle color differences observed by Voyager. The sources of these color differences are less certain, and the explanations given should be considered hypothetical until verified by independent means. Continued analysis and synthesis of Voyager data sets and the addition of supportive Earth-based observational and theoretical studies may do much to erase the uncertainties associated with the nature of the atmosphere of this strange planet.

NOTES AND REFERENCES

[1] Cameron, A. G. W. (1973) Abundances of the elements in the solar system. *Space Science Reviews*, **15**, 121–146.

[2] Heasley, J., Milkey, R. (1978) Structure and spectrum of quiescent prominences, III, Applications of theoretical models in helium abundance determinations. *Astrophysical Journal*, **221**, 677–688.

[3] Gautier, D., Conrath, B., Flasar, M., Hanel, R., Kunde, V., Chedin, A., Scott, N. (1981) The helium abundance on Jupiter from Voyager, *Journal of Geophysical Research*, **86**, 8713–8720.

[4] Conrath, B. J., Gautier, D., Hanel, R. A., Hornstein, J. S. (1984) The helium abundance of Saturn from Voyager measurements. *Astrophysical Journal*, **282**, 807–815.

[5] Nellis, W. J., Ross, M., Mitchell, A. C., van Thiel, M., Young, D. A., Ree, F. H., Trainor, R. J. (1983) Equation of state of molecular hydrogen and deuterium from shock-wave experiments to 700 kbar, *Physical Review*, **A, 27**, 608–611.

[6] This mechanism for helium 'enrichment' was suggested by Fegly, B., Prinn, R. G. (1986) Chemical models of the deep atmosphere of Uranus. *Astrophysical Journal*, **307**, 852–865, and by Pollack, J. B., Podolak, M., Bodenheimer, P., Christofferson, B. (1986) Planetesimal dissolution in the envelopes of the forming giant planets. *Icarus*, **67**, 409–443.

[7] MacFarlane, J. J., Hubbard, W. B. (1982) Internal structure of Uranus. In Hunt, G. (ed.) *Uranus and the Outer Planets*, Cambridge University Press, pp. 111–124.

[8] Conrath, B., Gautier, D., Hanel, R., Lindal, G., Marten, A. (1987) The helium abundance of Uranus from Voyager measurements. *Journal of Geophysical Research*, **92**, 15 003–15 010.

[9] Lindal, G. F., Lyons, J. R., Sweetnam, D. N., Eshleman, V. R., Hinson, D. P., Tyler, G. L. (1987) The atmosphere of Uranus: Results of radio occultation measurements with Voyager 2. *Journal of Geophysical Research*, **92**, 14 987–15 001.

[10] Herbert, F., Sandel, B. R., Yelle, R. V., Holberg, J. B., Broadfoot, A. L., Shemansky, D. E., Atreya, S. K., Romani, P. N. (1987) The upper atmosphere of Uranus: EUV occultations observed by Voyager 2. *Journal of Geophysical Research*, **92**, 15 093–15 109.

[11] Atreya, S. K., Sandel, B. R., Romani, P. N. (1989) Photochemistry and vertical mixing. In Bergstralh, J. T., Miner, E. D. (eds.) *Uranus*, University of Arizona Press, Tucson (in preparation).

[12] Willson, R. C., Duncan, C. H., Geist, J. (1980) Direct measurement of solar luminosity variation. *Science*, **207**, 177–179.

[13] Willson, R. C. (1989) private communication.

[14] Pearl, J. C., Conrath, B. J., Hanel, R. A., Pirraglia, J. A., Coustenis, A. (1990)

The albedo, effective temperature, and energy balance of Uranus, as determined from Voyager IRIS data. *Icarus*, **84**, 12–28.

[15] The Stephan–Boltzmann equation relates the flux energy per unit area being radiated by a body to the absolute temperature of the body. The equation is E $(W/m^2) = 5.66956 \times 10^{-8} T^4$, where T is the absolute temperature in Kelvins.

[16] The effective temperature of Uranus is that temperature a perfectly radiating body (i.e., a 'blackbody') would have to possess to radiate the same total thermal flux as that observed emerging from the Uranian atmosphere.

[17] Conrath, B. J., Pearl, J. C., Appleby, J. F., Lindal, G. F., Orton, G. S., Bezard, B. (1989) Thermal structure and energy balance of Uranus. In Bergstralh, J. T., Miner, E. D. (eds.) *Uranus*, University of Arizona Press, Tucson, in preparation.

[18] Table 9.2 is adapted from Lindal, G. F., Lyons, J. R., Sweetnam, D. N., Eshleman, V. R., Hinson, D. P., Tyler, G. L. (1987) The atmosphere of Uranus: results of radio occultation measurements with Voyager 2. *Journal of Geophysical Research*, **92**, 14 987–15 001.

[19] West, R. A., Pollack, J. B., Baines, K. H. (1989) Clouds and aerosols in the uranian atmosphere. In Bergstralh, J. T., Miner, E. D. (eds.) *Uranus*, The University of Arizona Press, Tucson, in preparation.

[20] Allison, M., Beebe, R. F., Conrath, B. J., Hinson, D. P., Ingersoll, A. P. (1989) Uranus atmospheric dynamics and circulation. In Bergstralh, J. T, Miner, E. D. (eds.) *Uranus*, The University of Arizona Press, Tucson, in preparation.

[21] The Coroilis force, named for French civil engineer Gaspard Coriolis, is technically not a force but rather a reaction. Because of its inertia, an object put in motion across the face of a rotating planet or other massive body will tend to veer to the right or left to preserve its initial direction and speed. On Earth, the Coriolis force causes winds and ocean currents to veer to the right in the northern hemisphere and to the left in the southern hemisphere.

BIBLIOGRAPHY

Bergstralh, J. T., Miner, E. D. (1989) *Uranus*, The University of Arizona Press, in preparation (see atmosphere chapters by Strobel *et al.*, Atreya *et al.*, Fegley *et al.*, Conrath *et al.*, Allison *et al.*, and West *et al.*; also the references cited in each chapter).

Hunt, G. (1986) Voyager 2 investigates the atmosphere of Uranus. *The Planetary Report*, **VI**, No. 6, pp. 14–15.

Ingersoll, A. P. (1987) Uranus. *Scientific American*, **256**, 38–45.

10

The magnetosphere of Uranus

10.1 STRENGTH AND ORIENTATION OF THE MAGNETIC FIELD

Some of the most eagerly awaited results from Voyager 2 in late 1985 were measurements of the Uranian magnetic and radiation fields. Six of Voyager's 11 investigation teams used instruments which obtained their Uranus data only while they were immersed in the environment they were measuring. These investigations include the cosmic rays (CRS), low energy charged particles (LECP), magnetometer (MAG), plasma science (PLS), plasma wave science (PWS), and planetary radio astronomy (PRA). Such instruments are generally described as '*in situ*' instruments as opposed to the 'remote sensing' instruments (imaging, photopolarimetry, ultraviolet and infrared spectrometry, and radio science) which did their studies from afar. Prior to the Uranus encounter, ground-based observers already knew much about the planet, its rings, and its satellite system. Until Voyager 2 crossed the boundary between the solar wind and the magnetosphere of Uranus, the nature of the magnetosphere belonged to the realms of theory and conjecture. The Voyager 2 encounter with Uranus forever changed that situation, and it did so with great flair and aplomb!

The first to detect the magnetic field was expected to be PRA. The interaction of the solar wind with a rotating planetary magnetic field typically generates radio waves at frequencies detectable by PRA. The magnetic field of Saturn was detected in this fashion more than nine months before the Voyager 1 encounter. The long time base of the Saturn measurements made it possible to determine with great precision the rotation rate of Saturn's magnetic field, which theoretically matches the rotation of Saturn's interior. Similar conditions were anticipated as Voyager 2 approached Uranus. The predicted strength of the magnetic field, based on the 'Magnetic Bode's Law' and other considerations (see Chapter 5), led PRA investigators to believe that similar radio emissions from Uranus would be detected several months before Voyager 2's January 1986 passage through the field. When none had been detected by late October 1985 a paper on implications of non-detection of the radio emissions

was presented at the annual meeting of the Division for Planetary Sciences [1]. There
was speculation that the low level of internal heat might imply that Uranus possessed
no magnetic field.

Radio emissions from Uranus were first detected only five days before Voyager
2's closest approach (C/A) to the planet, first by PWS near a frequency of 56 kHz and
then by PRA near 59 kHz. Voyager entered the bowshock wave in front of the
planet's magnetic field 10.5 h before Uranus C/A and entered the magnetic field 2.6 h
later. Because the solar wind stretches Uranus's magnetic field 'downwind' of the
planet, Voyager 2 did not leave the magnetosphere until 37.2 h after C/A; the final
bowshock crossing was nearly three days later. Times and distances of each of these
events are given in Table 10.1. The multiple crossings of the downstream bowshock

Table 10.1 — Voyager crossings of the bowshock and magnetopause

Boundary	Time from C/A (h)	Range from Uranus (km)
Bowshock (in)	−10.5	604 000
Magnetopause (in)	−7.9	461 000
Closest approach	0.0	107 000
Magnetopause (out)	+37.2	2 040 000
Bowshock (out)	+76.7	4 155 000
Bowshock (in)	+77.0	4 173 000
Bowshock (out)	+91.0	4 915 000
Bowshock (in)	+91.1	4 920 000
Bowshock (out)	+92.4	4 990 000
Bowshock (in)	+99.5	5 366 000
Bowshock (out)	+108.1	5 829 000

dramatically testify to the large fluctuations in size and position of this boundary.
Although some of this variability is caused by the 17.24-h rotation of Uranus rotation
period Uranus and its magnetic field, much of it is due to changes in the 'pressure' of
the solar wind. The outer edge of the magnetic field, known as the magnetopause,
undergoes similar but less dramatic changes in size.

As Voyager 2 traversed the magnetic field of Uranus, it became immediately
apparent from MAG data that the north and south magnetic poles were nowhere
near the rotation ('planetographic') poles. The dipole axis of the magnetic field was
found to be tilted 58.6° from the rotation axis of the planet. Furthermore, the
magnetic field was not centered in the planet but was offset about 8000 km, or nearly
one-third of the distance to the cloud tops, toward the dark north polar region (see
Fig. 10.1). This offset is emphasized by the difference in the latitudes where the
magnetic poles pierce the 1-bar pressure level in the atmosphere: 15.2° south latitude
and 44.2° north latitude! For comparison, Earth's magnetic field is tilted 11.4° from
the rotation axis and offset 462 km (1/14 of Earth's radius) in the general direction of
the Mariana Islands in Micronesia (148° E longitude, 18° N latitude); the magnetic

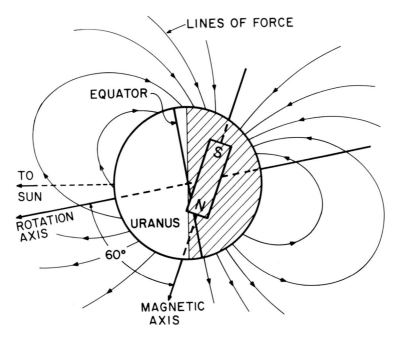

Fig. 10.1 — The simplest representation of the magnetic field is a magnetic dipole tilted 58.6°
from the rotation axis and offset 0.3 Uranian radii toward the dark north polar region.
(260–1829B)

poles pierce Earth's surface at latitudes of 82° N and 75° S. A comparison of some of
the relevant characteristics of the magnetic fields of Earth, Jupiter, Saturn, and
Uranus is given in Table 10.2.

Table 10.2 — Comparison of planetary magnetic fields

Characteristic	Earth	Jupiter	Saturn	Uranus
Equatorial radius (km)	6378	71492	60268	25559
Tilt of equator to orbit	23.45°	3.08°	26.73°	97.92°
Magnetic dipole tilt	11.4°	9.6°	0.0°	58.6°
Dipole offset (km)	460	7000	3000	8000
Dipole offset (planet radii)	0.07°	0.1	0.05	0.3
Typical sunward size of magnetosphere, (10^6 km)	0.06	4.6	1.2	0.46
Typical sunward size of magnetosphere (planet radii)	10	65	20	18
Typical surface magnetic field at equator (Oe)	0.31	4.28	0.21	0.25
Dipole moment (10^{12} Oe km^3)	0.08	1560	46	4.2
Rotation period (h)	23.934	9.925	10.656	17.24

The magnetic field generated in each of these planets is similar in nature to the field surrounding a bar magnet. As with bar magnets the magnetic fields of the planets have north and south magnetic poles. Compasses work on the principal that a small bar magnet on a freely turning spindle will align itself so that its north pole points in the direction of the planet's south magnetic pole. For that reason, the pole of Earth's magnetic field closest to the north geographic pole is called a south magnetic pole. The relationship is reversed for Jupiter, whose north magnetic pole lies close to its north planetographic pole. Saturn emulates Jupiter rather than Earth. The north magnetic pole of Uranus lies in the same hemisphere as its positive rotation pole, making it more like Jupiter and Saturn than like Earth. It should be remembered, however, that because the rotation pole of Uranus is tipped 98° it actually points slightly south of the plane of Uranus's orbit. The International Astronomical Union therefore designates the hemisphere containing the positive rotation pole as the *southern* hemisphere and the rotation as retrograde; the IAU convention is used in this book.

The measured magnetic field of Uranus is more complex than that of a simple magnetic dipole. In a fashion analogous to that of spherical harmonic representation of the gravity field of the planet (discussed in Chapters 5 and 8), non-dipolar magnetic field structure, including the offset, can be represented by spherical harmonics, including dipole, quadrupole, and octupole terms. The Uranus data permit relatively accurate determination of the dipole and quadrupole terms, but the data are insufficient to meaningfully constrain the octupole terms. The most widely used model of the magnetic field is the Q3 model [2], which includes the three dipole and five quadrupole coefficients to describe the internal magnetic field of Uranus. Differences between the Q3 model and the actual measurements for the 6-h period when the spacecraft was within 200 000 km of the center of Uranus were less than 1%. Fig. 10.2 is a contour map of magnetic field intensity in oersteds at the 1-bar pressure level as predicted from the Q3 model. Also shown in the figure are the points where the magnetic poles would pierce the surface, and the predicted areas of ultraviolet auroral activity.

10.2 IMPLICATIONS OF THE TILT AND OFFSET OF THE MAGNETIC FIELD

One very interesting question arose as scientists began to comprehend the unorthodox nature of the Uranian magnetic field. Was this the normal configuration of the field, or did Voyager just happen to catch it in the process of reorientation from alignment with the rotational pole in one direction to alignment in the opposite direction? Evidence from fossil magnetism in iron-bearing rock at the bottom of the Atlantic Ocean shows that Earth's magnetic field has 'flipped' 180° nine times in the last four million years. The Sun's magnetic field also reverses, but on a much shorter time scale. A solar cycle is the time period between such reversals and is only 11 years in length.

Although the solar magnetic field has been carefully observed during these reversals, the processes occurring in such reversals are undoubtedly much different for the Sun than for the Earth or for Uranus. The Sun's magnetic field is strongest in localized areas in the Sun's visible atmosphere where sunspots and bright solar flares

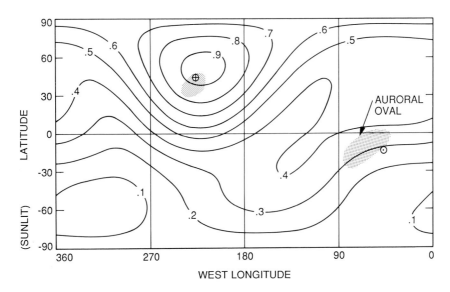

Fig. 10.2 — Contour map of magnetic field intensity in oersteds at the surface (1-bar pressure level) of Uranus as predicted from the Q3 model. Also shown in the figure are the magnetic equator(s), where magnetic field lines are parallel to the surface, and locations where the magnetic poles would pierce the surface.

occur. The solar magnetosphere reaches enormous proportions as the high-velocity stream of charged particles known as the solar wind drags the magnetic field outward beyond the orbit of Neptune.

Planetary magnetic fields are thought to be generated by the motions of electrically conducting fluids in the cores of the planets. Molten iron in the Earth's core is electrically conducting and probably constitutes the source of the terrestrial magnetic field. Radioactive decay may provide the energy necessary to drive such a dynamo, although the details of that process are still rather sketchy. Earth's magnetic field is drifting slowly westward, indicative of some differential motion of the core relative to the surface. Scientists have never had the opportunity to observe Earth's magnetic field during a reversal, but detailed studies of fossil magnetism in sea-floor rocks have provided some information on the strength and orientation of the field during such reversals. This study of Earth's magnetic field at past epochs as preserved in rocks whose ages are known is called paleomagnetism and is a relatively well-developed science. Paleomagnetists find evidence that Earth's magnetic field during reversals possesses a large dipole tilt and a large quadrupole moment. They also find that the field strength is about 25% of normal [3].

Uranus's magnetic field is undoubtedly generated by an active internal dynamo similar in nature to Earth's. It is not associated with a molten iron core as in the Earth's interior, but it must be generated in electrically conducting materials in the outer portions of extended molten ice/rock core of Uranus. The energy source which drives the Uranian magnetic dynamo is poorly understood; it could be heat generated by the gravitational sinking of heavier materials or it could be the result of heating from the decay of radioactive materials.

If reversals in the magnetic fields of Uranus and Earth have similar time scales, the chances are less than one in a thousand that a reversal was in progress at the time of the Voyager encounter. If, on the other hand, Uranus reverses its magnetic polarity much more frequently, the probability of encountering Uranus during a reversal may increase significantly. One characteristic of Uranus that makes it different from Earth and from the other giant planets is the large tilt of its rotation axis. The resulting preferential heating of the polar regions of the planet (see Chapter 9) may lead to a thermal bulge at the poles which is superimposed on the planetary oblateness (polar flattening) caused by rotation [4]. This thermal 'prolateness' is unstable and results in motion of the polar fluids toward the equator. Because of the fluid nature of Uranus such motion takes place over relatively short time periods (perhaps thousands of years), and the magnetic field lines may be effectively frozen in the polar fluid and transported toward the equator with it. The result would be frequent (perhaps continuous) changes in the magnetic dipole tilt. If Uranus emulates Earth during magnetic reversals, the present magnetic field strength of Uranus may be only 25% as large as when the magnetic dipole is aligned with the rotation axis.

One important effect of the large dipole tilt is correspondingly large changes in the magnetic field orientation as the planet rotates on its axis and as it revolves around the Sun (see Fig. 10.3). At the time of the Voyager 2 encounter, the south pole of Uranus was very nearly pointed at the Sun. The solar wind interaction with the magnetic field stretches the field downwind of the planet into an extended tail. The rotation of the planet then twists the magnetic tail into a very elongated corkscrew (helical) shape. The tail diameter is about 2 150 000 km and successive turns of the magnetic helix are separated by about 70 000 000 km.

The orbital motion of Uranus around the Sun changes the tail configuration dramatically. At southern summer solstice (near the time of the 1986 Voyager encounter) the angle β between the Sun–Uranus line and the magnetic dipole varied during a Uranus day from 113° to 129°. By about the year 2000 the diurnal variation of β will be from 63° to 180°, and about 14 years later the variation will be from 0° to 117°. Over an entire Uranus year, the variations cover the entire 180° range of values possible for β. For comparison, β remains between 55° and 125° for Earth, between 77° and 103° for Jupiter, and between 63° and 117° for Saturn.

10.3   AURORAE ('NORTHERN LIGHTS') ON URANUS

Charged particles spiralling down Earth's magnetic field lines collide with neutral atoms in the arctic atmosphere. If the rain of charge particles is intense enough both visible and ultraviolet light will be emitted in quantities sufficient for remote detection. These emissions constitute an aurora, commonly called the northern (or southern) lights. During active periods on the Sun the solar wind intensity increases substantially, and the auroral displays are brighter and often extend to much lower latitudes.

Aurora were observed by Voyager both at Jupiter and at Saturn. Several years before the Voyager encounter the International Ultraviolet Explorer (IUE) was trained on Uranus to look for evidence of an ultraviolet aurora there. A measurable excess of ultraviolet radiation convinced astronomers that Uranus also had aurorae

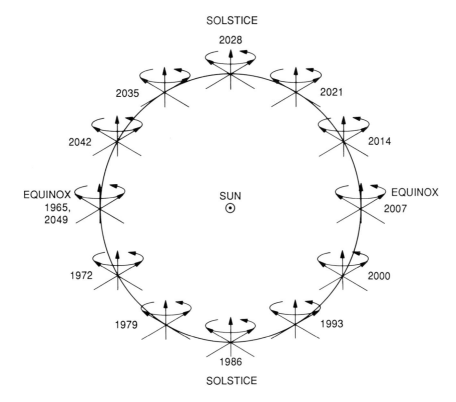

Fig. 10.3 — Uranus's magnetic field orientation with planet rotation and with orbital position. The magnetic axis will rotate through the sun at four separate orbital positions.

and therefore possessed both a magnetic field and radiation belts. It is now known that their conclusions, although correct, were not justified by the IUE data. The ultraviolet emissions detected by IUE were due instead to the dayglow phenomenon discussed in Chapter 9.

Jupiter, Saturn, and Earth aurorae all occur near the poles of these planets, and high-resolution ultraviolet auroral studies by Voyager were accordingly planned for the polar regions of Uranus. These searches completely missed any southern (daylit) hemisphere aurora, since the closest magnetic pole was at 15.2° south latitude. The bright ultraviolet dayglow also hampered searches for aurorae on the sunlit hemisphere. Darkside studies by Voyager did reveal ultraviolet auroral emissions near the +44.2° north latitude magnetic pole, however. The auroral region was 15° to 20° in diameter and is believed to be excited by electrons with energies of about 10 000 eV. Such electrons were seen in abundance by other Voyager investigations, though there is evidence that the population of such electrons (or appropriate ions) along the magnetic field lines which feed into a 20°-diameter auroral zone are insufficient to account for the observed auroral intensity [5]. Future analysis of the Voyager data may help to resolve this seeming inconsistency.

10.4 RADIO EMISSIONS FROM URANUS

Predictions of an early detection by Voyager 2 of Uranian radio emissions were based on an assumed beaming pattern which would sweep the radio beam across the direction of the approaching spacecraft each rotation of the planet. While the total amount of radiated power was correctly estimated in these predictions, the predominantly nightside anti-sunward beaming of the radio waves was not anticipated. It was this unusual beaming pattern which prevented detection of Uranus kilometric radiation (UKR) until just five days before Uranus closest approach. The radiation was first detected by PWS at 56 kHz (5.4-km wavelength) and then by PRA at 59 kHz (5.1-km wavelength). Based on the behavior of terrestrial radio emissions there had been a prediction [6] that UKR emissions would come predominantly from the night side of Uranus. The higher magnetic field on the dark side (due to the dipole offset in that direction) was also a contributor to the marked contrast between dayside and nightside radio emissions.

The dominant components of UKR are inferred from Voyager data to emanate from a region above the darkside magnetic pole at 44.2° north latitude. As the planet rotates, the radio signal received by PRA waxes and wanes due to the limited angular extent of the radio beam. Scientists carefully monitored these variations in UKR intensity and concluded that the magnetic field (and probably the bulk of the planet) rotates once each 17.24±0.01 h. As intense as were the nightside radio emissions, Voyager could no longer detect them by 24 February 1986 only 31 days after Uranus closest approach.

Radio emissions at Earth and Saturn are driven and controlled by the solar wind. Although some of the Uranian radio emissions are probably driven by the solar wind, they do not appear to be strongly dependent on the characteristics of the solar wind. Uranian emissions mimic more closely the complex Jupiter radio emissions which also exhibit relatively weak solar wind control. Voyager scientists [7] have identified six classes of UKR; some characteristics of each are noted in Table 10.3.

Two other types of radio emissions were detected near Uranus. The first was associated with tiny particles in the equatorial plane of the planet and is thought to be due to particles in an extended ring, which, striking the spacecraft, are vaporized and ionized, giving off radio emissions in the process. These will be discussed in more detail later. Lightning discharges in the atmosphere of Uranus are believed to generate the other type of emission and have been named Uranus Electrostatic Discharge (UED). The bursts are extremely short-lived and were seen at much higher frequencies than UKR.

Lightning discharges were sporadic and displayed no apparent periodicity, so no source location within the atmosphere could be inferred. About 140 UED events were identified within about 12 h of Uranus closest approach. No lightning bolts were identified in Voyager images, and radio whistlers associated with terrestrial and Jovian lightning were not detected at Uranus. Such whistlers are generated as massive discharges of electrons spiral down magnetic field lines, spinning at higher and higher frequencies as they fall toward the atmosphere.

10.5 THE PLASMA ENVIRONMENT OF URANUS

A plasma is a collection of charged particles, generally containing about equal numbers of positive ions and electrons. Plasmas exhibit some of the characteristics of

Table 10.3 — Types of Uranus radio emission

Radio emission characteristics	Frequency range (kHz)	Frequency peak (kHz)	Comments
Narrowband, bursty, 1–5 kHz bandwidth	15–120	60	First to be detected; source above dayside magnetic pole (?)
Narrowband, smooth, nearly constant	20–350	60	Possibly beamed from above the magnetic equator (?)
Dayside, smooth, weak, broadband	100–300	150	Appeared only once; source may near day side magnetic pole (?)
Broadband, smooth, most intense	100–850	400	Used to measure rotation; source above darkside magnetic pole
Broadband, bursty, intense	60–750	400	Follows 'B-smooth' in time; same darkside magnetic pole source (?)
Narrowband, bursty, very low frequency	Near 5	5	Only seen by PWS while outbound, source near Miranda (?)

a gas but differ from a gas in that they are good conductors of electricity and are affected by magnetic fields. A 'cold' plasma is one whose particle energies are small enough that electromagnetic forces constrain the plasma to move with the magnetic field; a 'hot' plasma is energetic enough to escape such constraints. If the hot plasma is sufficiently dense it can distort the magnetic field through which it flows. The solar wind is an example of a hot and relatively dense plasma which distorts the Sun's magnetic field and stretches it into an enormous magnetic bubble (the heliosphere) which envelops all of the known planets of the Solar System.

Voyager's plasma science investigation (PLS) found that a low-density plasma fills the Uranian magnetosphere. It consists of protons and electrons, with essentially no heavier components. Its composition essentially rules out the possibility that the Uranian moons serve as a primary source for the plasma. The regions through which the spacecraft passed were populated primarily with a cold plasma with maximum number densities near 3/cm^3. The temperatures of the cold plasma amount to a few tens of electron-volts (a few hundred thousand Kelvins). A hot plasma with much lower number densities is confined to regions of the magnetosphere more distant than an L-shell of 5 [8]. Temperatures of the hot plasma are a few thousand electron-volts (a few tens of millions of Kelvins).

The inner magnetosphere measurements can be divided into five distinct regions along the spacecraft trajectory (see Fig. 10.4). Between $L=9$ and $L=6.6$ inbound

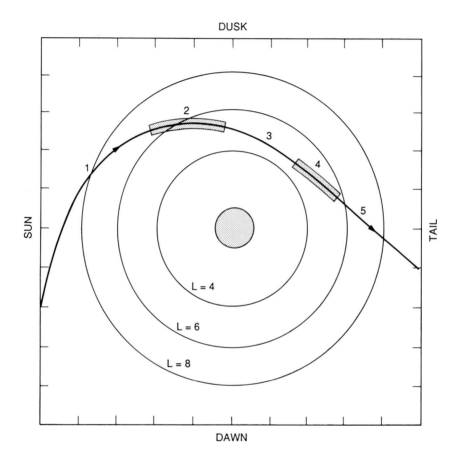

Fig 10.4 — The inner magnetosphere plasma measurements can be divided into five distinct regions along the spacecraft trajectory described in the text. These regions are shown in a magnetic coordinate system which fixes the direction of the Sun and the magnetotail and represents L-shell value as a radial distance from the center of the planet.

only the hot proton component was seen. The second region extends inward to $L=5.3$. In this region the hot protons are joined by protons with intermediate and cooler temperatures and by cold and hot electron plasmas. From $L=5.3$ inbound through the minimum $L=4.6$ and outbound to $L=4.8$ the hot and intermediate plasmas are absent and only the cold plasmas remain. Outbound from $L=4.8$ to

$L=5.6$ the hot electrons and hot protons reappear, all at somewhat higher temperatures than for the inbound hot plasma. At this point in the trajectory the spacecraft passes into the shadow of Uranus. There is evidence that the spacecraft became highly charged (to about -400 V) during this period, repelling electrons of lower energies so they did not reach the PLS detectors. The effect on the protons was just the opposite as the spacecraft voltage accelerated both cool and hot protons toward the detectors, imparting enough energy to them to make the cool protons appear warmer and raising the hot proton energies beyond the detection limit of PLS. The hot electrons were little effected by spacecraft charging, but began to disappear at these L-shell distances. Once the spacecraft exited the shadow of Uranus the spacecraft charge slowly dissipated. The hot protons persisted until about $L=16$, after which the densities of both positively and negatively charged plasmas dropped below PLS detection levels and the spacecraft entered the magnetic tail of Uranus.

During Voyager's traversal of the downstream magnetosphere, PLS repeatedly detected enhancements of the proton and electron fluxes. These were due to a plasma sheet in the magnetotail of Uranus. The plasma sheet is a relatively thin (about 10 to 15 Uranus radii thick) sheet of charged particles which populates the magnetic equatorial region above the planet's dark side. At distances beyond about 15 Uranus radii behind the planet the pressure of the solar wind bends the sheet in a more nearly anti-solar direction. This whole structure rotates with the planet every 17.24 h. As Voyager receded from the planet the plasma sheet swept through the spacecraft position four times before Voyager crossed the magnetopause (the edge of the Uranian magnetic field). The first three of these passages occurred at times predicted, while the last was significantly displaced from its predicted location, possibly because the spacecraft was close to the magnetopause, where large distortions can occur. The three-dimensional model is depicted with a superimposed Voyager trajectory in Fig. 10.5. The trajectory appears helical because the coordinate system is fixed with respect to the rotating Uranian magnetic field.

10.6 ENERGETIC CHARGED PARTICLES AT URANUS

Closely related to the plasma content of the Uranus magnetosphere are the charged particle measurements from the Low Energy Charged Particle (LECP) and Cosmic Ray (CRS) investigations. The PLS measured charged particles in the range of energies from 10 eV to 5950 eV. The combined range of LECP and CRS is 22 000 eV (22 keV) to 10 000 000 eV (10 MeV) for electrons and 28 keV to several hundred MeV for ions. Charged particles in these energy ranges are not constrained to remain within the magnetosphere but can 'leak' out into the solar wind. Such particles move most freely along magnetic field lines, whether within the Uranus magnetic field or exterior to it in the solar magnetic field.

Energetic electron number densities dropped to interplanetary background levels outside the magnetopause, but energetic protons from Uranus were seen more than 40 Uranus radii upstream (toward the Sun) from the planet and they may extend as far as 500 Uranus radii (i.e., nearly 13 000 000 km) upstream. As the spacecraft crossed the inbound bowshock and entered the magnetosheath (the region between the bowshock and the magnetosphere) the numbers and energies of detected protons began to increase. Proton numbers increased rapidly as the spacecraft crossed into

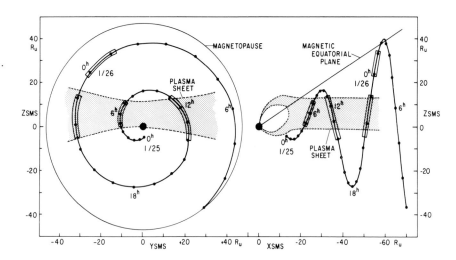

Fig. 10.5 — A three-dimensional model depicting the Voyager trajectory through the magnetotail of Uranus. Shaded regions are the measured times of plasma sheet crossing. The trajectory appears helical because the coordinate system is fixed with respect to the rotating Uranus magnetic field.

the magnetosphere and energetic electrons joined the population. Near the L-shell distances of the satellite orbits of Miranda, Ariel, and Umbriel there are noticeable decreases in the numbers of the high-energy protons and electrons. These dips are an indication that some of the protons and electrons are being swept up by these satellites as they circle Uranus. The more distant Titania and Oberon were outside the most intense regions of energetic charged particles, especially on the outbound leg of the trajectory where the numbers of both the high-energy and the lower-energy charged particles dropped to background levels well within the L-shells of Titania and Oberon.

Energetic proton bombardment of the satellite surfaces and the ring particles may cause radiation damage to the surface materials of these bodies [9]. The most noticeable effect is likely to be a darkening. The ring particles and the surfaces of satellites interior to the orbit of Miranda are very dark, reflecting only 7% or less of the sunlight incident upon them. Because of the inclined magnetic axis of Uranus, magnetospheric particles are swept across the satellite and ring surfaces from a variety of directions as they circle the planet. Large amounts of methane (CH_4) are believed to be present on the surfaces or in the interiors of these bodies. Proton bombardment effectively splits these methane molecules and much of the hydrogen (H_2) escapes and leaves a black residue of carbon (C) behind. In less than 100 000 years, an extremely short geologic time period, surfaces can be completely blackened to a depth of a millimetre or more.

Miranda orbits Uranus near an L-shell of 5. That distance marked an inner boundary for protons with several-keV energies as measured by PLS. The higher-energy protons and electrons measured by LECP and CRS were instead enhanced inside Miranda's magnetic distance, reaching their maximum number densities at the minimum L-shell (approximately 4.5) reached by Voyager 2. This is seen [10] as a possible indication that Uranus possesses a region of dense plasma (a 'plasma-sphere') inside $L=5$. The availability of high-energy protons for surface darkening processes therefore seems assured.

As with the PLS measurements, the main constituents of the charged particle population were the protons (H^+) and the electrons (e^-). A small amount of ionized hydrogen gas (H_2^+) may also have been present, but no helium (He) or any other heavier ions were detected, in marked contrast to the radiation environments within the magnetospheres of Jupiter and Saturn. It seems evident that the primary source of these particles is the extended hydrogen atmosphere responsible for the ultraviolet dayglow discussed in Chapter 9.

Enhancements in the numbers of energetic protons were seen at and near the times of the four plasma sheet crossings in the Uranian magnetotail. The most intense was the fourth, whose position did not match the model depicted in Fig. 10.5, possibly because of distortions near the outbound magnetopause. Enhanced energetic electron numbers were also seen at the first two plasma sheet crossings, but it is obvious that the number of energetic electrons decreases rapidly with distance down the tail. The second crossing shows electron number densities less than 10% those seen in the first crossing, and the third and fourth crossings are not seen in the electron data.

10.7 WAVE-PARTICLE INTERACTIONS IN THE MAGNETOSPHERE

Charged particles interact in a variety of ways with a magnetic field. A flow of charged particles (i.e., an electrical current) creates its own magnetic field, thus distorting any pre-existing field. Any charged particle moving through a magnetic field will experience a force which tends to resist motion across field lines but allows relatively free motion along field lines. These interrelationships are described in four basic equations known as Maxwell's equations, named for nineteenth-century Scottish physicist and mathematician James Clerk Maxwell.

As charged particles are accelerated or decelerated they experience an increase or decrease in energy. The increase in energy comes from the absorption of electromagnetic energy in the form of waves; the decrease in energy results in the emission of electromagnetic waves. These electromagnetic waves are often termed light waves, although they can be at any wavelength. The highest energy waves are known as gamma rays. In order of decreasing energy, other forms are known as X-rays, ultraviolet light, visible light, infrared light, microwaves, and radio waves. The Voyager Plasma Wave investigation (PWS) detects some of the lowest energy radio waves, whose frequencies range from 10 Hz (cycles per second) to 56200 Hz. Interestingly, sound waves at these frequencies span the range of sensitivity of the human ear, and one of the methods used by scientists to analyze PWS data is to convert the data from radio waves to sounds of the same frequency and 'listen' to the data. The 'sounds' of Jupiter, Saturn, and Uranus may be familiar to many readers and were produced in this fashion.

One of the more energetic interactions in a magnetic field are those which generate the ultraviolet auroral phenomena seen on Earth and other planets. Many phenomena in the magnetospheres of Solar System planets and the Sun involve low-energy interactions that result in emission or absorption of radio waves in the PWS frequency range. Such radio waves interact readily with charged particles and do not generally propagate over large distances. Hence PWS data can be used to study nearby interactions between the charged particles and the magnetic field through which the spacecraft is passing. 'Nearby' in this case means from a few meters to a few millions of kilometers, depending on the frequency being monitored and the type of charged-particle interactions present.

Most of the intense radio emissions detected by PWS were seen in the inner magnetosphere during a 7-h period surrounding closest approach to the planet. The different kinds of emissions are often named for the nature of the sounds represented by the transformed radio signals (e.g., hiss, chorus, chirp, whistlers, and noise).

The first Uranian radio emissions detected were seen in PWS data five days before closest approach to the planet. These emissions were probably due to the energetic upstream protons seen by LECP and CRS. The radio signals were the first convincing herald that Uranus was indeed the possessor of a magnetic field. A turbulent flow of charged particles near the inbound bowshock and in the magneto-sheath also gave rise to radio emissions in the PWS frequency range. However, once the spacecraft entered the magnetosphere of Uranus, wave activity was very nearly absent until Voyager reached the inner portions of the magnetosphere within 250 000 km of the planet.

One of the most intense types of emission inside that distance was whistler activity. A whistler is so called because the frequency of the emission increases with time (like a whistle whose pitch gets higher as time goes by). Whistlers are thought to be caused by high-speed electron motions with the upper atmosphere. Some of the whistlers may be associated with auroral activity, while others are probably the result of lightning-like discharges. When these rising frequencies occur rapidly enough to overlap with one or more prior whistlers the emissions are appropriately called chorus. Audio recordings of these events sometimes sound like chirps from a flock of birds.

Emissions at 562 Hz (a frequency which in musical sound corresponds approximately to an octave above middle C) are greatly enhanced near $L=7.5$ during both the inbound and outbound parts of the trajectory. There exists, however, a marked asymmetry between the two periods; the outbound emissions are about a factor of 30 stronger than the inbound emissions. The magnetic latitude was 33° during the inbound event and 13° during the outbound event. It is not known whether the asymmetry is related to the differences in magnetic latitude, to dayside/nightside differences, or to temporal changes in the emission intensity.

Voyager 2 crossed the equatorial plane of Uranus about 43 min before closest approach to the planet. The radial distance from the center of the planet at that time was about 115 400 km, more than twice the distance of the epsilon ring, outermost of the visible rings of Uranus. Voyager remote sensing instruments found no evidence for ring material external to the epsilon ring. Nevertheless, PWS data indicate that the spacecraft passed through a swarm of particles nearly 3500 km in thickness about that time. The particles are probably about a micrometer in size and are believed to be part of an extended dusty ring system not detectable in other Voyager instrumentation. As the particles strike the spacecraft they are instantly vaporized and ionized by the impact, thereby releasing a cloud of charged particles, some of which strike the PWS antenna. This causes a voltage impulse which is detectable in PWS data. PWS scientists estimate [11] that the maximum particle number density was between one and two particles in each 1000 m^3 (1000 m^3 is about the size of a small theater). Because the spacecraft was moving at a speed of about 18 km/s even that small number density resulted in a peak rate of 54 impacts/s. When the PWS data is displayed as audio frequencies, it sounds much like hail striking a tin roof.

The spacecraft crossed the magnetic equator only once when the spacecraft was within $L=30$. That crossing occurred about 4.8 h before closest approach when the L-shell value was approximately 12. A slight enhancement in radio emission intensity seen at that time has been interpreted by PWS scientists [12] as the best indicator of the magnetic equator location. It has also led them to conclude that much more intense emissions would have been detected if the spacecraft had crossed the magnetic equator inside of $L=8$.

Surprisingly, no wave activity was detected during the outbound crossings of the plasma sheet. Scientists can only speculate [13] that this unexpected circumstance is related to the high tilt of the magnetic axis with respect to the rotation axis and to the Sun-pointed orientation of the rotation axis at the time of the Voyager 2 encounter. Possibly such orientations lead to an 'open' magnetosphere and a reduction in the amounts of trapped radiation in the magnetotail relative to Earth, Jupiter, and Saturn.

NOTES AND REFERENCES

[1] Kaiser, M. L., Desch, M. D. (1985) Voyager search for uranian radio emissions. *Bulletin of the American Astronomical Society*, **17**, 743 (abstract only).

[2] Connerney, J. E. P., Acuña, M. H., Ness, N. F. (1987) The magnetic field of Uranus. *Journal of Geophysical Research*, **92**, 15329–15336.

[3] Merrill, R. T., McElhinny, M. W. (1983) *The Earth's Magnetic Field: Its History, Origin and Planetary Perspective*, Academic Press, New York, 401 pages. See Chapter 5: 'Reversals of the Earth's magnetic field', pp. 135–168.

[4] Podolak, M., Hubbard, W. B., Stevenson, D. J. (1989) Models of Uranus' Interior and Magnetic Field. In Bergstralh, J. T., Miner, E. D. (eds.) *Uranus*, The University of Arizona Press, Tucson, in preparation.

[5] Cheng, A. F., Krimigis, S. M., Lanzerotti, L. J. (1989) Energetic particles at Uranus. In Bergstralh, J. T., Miner, E. D. (eds.) *Uranus*, The University of Arizona Press, Tucson, in preparation.

[6] Curtis, S. A. (1985) Possible night side source dominance in nonthermal radio emissions from Uranus. *Nature*, **318**, 47.

[7] Desch, M. D., Kaiser, M. L., Zarka, P., Lecacheux, A., Leblanc, Y., Aubier, M., Ortega-Molina, A. (1989) Uranus as a radio source. In Bergstralh, J. T., Miner, E. D. (eds.) *Uranus*, The University of Arizona Press, Tucson, in preparation.

[8] An L-shell of 5 is one whose surface is defined by all magnetic field lines which cross the magnetic equator at a distance of 5 planetary radii from the magnetic dipole axis. The radial distance, r, of the L-shell at other magnetic latitudes (for a purely dipolar magnetic field) is given by $r=L \cos^2 (\text{lat})$, where $L=5$ in the example given and r is in units of planetary radii.

[9] Cheng, A. F., Haff, P. K., Johnson, R. E., Lanzerotti, L. J. (1986) Interactions of planetary magnetospheres with icy satellite surfaces. In Burns, J. A., Matthews, M. S. (eds.) *Satellites*, The University of Arizona Press, Tucson, pp. 403–436.

[10] Cheng, A. F., Krimigis, S. M., Lanzerotti, L. J. (1989) Energetic particles at Uranus. In Bergstralh, J. T., Miner, E. D. (eds.) *Uranus*, The University of Arizona Press, Tucson, in preparation.

[11] Gurnett, D. A., Kurth, W. S., Scarf, F. L., Burns, J. A., Cuzzi, J. N., Grün, E. (1987) Micron-sized particle impacts detected near Uranus by the Voyager 2 plasma wave instrument. *Journal of Geophysical Research.* **92**, 14959–14968.

[12] Kurth, W. S., Barbosa, D. D., Gurnett, D. A., Scarf, F. L. (1987) Electrostatic waves in the magnetosphere of Uranus. *Journal of Geophysical Research*, **92**, 15225–15233.

[13] Scarf, F. L., Gurnett, D. A., Kurth, W. S., Coroniti, F. V., Kennel, C. F., Poynter, R. L. (1987) Plasma wave measurements in the magnetosphere of Uranus. *Journal of Geophysical Research*, **92**, 15217–15224.

BIBLIOGRAPHY

Bergstralh, J. T., Miner, E. D. (1989) *Uranus*, The University of Arizona Press, in preparation (see magnetosphere chapters by Ness *et al.*, Belcher *et al.*, Cheng *et al.*, Desch *et al.*, and Kurth *et al.*; also the references cited in each chapter).

Journal of Geophysical Research, **92**, No. A13 (December 30, 1987). This special Voyager issue contains 45 articles, 23 of which deal with the Uranian magnetic field, its associated radiation field, and radio emissions generated by interactions between the two.

Ness, N. F. (1986) The magnetosphere of Uranus. *The Planetary Report*, **VI**, No. 6, pp. 8–10.

11

The rings of Uranus

11.1 STRUCTURE OF THE RINGS

Each planetary ring system has its own unique characteristics. Jupiter's ring is devoid of water ice, probably composed of silicate dust particles with average sizes near a micrometer. The ring is confined to a region relatively near the planet and shows little discernible internal structure. It is likely a product of the tidal and collisional forces acting on the tiny satellites Metis and Adrastea. Saturn possesses the most extensive and most easily observed ring. Saturn's ring particles range from microscopic size upward to small satellites tens of meters or more in diameter; its composition may be dominated by water ice. The Uranus rings are very dark, consist primarily of ten relatively narrow strands, and are composed of black carbonaceous material.

Reasons for the differences in these three planetary ring systems may be intimately related to the processes which formed them. There is a great likelihood that the Neptune ring system bears little resemblance to any of its three predecessors. One marvels at Nature's propensity for building rings in such a wide variety of environments. Nothing in the collected experience of astronomers and planetologists would have permitted accurate predictions of the basic structure of a newly discovered planetary ring system on the basis of previously studied ring systems. Nor is it possible to assess the total character of any ring system from a limited data set.

In spite of these diversities, the same laws of motion are in operation in each of the known ring systems. Each derives its major control from the gravity of the central planet, responds to secondary forces caused by the gravity of nearby satellites or other ring particles, and experiences small retarding forces which tend to cause the particles to spiral toward the planet atmosphere. Each is constrained to orbit close to the equator of these oblate giants, appears to be much younger than its planet, and requires ongoing processes to maintain its very existence.

11.1.1 Newly discovered rings

In Chapter 4 the story of the serendipitous discovery of the Uranian rings was detailed. A careful analysis of the original data sets revealed the presence of nine narrow rings named (in order of increasing distance from Uranus) 6, 5, 4, alpha, beta, eta, gamma, delta, and epsilon. Subsequent stellar occultation observations from Earth expanded the original data set and permitted scientists to study the azimuthal and temporal variations of each ring. But the data consisted of a collection of one-dimensional profiles of the radial structure of the rings. It is fortunate that the discoveries occurred at a time when the pole of Uranus was pointed in the general direction of Earth, because this provided a maximally open ring system whose radial characteristics were then easier to probe. This geometry also facilitated studies of ring eccentricity, width variations, and precession rates. Studies of ring inclinations were more difficult, but reasonably precise estimates for all these quantities were available early in the Voyager sequence planning period.

Attempts to obtain good two-dimensional images of the ring system from Earth were stymied by the blackness of the rings, the proximity of the bright planet, the low solar lighting levels, and the enormous distance to Uranus. That deficit was erased by Voyager's first detection of the epsilon ring in late November 1985 from a distance of 72 300 000 km. Fig. 11.1 is one of the better images of the ring system, taken the day before Uranus closest approach from a distance of 1 120 000 km. A previously unknown narrow ring may be seen between the delta and epsilon rings [1]. The ring has been given a provisional name of 1986U1R, which it will retain until the International Astronomical Union eventually decides upon a new nomenclature system for the Uranian rings that is internally consistent.

Several images were shuttered near the time Voyager passed through the equatorial plane of Uranus. One of these (Fig. 11.2) was processed to remove image smear and again showed 1986U1R. In another (Fig. 11.3) the smear remains, but a faint ring about 2500 km in width, temporarily named 1986U2R, may be seen between the planet and the innermost of the narrow rings. Just minutes later the rings were seen silhouetted against the much brighter background of the Uranian atmosphere (Fig. 11.4).

The highest resolution was obtained with stellar occultation measurements similar to those done from Earth but utilizing the photopolarimeter (PPS) and the ultraviolet spectrometer (UVS). PPS measurements of the brightness of the star sigma Sagittarii (sometimes known as 'Nunki') provided data on the optical thickness of the epsilon, 1986U1R, and delta rings with separations of only 10 m in radial distance from the center of the planet. A similar occultation cut of all of the known rings made use of the light of the star beta Persei ('Algol') and provided radial resolution of about 100 m. In addition to providing multiple measurements of the radial profiles of the ten known narrow rings of Uranus the PPS occultation data also provided evidence for the existence of several other rings or partial rings ('ring arcs') [2]. No new rings (other than 1986U1R) were seen in the UVS data. Radio science (RSS) measurements from Earth of the occulted spacecraft radio signal yielded 50-m radial resolution for the nine previously known rings, but the new rings seen by UVS, PPS, and ISS were not detected, possibly because particle sizes in those rings were too small to absorb radio waves. Pertinent data on all the rings (except the dusty ring shown in Fig. 11.5 and discussed below) are collected in Table 11.1.

11.1.2 Appearance

Galileo first sighted the 'cup handles' on either side of Saturn that later were to be identified as rings. From that early seventeenth century observation until the

Fig. 11.1 —This image of the ring system was taken the day before Uranus closest approach
from a distance of 1 120 000 km. A previously unknown narrow ring, temporarily designated
1986U1R, is barely visible between the outer two rings (the delta and epsilon rings) discovered
from Earth. Note the relatively even spacing of the three inner rings (rings 4, 5, and 6).
(P-29507)

present, rings have offered challenges for dynamicists. At first it was the determination that these 'appendages' were actually circling the planet in Keplerian orbits
[3]. Then it was determined that the rings were swarms of individual particles
orchestrated by gravitational forces to march around the planet in an extremely
narrow latitude band. Some attempts were made to associate major divisions in the
rings with the gravitational forces of known satellites. Observations repeatedly
showed that Saturn's rings were behaving in theoretically 'impossible' fashion; each
new discovery led to re-examination and revision of prior theories.

   The discovery of Uranus's rings in 1977 rekindled the fires of wonder for those
who had finally become comfortable with the broad and basically continuous rings of

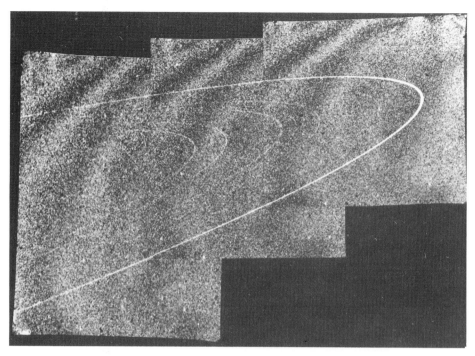

Fig. 11.2 — The outer portion of the rings was shuttered near the time Voyager passed through the equatorial plane of Uranus. This image was processed to remove image smear; 1986U1R is visible as a short arc between the delta and epsilon rings near the upper right. (260-1789)

Saturn. Uranus possessed rings that could not be directly observed through telescopes. Blockage of starlight confirmed that they were narrow and dark, not wide and bright as rings should be. As ring particles circle the planet, those closer to the planet orbit in slightly less time than their more distant neighbors. Frequent jostling and nudging should spread the rings out into wider and wider sheets. Profiles of these strange Uranian rings confirmed that the particles tended to spread outward toward the edges of each ring, but were then prevented from going farther, almost as if they were piling up against an impenetrable wall (see Fig. 11.6).

To understand this phenomenon, theoreticians again sought a gravitational explanation. Perhaps small satellites orbit both inside and outside each ring and provide the gentle but repeated tugs that served as walls for the spreading particles [4]. It was with no small amount of satisfaction that Voyager scientists noted Prometheus and Pandora flanking either side of Saturn's narrow F ring (Fig. 11.7) and Atlas just outside Saturn's abrupt A-ring edge (Fig. 11.8). All that remained was to verify that similar mechanisms were at work in the Uranian ring system.

Among the ten satellites of Uranus discovered in Voyager 2 imaging, two (Cordelia and Ophelia) were found surrounding the epsilon ring, the most distant and densely populated of the Uranian rings (Fig. 11.9). Continued searches of the images failed to show evidence for shepherding satellites surrounding any of the other rings. If such satellites exist, and as yet no better theory of Uranian ring

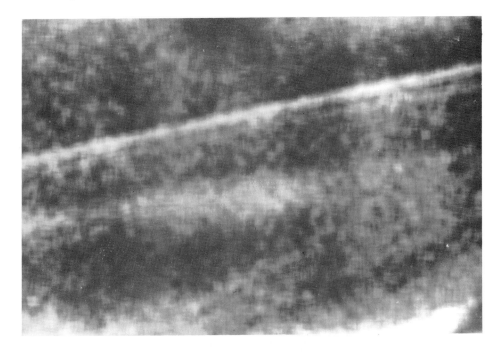

Fig. 11.3 — (a) The inner portion of the rings was also shuttered shortly after the image of Fig. 11.2. The image is badly smeared by spacecraft motion during the exposure, but a faint ring about 2500 km in width, temporarily designated 1986U2R, may be seen between the planet and the innermost of the narrow rings (ring 6). (P-29540) (b) A sketch of the ring system is shown for comparison. (260-1776B)

Fig. 11.4(a) — Just minutes after Voyager crossed the equatorial plane of Uranus the rings were seen silhouetted against the much brighter background of the Uranian atmosphere. Note the uneven relative spacing of the 4, 5, and 6 rings due to their different inclinations. (260-1653)

confinement has been proffered, they are either too small or too dark to be seen in the Voyager images. More will be said about shepherding mechanisms later in this chapter.

Two of the rings have diffuse detached companion rings. The delta ring companion is about 10 km wide and is planetward of the narrow delta ring. The eta ring has a companion ring of 55 km width external to its narrow component. Most of the narrow rings have relatively abrupt and well-defined edges, especially in the highest resolution Voyager data. The epsilon and gamma rings are the only two with sharp edges at both their outer and inner boundaries. The outer edges of 1986U1R, delta, 4, 5, and 6 appear sharper than their inner edges, while the reverse is true for the eta ring. The beta ring is not sharp at either edge; the alpha ring is relatively sharp at both

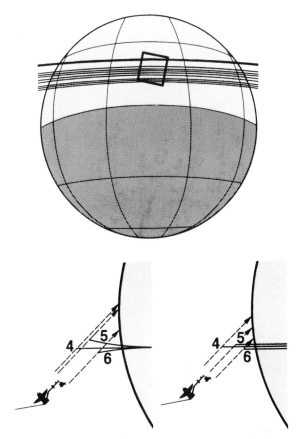

Fig. 11.4(b) — The sketch displays graphically how the apparent spacing of the inclined rings is changed by the nearly edge-on viewing. (260-1777)

edges near its periapsis (closest point to the planet), but less sharp at both edges near apoapsis (farthest point from the planet).

The ring system appears more nearly continuous in forward-scattered light. It was anticipated that tiny particles in the ring system might make the rings much brighter when viewed with backlighting from the Sun. In fact, of the more than 200 ring images taken with solar backlighting, the only image to reveal any ring material is the one displayed in Fig. 11.5. The concentration of dust-sized grains in the ring is much smaller than had been anticipated, and only the single 96-s exposure at the highest possible solar phase angle (172.5°) had both long enough exposure and high enough phase angle to bring out these tenuous dust bands. The apparent reason for the dearth of dust is atmospheric drag forces caused by the extended atomic hydrogen exosphere. Lifetimes of dust in the Uranian rings [5] are extremely short: less than a year at the inner edge of the rings and only about 1000 years at the outer edge of the rings. The dust is apparently replenished by interparticle collisions within the rings, by micrometeoroid impacts from space, and by tidal breakup of nearby satellites or large ring particles.

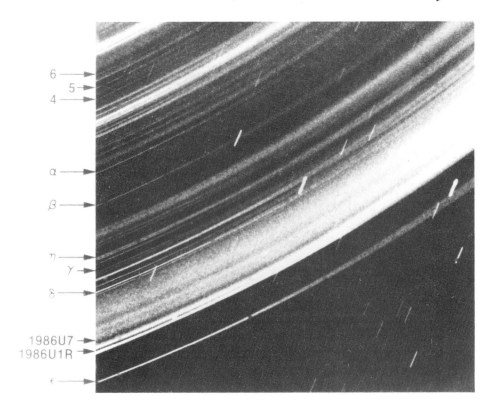

Fig. 11.5 — During passage through Uranus's shadow, this wide-angle 96-s exposure of the ring region was shuttered. Although the rings previously seen are identifiable, 1986U1R is far brighter than the rest and most of the features are not seen in any other images or non-imaging data sets. (260-1776A)

Table 11.1 — Uranus ring data from Voyager

Ring name	Distance (km)	Width (km)	Equivalent depth: PPS	Equivalent depth: RSS	Discovery year
1986U2R	37 000–39 500	~2500	—	—	1986
6	41 837.2(0.3)	1–3	0.47(0.17)	0.47(0.04)	1977
5	42 234.8(0.3)	2–3	0.93(0.19)	1.00(0.04)	1977
4	42 570.9(0.3)	2–3	0.62(0.19)	0.73(0.04)	1977
Alpha	44 718.4(0.2)	4–13	3.05(0.60)	3.13(0.11)	1977
Beta	45 661.0(0.1)	7–12	1.95(0.61)	2.08(0.06)	1977
Eta	47 175.9(0.1)	1–2	0.62(0.16)	0.48(0.04)	1977
Gamma	47 626.9(0.3)	1–4	1.5–3.4	3.4–4.8	1977
Delta	48 300.1(0.1)	3–7	2.34(0.34)	2.1–2.7	1977
1986U1R	50 023.9(0.3)	2–3	0.17(0.08)	—	1986
Epsilon	51 149.3(0.1)	20–95	47.2(5.5)	48.1(0.4)	1977

Numbers in parentheses represent the statistical error (standard deviation) in the distance or equivalent depth. RSS equivalent depths have been reduced by a factor of two to account for an expected apparent doubling of absorption (relative to equivalent widths from stellar occultations) caused by the geometry of the experiment. Tabulated data do not include the diffuse companions of the eta and delta rings.

Fig. 11.6 — Computer processing of one-dimensional radio science data has been used to produce this two-dimensional view of the epsilon ring showing the many internal details, including very abrupt termination of ring material at the two edges. (Courtesy G. L. Tyler)

The dust bands of Fig. 11.5 are probably due to the presence of large ring particles which serve as sources of the dust. Gravitational interaction with these particles may retard the planetward motion of the dust particles temporarily, giving rise to the apparent structure. The radial positions of Cordelia and the ten narrow rings seen in backscatter are labelled in the figure. Although each of the ten rings can be seen, only the newly discovered 1986U1R is prominent, and there exist a large number of narrow dust rings which are visible neither in the approach images nor in any of the stellar or radio occultation data sets. This wide-angle image was shuttered during the radio science atmospheric occultation experiment. The spacecraft was being continuously maneuvered to keep the antenna pointed at the nearest edge of the planet along a path from the spacecraft to Earth. A location in the rings was found for which

Fig. 11.7 — Prometheus and Pandora are seen on either side of Saturn's narrow F ring in this
Voyager 2 image. These satellites are believed to confine material within the F ring and to be
responsible for the 'braided strands' appearance of parts of the ring. (260-1135A)

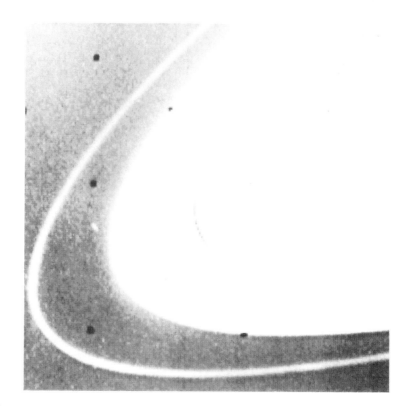

Fig. 11.8 —Saturn's satellite Atlas is close to the abrupt A-ring edge. Although originally
thought to gravitationally confine the outer boundary of the A ring (hence its name, Atlas), it is
now considered too small and too distant to accomplish that task. (P-23070)

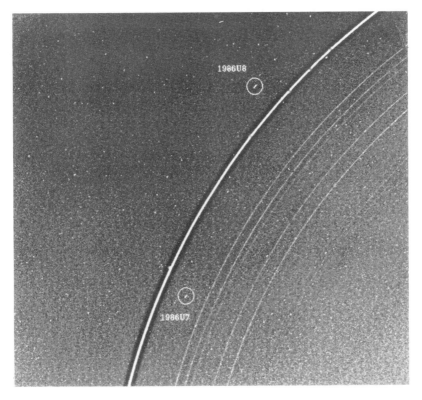

Fig. 11.9—Among the ten satellites of Uranus discovered in Voyager 2 imaging, two (Cordelia
and Ophelia) were found surrounding the epsilon ring, the most distant and densely populated
of the Uranian rings. (P-29466)

the spacecraft rotation would very nearly cancel image smear due to the relative
motions of Voyager and the rings. The success of this technique is witnessed by the
sharpness of the ring image at the left edge of the figure, even though background
stars and the rings at the right edge of the figure are badly smeared.

11.1.3 Sizes and shapes

Non-zero inclinations have been measured for most of the narrow rings of Uranus.
Most also have measurable eccentricities (departures from circular shape). Particles
in the epsilon ring, for example, are about 800 km closer to the planet at their
periapsis (closest point) than at their apoapsis (farthest point). The most highly
inclined rings are 4, 5, and 6, which are tilted from Uranus's equatorial plane by only
0.032°, 0.054°, and 0.062°, respectively. Even this small tilt was easily discerned in
Voyager imaging, as may be seen by the differences in their relative spacings in Figs
11.1 and 11.4(a). The effect of these small tilts is displayed in the sketches at the
bottom of Fig. 11.4(b). Orbital characteristics of each of the rings are given in Table
11.2. Most of the tabulated data was adapted from French *et al.* [6] Orbital periods
were not directly observed, but are calculated from the mean distance of the ring

Table 11.2 — Uranus ring orbital data

Ring name	Semimajor axis (km)	Semimajor axis (R_U)	Orbit Eccentricity	Inclination (°)	Period (h)
1986U2R	37 000–39 500	1.450–1.550	~0.0	~0.0	5.1–5.7
6	41 837.2(0.3)	1.63689(1)	0.001013(4)	0.0616(10)	6.1988
5	42 234.8(0.3)	1.65244(1)	0.001899(5)	0.0536(13)	6.2875
4	42 570.9(0.3)	1.66559(1)	0.001059(4)	0.0323(06)	6.3628
Alpha	44 718.4(0.2)	1.74962(1)	0.000761(4)	0.0152(06)	6.8508
Beta	45 661.0(0.1)	1.78650(1)	0.000442(3)	0.0051(06)	7.0688
Eta	47 175.9(0.1)	1.84577(1)	0.000004(3)	0.0011(08)	7.4239
Gamma	47 626.9(0.3)	1.86341(1)	0.000109(7)	0.0015(18)	7.5307
Delta	48 300.1(0.1)	1.88975(0)	0.000004(2)	0.0011(04)	7.6911
1986U1R	50 023.9(0.3)	1.95719(1)	~0.0	~0.0	8.1069
Epsilon	51 149.3(0.1)	2.00123(1)	0.007936(5)	0.0002(08)	8.3823

Numbers in parentheses refer to the statistical uncertainty (standard deviation) in the last digit of the orbit semimajor axes (mean distance from the planet for an elliptical orbit, or the orbit radius for a circular orbit), orbit eccentricities, and orbit inclinations.
R_U, radius of Uranus.

from the center of the planet (i.e. the 'semimajor axis' of the orbit) and from the Uranus mass and gravity harmonics [7].

The four widest rings (alpha, beta, delta, and epsilon) vary in width in a predictable fashion. Each reaches a maximum width near apoapsis and is narrowest near periapsis. The total amount of width variation observed for each of the rings is given in Table 11.1. The eta ring has a diffuse outer component about 55 km in width; the delta ring has a similar 10-km diffuse inner component. Data in the tables are exclusive of these diffuse components.

11.1.4 Optical thicknesses

Starlight (or a radio beam) passing through a ring is diminished in intensity by an amount that depends on the number density and sizes of the ring particles. A convenient measure of the amount of absorption is the optical thickness, t, often called the optical depth or opacity, of each portion of the ring. Optical thickness is defined as the natural logarithm of the ratio of incident light, I_0, to transmitted light, I; in mathematical terms, $t = \ln(I_0/I)$. If none of the light is absorbed, the optical thickness is zero. If the ring is opaque and transmits no light, the optical thickness is infinite. An optical thickness of unity means that only 36.8% ($e^{-1} = 1/2.718$) of the light is transmitted.

The optical thickness varies as the beam of starlight (or radio waves) passes radially across each ring. Optical thicknesses as high as 3 were observed by Voyager at several places in the epsilon ring. Even higher optical thicknesses were observed in the gamma and delta rings. Typical maximum optical thicknesses of the other classical rings is near unity. 1986U2R is estimated to have optical thicknesses in the range of 0.0001 to 0.001, and the dust bands of Fig. 11.5 have still smaller optical thicknesses.

One question of interest is whether the integrated optical thickness across any variable-width ring is a constant. This would be an indication that the number of particles per second passing any point of the ring is constant. This integrated optical thickness is sometimes called the 'equivalent depth' of a ring. Equivalent depths determined independently from PPS stellar occultation data and from RSS radio occultation data are given in Table 11.2. Most of the rings seem to have constant equivalent depths to within the stated statistical uncertainties.

The gamma and delta rings may be exceptions; their variations are much larger (especially for the RSS data) than the uncertainties of each equivalent depth measurement. Each of these two rings exhibit width variations, and where they are narrowest the optical thicknesses exceed 3 in the PPS data and 6 in the radio data. Equivalent depth remains relatively constant only if there is no mutual shadowing by the particles, i.e. the starlight shadow cast by one particle does not fall on any other particles. It is likely that these conditions are not met for the narrowest portions of the gamma and delta rings. The effect would reduce the equivalent depth when these rings are narrowest, and that is precisely what the data show.

An interesting but perhaps unrelated characteristic of the gamma and delta rings is that they are the only ones whose orbits are not described by simple ellipses. Instead, they appear to undergo periodic radial oscillations with amplitudes of 5 km and 3 km respectively. In the case of the gamma ring, this ±5 km radial oscillation can be compared to 'breathing' of the ring and is superimposed on the ±5 km orbital distance variations caused by the elliptical shape of the orbit. The periodicity of the two motions is very nearly equal (7.5421-h 'breathing' period versus 7.5307-h orbital period). The average motion of delta ring particles is almost a perfect circle, but the 3 km 'breathing' pattern is more complex and has a period of 15.3595-h, nearly double that of the 7.6911-h orbital period.

A factor of two in equivalent depth between PPS and RSS values for all the rings is well understood and is due to the difference in the nature of the two experiments. In Table 11.1, RSS values have been divided by two, and with the possible exception of the gamma and delta rings they match well with the equivalent depth values derived from PPS data.

11.2 PHYSICAL PROPERTIES OF THE RINGS

The ring data collected by Voyager span the electromagnetic spectrum from ultraviolet to radio wavelengths. Data have also been obtained in a variety of viewing and illumination geometries. An intercomparison of the data sets provides much information of relevance in studies of the properties of the individual ring particles. Voyager results are also descriptive of the environment experienced by the ring particles and are therefore useful for theoretical studies of ring evolution. By early 1990 scientists were only beginning to digest the quantities of data returned by Voyager 2, but some interesting facts had already come to light.

11.2.1 Composition of ring particles

No high-resolution ultraviolet, visible, or infrared spectra of the Uranian rings were obtained by Voyager. The rings were too dark, too narrow, and too tenuous to permit the collection of such data. Reflection spectra of solid particles are not always good indicators of particle composition anyway, either because surface materials generally have no sharp compositional absorptions or because other physical characteristics (particle size, compaction, temperature, etc.) complicate identification.

Voyager images of the rings contain information in several colors. A careful analysis and comparison of the images shows the rings to be uniformly dark at all colors. The only cosmically abundant material which matches both the low reflectivity and the uniformly gray color of the Uranian rings is carbon. It is perhaps noteworthy that the Uranian satellites also have neutral colors, but their reflection spectra indicate that water ice is present on their surfaces. No indication of water ice is seen in Earth-based spectra of the rings. The absence of water ice on the ring particle surfaces may partially explain their much lower reflectivity relative to the satellites.

The source of the carbon is very possibly decomposition of methane (CH_4) ice through bombardment by energetic protons (see Chapter 10). Methane is known to be abundant in the outer Solar System, and energetic proton fluxes are sufficiently high to blacken most exposed methane in the inner magnetosphere. It is also possible that part or all of the carbon is in the form of originally elemental carbon from the breakup of carbonaceous meteors or asteroids captured into orbit around Uranus.

11.2.2 Reflectivity of ring particles

The low visibility of the rings against the background sky is evident both from Voyager and from Earth-based observations. Reflectivity of the epsilon ring has been estimated to be only 1.4±0.4%, making it one of the darkest natural structures known [5]. The remaining rings have similarly low reflectivities. The rings are not filled with particles, but they are likely several tens of meters in thickness, with about 1% of the volume occupied by particles. A flat sheet composed of the same material as the ring particles would have a reflectivity of about 3.2%. For comparison, the bright A and B rings of Saturn have reflectivities closer to 60%, and the less prominent C ring and Cassini Division reflect 20% to 30% of the sunlight incident upon them [8]. The optical depth of Jupiter's ring is too small (~0.00005) to provide direct information on the absolute reflectivity of the individual ring particles, but Earth-based measurements of the spectral variations in the near infrared [9] have shown the ring to have a spectral slope similar to lunar soil, which has a reflectivity of about 11%.

11.2.3 Size distribution of ring particles

Particle sizes in Jupiter's main ring are clustered near a micrometer; the halo surrounding the main ring is composed primarily of sub-micrometer particles [9]. Saturn's rings have a wide distribution of sizes, ranging in size from micrometers to many tens of meters. Much of the information on particle sizes comes from observed differences in the optical thickness at different wavelengths. Particles tend to absorb radiation at wavelengths equal to or smaller than the particle size and to be transparent to radiation at wavelengths larger than the particle size.

The nine Uranian rings discovered from Earth seem to have nearly identical optical thicknesses at all measuring wavelengths from the ultraviolet to the radio regime. This is also true of the diffuse companion rings of the delta and eta rings. The implications are obvious: the vast majority of the ring mass must be in particles with sizes comparable to or larger than the longest measuring wavelength (13 cm). Models consistent with the Voyager data have been constructed to try to place bounds on the particle sizes [5]. In the epsilon ring, these models lead to the conclusion that the effective ring particle size is near 70 cm and the ring thickness is more than 30 m. Although 30 m is small compared to the width of the epsilon ring, such a large thickness does definitely suggest that the epsilon ring is not composed of a monolayer of ring particles. It is unlikely that all the ring particles are the same size; a distribution of sizes is more probable, but there is little information available to constrain that distribution. Nevertheless, the data seem to suggest that most of the particles are centimeter size or larger.

One notable exception is the newly discovered ring 1986U1R. It is seen in neither the Earth-based telescopic data nor the Voyager RSS data. Most of the telescopic data utilize infrared data at a wavelength of 2.2 μm. UVS data at 0.11 μm show an equivalent depth of about 0.31 km, significantly larger than the equivalent depth of about 0.13 km derived from PPS data at 0.27 μm. This suggests that more than 60% of the particles in 1986U1R are less than 0.3 μm in radius and that the ring has very few particles of centimeter size or larger.

This conclusion is also supported by the image of the Uranus dust bands introduced earlier (Fig. 11.5). The 172.5° phase angle means that the cameras were pointed only 7.5° away from the Sun at the time of the picture. Such a geometry favors dust-sized particles, which scatter sunlight predominantly in a direction away from the light source. A similar effect occurs when an automobile driver experiences extreme glare from a dusty windshield while driving toward the rising or setting Sun but has no difficulty seeing through the windshield near midday or after turning the automobile toward a different direction. 1986U1R is the brightest feature in Fig. 11.5. Considering its relatively low optical thickness, micrometer and sub-micrometer particles must completely dominate its population.

The diffuse companions of the delta and eta rings are not visible, implying that their dust content is very low. Most of the dusty rings in the image were not detected in approach images, nor were they seen in the stellar or radio occultation data sets. It is estimated [10] that the fractional area in and around the various known rings ranges from 0.03% to 0.2%. Cordelia orbits just inside 1986U1R, and the brightest dust band (other than 1986U1R) lies just inside Cordelia's orbit. The optical thickness of this previously unknown dust ring is estimated to be about 0.00001; the optical thicknesses of the other dusty rings are even smaller. These dust bands are similar in appearance to Saturn's D ring (Fig. 11.10), which was imaged under similar lighting conditions; there is no satisfactory dynamical explanation for the division of these two rings into multiple well-defined bands of material.

During Voyager's passage through the plane of the Uranus rings the Plasma Wave investigation (PWS) and Planetary Radio Astronomy (PRA) investigations detected particle impacts on the spacecraft (see Chapter 10). Analysis of the PWS and PRA signals generated by these impacts has led to the conclusion that most of the particles were a few micrometers in radius. At the crossing distance of 115 400 km from the center of the planet, this swarm of dust motes occupied a disk several thousand kilometers in thickness. The maximum number of particles striking the spacecraft was just over 50 per second, but the number was sharply peaked near the ring plane, dropping to 20 per second about 150 km (7 s) on either side of the peak.

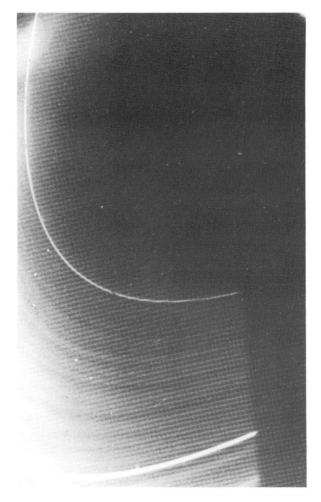

Fig. 11.10 — The closest analogy to the dusty rings of Uranus may be Saturn's D ring, which was imaged by Voyager 2 under similar lighting conditions. (P-23967)

11.3 INTERACTION WITH SHEPHERDING SATELLITES

One of the primary science objectives of the Voyager encounter was to study the interactions between ring particles and the shepherding satellites most scientists expected to find between the rings. Such interactions are responsible for some of the major features of Saturn's rings, such as the outer edges of the A and B rings and a large number of smaller features spread throughout the rings.

Small satellites are also thought to be the source of new ring material as material from their surfaces is freed into the ring plane by micrometeoroid impact or by tidal breakup due to the difference in Uranus's gravitational pull between the Uranus-facing side and the opposite side. This latter effect is related to the Roche limit, the minimum distance from the planet's center a solid object can reach before these uneven tidal forces become stronger than the cohesiveness of the object. Sometimes the Roche limit is described from a different point of view as the distance beyond which ring particles could coalesce to form stable larger bodies. The Roche limit is understandably somewhat imprecise; it also varies depending on the density of the material, its cohesiveness, and the size of the particles being considered. Generally, rings are assumed to form inside the Roche limit for objects large enough to be detected from imaging observations. That limit at Uranus is outside the epsilon ring for objects with density near 2 g/cm^3. Larger moonlets may survive for a time at closer ranges if their construction is of higher density or is more consolidated. Ophelia's orbit is very near the Roche limit for moonlets of density 2; Cordelia is well within that limit. These two moonlets have estimated diameters of 25 to 30 km and must therefore be denser than 2 g/cm^3 or stiffer in construction than was assumed for a Roche limit near 2.1 radii from the center of the planet. Other objects of smaller size must also exist within the Roche limit. Objects with diameters of 5 to 10 km could provide the gravitational forces which confine some of the narrow Uranian rings. Still smaller objects with diameters of 100 m to 1 km may serve as sources for the replenishment of ring particles.

11.3.1 Resonances within the rings

Once the moonlets have been broken up into smaller pieces the likelihood of inter-particle collisions increases and the process of ring formation is accelerated. Eventually these repeated collisions cause the ring particles to form circular or nearly circular orbits around the planet, their speeds relative to nearby particles drop drastically, and further breakup from inter-particle collisions is inhibited. The main forces that now act to change the ring structure are the weak but repeated tugs of nearby moonlets or more distant but much larger satellites. The primary vehicle through which these forces are felt is orbital resonances.

Particles near the outer edge of Saturn's B ring circle the planet once every 11.3 h. Saturn's satellite Mimas has an orbital period of 22.6 h. Outer B-ring particles circle Saturn twice each time Mimas completes one orbit. Each of the ring particles at that distance will encounter Mimas at the same part of their orbit every second time around the planet. These repeated encounters will tend to slow down the ring particles in much the same way that tides on Earth tend to lag behind Earth's rotation due to the retarding gravitational pull of the Moon. Because of their slightly changed speed the chances of collisions between particles increases, and the lower speed is partly transmitted to nearby particles. The resulting loss of energy causes the particles to drop to slightly lower, less energetic orbits, and Mimas has succeeded in effectively repelling outer B-ring particles back toward the planet. Given sufficient time for this process to continue, a gap eventually forms at the outer edge of the B ring and the ring develops a sharp outer boundary.

If the satellite is smaller but closer to the ring particles, the orbital periods more nearly match, and instead of the 2:1 resonance described above for Mimas and the outer edge of the B ring, less frequent alignments would cause a similar effect. For example, consider the effects of Cordelia and Ophelia on the epsilon ring (Fig. 11.9) [11]. Ophelia completes 13 orbits around Uranus in the same amount of time a particle at the outer edge of the epsilon ring completes 14 circuits of the planet. Similarly, particles at the inner edge of the epsilon ring are in a 24:25 resonance with

Cordelia; that is, Cordelia circles the planet 25 times in the same period that inner epsilon ring particles circle the planet 24 times. Ophelia effectively prevents particles at the outer edge of the epsilon ring from escaping; Cordelia performs the same task at the ring's inner edge. A 47:49 resonance with Cordelia near the middle of the epsilon ring may also account for that ring's relative minimum in optical thickness near its center and for the larger number of particles confined to the outer half of the ring (see Fig. 11.6).

The interaction is complicated somewhat by the fact that the orbit of Ophelia is slightly elliptical (non-circular). The average eccentricity (measure of the departure from circularity) of the epsilon ring is intermediate between the negligible eccentricity of Cordelia and the somewhat larger eccentricity of Ophelia; this may be related to the fact that the epsilon ring is widest at its apoapsis and narrowest at its periapsis.

As successful as resonance theory with known satellites may be in explaining some of the major features of the epsilon ring, it has thus far been singularly unsuccessful in explaining the finer details of the epsilon ring or in explaining most of the boundaries or other features in the smaller rings of Uranus. The 23:22 resonance of Cordelia falls very close to the sharp outer edge of the delta ring, and the 6:5 resonance of Ophelia is close to the outer edge of the gamma ring [11]. It is likely that other small satellites exist which control the boundaries of others of the rings, but this is by no means a certainty. Following the success of the Ophelia/Cordelia resonance studies, several other dynamicists have conducted studies to try to predict the locations of such satellites from the locations of sharp ring edges or other gravity-driven features within the rings. Another such study worthy of mention will be discussed in the next section.

11.3.2 Density waves

In addition to bounding the edge of a ring, satellite gravitational resonances can also cause tightly wound spiral density waves within a ring. It is this mechanism on a much grander scale which creates the spiral arms of a galaxy. Large numbers of such density waves have been found in the A ring of Saturn [10,12]. Because of the limited widths of the Uranian rings it was not initially expected that density waves would exist. One such density wave may have been found in the inner half of the delta ring [13]. Four 'windings' of this density wave are seen in each of the highest resolution profiles of the ring obtained by Voyager. The most likely location for the unseen satellite is between the gamma and the delta rings; assuming that the satellite has a diameter of about 10 km (still too small to have been seen in Voyager imaging), six possible locations were identified. Two of the possible locations are intriguing. If the satellite is located at a distance of 47 984.1 km from the center of Uranus it may possibly have additional resonance effects on the rings. The density wave would then be caused by a 101:102 resonance, the slightly stronger 102:103 resonance would fall very near the inner edge of the delta ring, and the 90:89 resonance might assist Ophelia in constraining the outer edge of the gamma ring. If the satellite is at a distance of 47 920.8 km, the delta density wave would be a 84:85 resonance, the 43:42 resonance would fall near the outer edge of the eta ring, and the 15:16 resonance would fall near the inner edge of 1986U1R.

11.3.3   Ring edges, ring thickness, and particle confinement

Density waves are more than just a curiosity and an indication of possible unseen satellites. They can also provide additional information about the rings themselves. For example, the delta ring density wave tells dynamicists that the physical thickness of the ring is between 7 and 20 m and that the amount of ring material contained in that physical thickness is between 5 and 10 g for every square centimeter. These numbers are comparable to numbers derived for parts of the A ring of Saturn from a study of the density waves seen there.

Another observational result of the Voyager ring observations is a direct correlation between the optical thickness near the edge of a ring and the abruptness of the ring edge. This is especially true of the outer edges of the gamma, delta, and epsilon rings. Because of the small gravitational effects the ring particles have on each other and the results of collisions between ring particles, unconfined rings will tend to spread in a radial direction both inward and outward. The congregating of particles near the outer edge of a ring is strong evidence that an active confinement mechanism is counteracting the spreading tendency. Orbital resonances with nearby satellites (discussed earlier in this chapter) are an effective means of providing such confinement. Too little is presently known about the rings of Uranus and the physics of interactions within and between rings to know whether other effective mechanisms are both possible and important in the Uranian ring system.

Studies of the possible particle sizes in the epsilon ring were discussed earlier in this chapter and lead to the conclusion that the ring is many particles thick. One model leads to the conclusion that the ring is about 30 m thick. The absence of large variations in equivalent depth with variations in physical width also provides evidence that the alpha, beta, gamma, and delta rings are many particles thick. If the dominant sizes for particles in these rings are tens of centimeters, these rings must also be several meters thick.

PWS and PRA data at ring-plane crossing show that the ring particles at distances near a distance of 115 400 km are spread into a centrally condensed disk of tiny particles that extends over thousands of kilometers of thickness. The E and G rings at Saturn are also much thicker than the main A, B, and C rings closer to the planet. The small particles of these 'etherial' rings may interact more easily with the planetary magnetic and radiation fields, and the resultant forces could possibly account for the vertical spreading of these rings.

11.4   AGE AND EVOLUTION OF THE RINGS

Uranus has an exosphere of atomic and molecular hydrogen which extends outward through the orbits of most of the rings (see Chapter 9). This hydrogen must be a significant source of atmospheric drag forces on the dust particles in the rings. Voyager UVS team members calculate [14] that the effectiveness of this drag is sufficient to sweep all particles a micrometer or smaller in radius out of the Uranian rings within 1000 years. Because of the greater hydrogen densities at lower altitudes, such particles, if not replenished by infall from greater distances, would be swept from the inner parts of the ring system in less than a year. This radial gradation of the drag forces may explain the larger amounts of dusty material in the outer portions of the rings near 1986U1R. Orbital lifetimes for larger particles are greater in direct proportion to their sizes. Centimeter-sized particles would disappear into the Uranus atmosphere in about 10 000 000 years, a geologically short interval, and meter-sized particles would disappear in a fraction of the apparent age of Uranus.

Other slower processes also contribute to the destruction of the rings. Micrometeoroid bombardment probably shatters larger fragments, making them more susceptible to drag forces. Another is Poynting-Robertson drag, in which an orbiting particle absorbs sunlight and re-emits the light in a preferentially forward direction in its orbit [15]. P–R drag, like atmospheric drag, is most effective on the smallest ring particles. Collisions between ring particles also serve to transfer energy from one to the other. The lower energy particle, if not inhibited by particle confinement mechanisms, moves inward toward the planet, while the more energetic particle moves outward where particle confinement is generally more efficient.

Particle confinement by means of satellite orbital resonances may retard the orbital decay of ring particles, but it is clear that the Uranian rings undergo relatively rapid evolution. Theoretical arguments [16] favor a development of rings from the breakup of satellites well after rather than concurrent with planet formation. That process would also favor formation of rings around the more massive planets, where forces leading to tidal breakup of satellites are stronger. The close association of small satellites with each of the ring systems and the seemingly continuous ring particle size distribution, from micrometer to meters and probably to kilometers in diameter, also lend support to the breakup theory.

Saturn's D ring and the dusty rings of Uranus both show definite banding in their structure. By analogy with the Jupiter ring, whose source is thought to be the satellites Metis and Adrastea, the location of these dust bands may be an indicator of the radial positions of small satellites or large ring particles still serving as sources for replenishment of ring material. This picture is one of a constantly changing ring system whose particle sources are slowly but surely being depleted. Although the argument made here is for creation of the dusty rings, the procedure must be effective for the nine rings of Uranus discovered in Earth-based observations. These rings may be the remnants of 'creation' events of the more distant past which have since been swept clean of most of the smallest particles. In such a scenario, 1986U1R and the other dusty rings are very young constructs and are not destined to endure for geologically long periods of time. Voyager has succeeded in obtaining snapshots in one brief instant of time of a rapidly evolving system.

NOTES AND REFERENCES

[1] Smith, B. A., Soderblom, L. A., Beebe, R., Bliss, D., Boyce, J. M., Brahic, A., Briggs, G. A., Brown, R. H., Collins, S. A., Cook, A. F. II, Croft, S. K., Cuzzi, J. N., Danielson, G. E., Davies, M. E., Dowling, T. E., Godfrey, D., Hansen, C. J., Harris, C., Hunt, G. E., Ingersoll, A. P., Johnson, T. V., Krauss, R. J., Masursky, H., Morrison, D., Owen, T., Plescia, J. B., Pollack, J. B., Porco, C. C., Rages, K., Sagan, C., Shoemaker, E. M., Sromovsky, L. A., Stoker, C., Strom, R. G., Suomi, V. E., Synnott, S. P., Terrile, R. J., Thomas, P., Thompson, W. R., Veverka, J. (1986) Voyager 2 in the uranian system: Imaging science results. *Science*, **233**, 43–64.

[2] Colwell, J. E., Horn, L. J., Lane, A. L., Esposito, L. W., Yanamandra-Fisher, P. A., Pilorz, S. H., Simmons, K. E., Morrison, M. D., Hord, C. W., Nelson, R. M., Wallis, B. D., West, R. A., Buratti, B. J. (1990) Voyager photopolari-meter observations of uranian ring occultations. *Icarus*, **83**, 102–125.

[3] Keplerian orbits are so named because German astronomer Johannes Kepler was the first to recognize that orbiting bodies trace ellipses (and not circles with second, third, and fourth-generation 'epicycles') about their central bodies.

[4] Goldreich, P., Tremaine, S. (1979) Towards a theory for the Uranus rings. *Nature*, **277**, 97–99.

[5] Esposito, L. W., Brahic, A., Burns, J. A., Marouf, E. A. (1989) Particle properties and processes in Uranus' rings. In Bergstralh, J. T., Miner, E. D. (eds.) *Uranus*, The University of Arizona Press, Tucson, in preparation.

[6] French, R. G., Nicholson, P. D., Porco, C. C., Marouf, E. A. (1989) Dynamics and structure of the uranian rings. In Bergstralh, J. T., Miner, E. D. (eds.) *Uranus*, The University of Arizona Press, Tucson, in preparation.

[7] The orbital periods of the ring particles were calculated from Equation (4) in Elliot, J. L., Nicholson, P. D. (1984) The rings of Uranus. In Greenberg, R., Brahic, A. (eds.) *Planetary Rings*, The University of Arizona Press, Tucson, pp. 25–72.

[8] Cuzzi, J. N., Lissauer, J. L., Esposito, L. W., Holberg, J. B., Marouf, E. A., Tyler, G. L., Boischot, A. (1984) Saturn's rings: properties and processes. In Greenberg, R., Brahic, A. (eds.) *Planetary Rings*, The University of Arizona Press, Tucson, pp. 73–199.

[9] Smith, B. A., Reitsema, H. J. (1980) CCD observations of Jupiter's ring and Amalthea. Presented at IAU Colloquium 57, Kona, Hawaii. Quoted in Burns, J. A., Showalter, M. R., Morfill, G. E. (1984) The etherial rings of Jupiter and Saturn. In Greenberg, R., Brahic, A. (eds.) *Planetary Rings*, The University of Arizona Press, Tucson, pp. 200-272.

[10] Ockert, M., Cuzzi, J. N., Porco, C. C., Johnson, T. V. (1987) Uranian ring photometry: results from Voyager 2. *Journal of Geophysical Research*, **92**, 14 969–14 978.

[11] Porco, C. C., Goldreich, P. (1987) Shepherding of the uranian rings. I. Kinematics. *Astronomical Journal*, **93**, 724–729.

[12] Shu, F. H. (1984) Waves in planetary rings. In Greenberg, R., Brahic, A. (eds.) *Planetary Rings*, The University of Arizona Press, Tucson, pp. 513–561.

[13] Horn, L. J., Yanamandra-Fisher, P. A., Esposito, L. W., Lane, A. L. (1988) Physical properties of the uranian delta ring from a possible density wave. *Icarus*, **76**, 485–492.

[14] Broadfoot, A. L., Herbert, F., Holberg, J. B., Hunten, D. M., Kumar, S., Sandel, B. R., Shemansky, D. E., Smith, G. R., Yelle, R. V., Strobel, D. F., Moos, H. W., Donahue, T. M., Atreya, S. K., Bertaux, J. L., Blamont, J. E., McConnell, J. C., Dessler, A. J., Linick, S., Springer, R. (1986) Ultraviolet spectrometer observations of Uranus. *Science*, **233**, 74–79.

[15] Poynting–Robertson drag is described in more detail by Mignard, F. (1984) Effects of radiation forces on dust particles in planetary rings. In Greenberg,

R., Brahic, A. (eds.) *Planetary Rings*, The University of Arizona Press, Tucson, pp. 333–366.

[16] Harris, A. W. (1984) The origin and evolution of planetary rings. In Greenberg, R., Brahic, A. (eds.) *Planetary Rings*, The University of Arizona Press, Tucson, pp. 641–659.

BIBLIOGRAPHY

Bergstralh, J. T., Miner, E. D. (1989) *Uranus*, The University of Arizona Press, in preparation (see rings chapters by French *et al.* and Esposito *et al.*; also the references cited in each chapter).

Cuzzi, J. N., Esposito, L. W. (1987) The rings of Uranus. *Scientific American*, **257**, No. 1, 52–66.

Greenberg, R., Brahic, A. (1984) *Planetary Rings*, The University of Arizona Press, Tucson, 784 pages, written prior to Voyager 2 Uranus encounter.

Porco, C. C. (1986) Voyager 2 and the uranian rings. *The Planetary Report*, **VI**, No. 6, 11–13.

12

The satellites of Uranus

12.1 DISCOVERY OF TEN ADDITIONAL SATELLITES

The Voyager mission has become a public legacy. The two Voyager spacecraft were built and launched in 1977 with promised funding and 'guaranteed' longevity for their basic four-year mission to study the Jupiter and Saturn systems. As Voyager 2 flew past Neptune in the summer of 1989, each of the two spacecraft had already logged 12 years in space! Those who worked with the spacecraft for most of those years learned to expect spectacular results from each of the 11 investigations when the spacecraft passed each new planet, and they were seldom disappointed in that expectation.

During the planetary encounters of Voyager it became customary to display images and some other data on monitors at the Jet Propulsion Laboratory almost as soon as they arrived. As might be expected, the images returned during an encounter period received the greatest attention. Most humans depend more heavily on their sense of sight than any other sense. Voyager 2 gave mankind the opportunity to see at close range worlds so remote that most of Earth's inhabitants were totally unaware of their existence.

Even in the most powerful telescopes, only the barest of details can be seen in the atmospheres of the giant planets, and their satellites are mere pinpoints of light. Now for the first time in history a remote robot spacecraft enabled man to see the far reaches of the outer Solar System. Those fortunate enough to have had access to the television monitors sat fascinated, often shunning both food and sleep to catch glimpses of the latest images. Such behavior was certainly not limited to avid amateurs; seasoned members of the Voyager Imaging Team often acted much like young children suddenly given free access to all the contents of a candy shop. The author of this book shared in those same feelings.

As each new picture appeared on the screen, watchers would scrutinize the image. It became an unofficial contest to see who could spot each new discovery first. ISS team members had the marked advantage of higher resolution screens and access to contrast enhancement and other software routines, and some of the more privileged watchers peered over their shoulders. (It was traditional on the Voyager project, because of the high degree of teamwork necessary for the success of the mission, to give credit for new discoveries to 'Voyager' rather than to any specific individual who happened to be the first to recognize the discovery. This book will not depart from that tradition. Hence the scramble to be the first to see any new discovery was due to a general desire to be a contributor to the overall success of the mission rather than to individual attempts aimed at self-aggrandizement. The Voyager team can be rightfully proud of the accomplishments of its mission, and major contributions to the success of the mission were made by each of those who took part. For lists of Voyager personnel and science team principal investigators, see Chapter 6. Individual science investigation team members and associates are listed as co-authors of each investigation's preliminary report in the special Uranus encounter issue of *Science*. [1])

The pictures of Uranus were moderately disappointing because of the absence of atmospheric details. The rings, whose presence and location were known before the encounter, were first detected in late November (Fig. 12.1). Initially only the epsilon

Fig. 12.1 — The Uranian rings, whose presence and location were known before the Voyager encounter, were first detected in late November. The picture is a computerized summation of six images taken at a distance of 72 000 000 km. The epsilon ring is prominent, but several of the narrower rings may also be seen in the original data. (P-29314)

ring was easily seen, but by mid to late December all nine of the previously known rings had been 'captured' in specially processed Voyager imaging data. In the meantime the attention of ISS team members turned to images of the five known satellites and to extensive searches for possible new satellites. Both of these efforts were to bear unprecedented fruits. Thanks to Voyager 2 the number of known satellites of Uranus was about to triple!

12.1.1 Discovery dates and images

The first success in the hunt for new characters in the Uranus plot came in the waning hours of 1985. A small but unmistakable satellite was orbiting the planet at a distance almost precisely midway between the orbit of Miranda and the epsilon ring. The date was 30 December 1985 and the official temporary designation was 1985U1, indicating that it was the first previously unknown satellite of Uranus discovered in the year 1985. Unofficially, ISS team members took their lead from the names of the outer two satellites, Oberon and Titania, who were characters from William Shakespeare's *A Midsummer Night's Dream*: the new family member was nicknamed 'Puck' for the tricksy household fairy who served King Oberon.

The early discovery of 1985U1 made it possible to slightly revise the encounter sequence to divert one recorded image reserved for Miranda to the new satellite. The image of Puck was successfully shuttered (see Fig. 12.2), although a problem with one of the tracking antennas at Canberra, Australia, resulted in the loss of the first playback of the image. Commands were quickly generated to replay the image before it was overwritten on the tape recorder by later data, and the second playback was successfully received.

The next two satellites were discovered on 3 January 1986 and were designated 1986U1 and 1986U2. The reader should understand that a 'discovery' occurred only when the same satellite was seen in two or more images shuttered at sufficiently separated times to enable Voyager Navigation Team Members to calculate an approximate orbit. 1986U3 was discovered even closer to the rings on 9 January 1986. Three more satellites, 1986U4, 1986U5, and 1986U6, were discovered on 13 January 1986. Five days later on 18 January, 1986U1, 1986U3, and 1986U4 were all captured in a single image which also included the outer rings of Uranus (Fig. 12.3).

One of the primary objectives of the satellite searches was the detection of ring shepherds (discussed in Chapter 10) which theoreticians were certain must be responsible for shaping the narrow rings of Uranus. On 20 January 1986 two small satellites were found flanking either side of the bright epsilon ring, and scientists were convinced that calculations would verify that the gravitational influences of 1986U7 and 1986U8 were indeed responsible for the radial confinement of epsilon ring particles. On 21 January the two were captured in a single image (Fig. 12.4), which also shows the nine rings first detected from Earth. The tenth Uranian satellite discovered by Voyager (1986U9) was announced that same day, just three days before the spacecraft reached its closest approach to the planet.

12.1.2 Naming process

On 22 January 1986 daily press conferences in the Jet Propulsion Laboratory's Theodore von Karman Auditorium began. Excitement was building rapidly as each new set of satellite pictures revealed new details. Voyager scientists and Flight

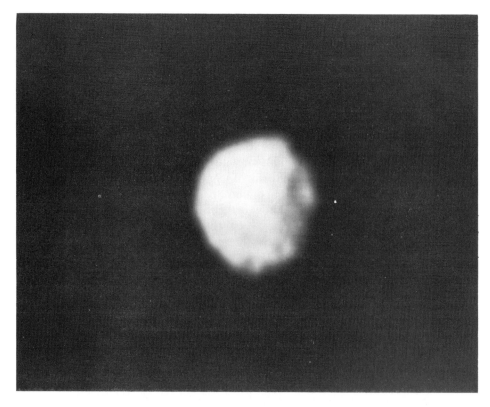

Fig. 12.2 — This image of Puck (1985U1) replaced a previously planned image of Miranda. Voyager 2 was 500 000 km from Puck and the resolution was about 10 km. (P-29519)

Science Office personnel were meeting each morning to compare their newest findings and to report on instrument health and data quality. Each afternoon in a science discussion meeting they informally talked about the interpretation and implications of their mutual findings and planned for the next day's press conference presentations. Just four days after the 24 January 1986 closest approach of Voyager 2 to Uranus plans were being made for the final press conference. The morning meeting had concluded a few minutes early to permit everyone to view NASA's companion feature to the Voyager Uranus encounter: the launch of the heralded 'Teacher in Space' mission of the Space Shuttle Challenger.

Challenger's crew was headed by Francis R. (Dick) Scobee. Other crew members included Michael J. Smith, Judith A. Resnick, Ronald E. McNair, Ellison S. Onizuka, and Gregory B. Jarvis. The seventh crew member was grade school teacher Christa McAuliffe. Voyager's encounter and the Teacher in Space mission were to be the first two major events in NASA's year of spectaculars. The launch of the Galileo spacecraft to Jupiter and of the Hubble Space Telescope into Earth orbit were to follow later in the year, potentially ending a nine-year hiatus in NASA launches of major interplanetary spacecraft. Nature itself was to be a part of the

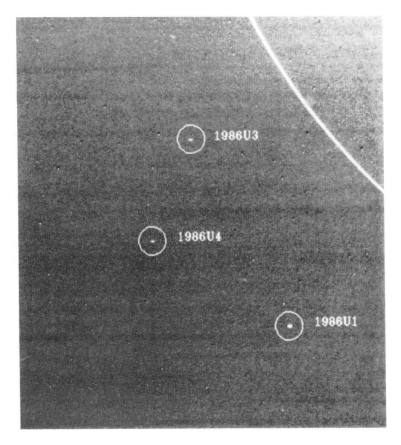

Fig. 12.3 — On January 18, 1986, Portia (1986U1), Juliet (1986U3), and Cressida (1986U4) were all captured in a single image which also included the outer rings of Uranus. (P-29465)

spectacle with the return of Halley's Comet, and spacecraft built by the European Space Agency, Russia, and Japan were closing in on that infrequent celestial visitor to the inner Solar System.

All the euphoria disappeared that sunny but bleak Tuesday morning as Voyager scientists and millions of others watched in stunned disbelief the explosion of the Challenger 74 s into its flight. The final press conference of the Voyager Uranus encounter scheduled for 10:00 am PST was postponed until the following day, and Voyager personnel cancelled all press interviews for the remainder of the day. Voyager joined the worldwide mourning for the loss of the seven Challenger astronauts.

Many suggested that a fitting memorial would be to name seven of the newly discovered Uranian satellites for the fallen astronauts. Many Voyager personnel openly expressed support for the idea. In the end, members of the Nomenclature Committee of the International Astronomical Union (IAU) chose to name several craters on the far side of the Moon for the seven astronauts. There was precedent for

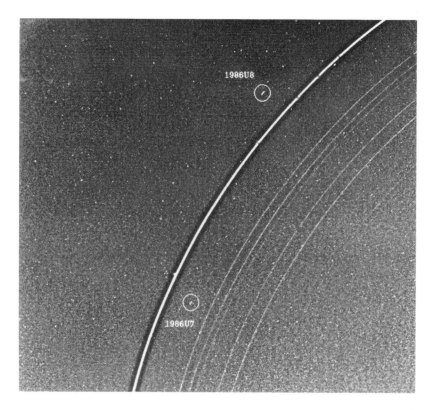

Fig. 12.4 — Cordelia (1986U7) and Ophelia (1986U8) are believed to be responsible for the
radial confinement of epsilon ring particles. The two were captured in a single image on January
21, 1986. The nine 'classical' rings of Uranus are also apparent. (P-29466)

this choice: three Apollo astronauts who died in a test of that spacecraft and Russian
cosmonauts who died in the line of duty had been accorded similar tributes.

Nine of the names chosen by the IAU were taken from the writings of
Shakespeare; the tenth (like Umbriel) is from Alexander Pope's *The Rape of the
Lock*. In order of increasing distance from the planet, the first satellite (1986U7) was
named Cordelia after the youngest of King Lear's three daughters in *King Lear*.
Ophelia (1986U8) was the daughter of Polonius, Lord Chamberlain to the King of
Denmark in *Hamlet*. Bianca (1986U9) is the daughter of Baptista and younger sister
of Katherine in *The Taming of the Shrew*. 1986U3 was named Juliet, who is both the
heroine of *Romeo and Juliet* and a lady beloved of Claudio in *Measure for Measure*.
Desdemona, wife of *Othello*, was the name chosen for 1986U6. Rosalind (1986U2)
was named for the daughter of the exiled duke in *As You Like It*. Portia (1986U1) is
the rich heiress who becomes the wife of Bassanio in *The Merchant of Venice*.
Cressida (1986U4) is the daughter of Calchus and the title character of *Troilus and
Cressida*. Pope's *The Rape of the Lock* is a poem about a lock of hair mischievously
cut from the head of the angered Belinda (1986U5). In deference to the ISS
scientists, the IAU retained the name Puck for 1985U1. All 15 of the satellites of

Uranus therefore derive their names from characters in classical English literature. This is in contrast to the planets and the satellites of Mars, Jupiter, Saturn, and Neptune, which all derive their names from Greek and Roman mythology.

12.1.3   Orbital and physical characteristics

Once discovered, many of the small satellites could then be located in earlier images. A careful search revealed 49 separate images of Puck, spanning a period of more than 35 days (46 orbits of Puck around Uranus). The other nine satellites were found in 13 to 38 images each, spanning time periods corresponding to 14 to 43 orbital periods. Careful analysis of these data [2] led to determination of their orbital periods to within a second. Their average distances from Uranus (i.e., the semimajor axes of their orbits) were then determined by two different methods. The first method utilized the positional measurements only and yielded accuracies of 18 to 39 km. The second used the measured orbital periods and calculated the semimajor axis of each orbit from the mass of Uranus and its gravity harmonics [3], and yielded accuracies of 1.4 km or less. Names, discovery dates, orbit sizes, and orbital periods are given in Table 12.1. Also tabulated are the estimated sizes and their uncertain-

Table 12.1 — The small satellites of Uranus

Satellite number/name	Discovery date	Semimajor axis (km)	Semimajor axis (R_U)	Period (h)	Radius (km)
U13/Cordelia	20 Jan 1986	49 751.7(0.2)	1.946 54(1)	8.0408(1)	13(2)
U14/Ophelia	20 Jan 1986	53 764.3(1.4)	2.103 54(6)	9.0338(3)	15(2)
U15/Bianca	21 Jan 1986	59 165.4(0.9)	2.314 86(4)	10.4299(2)	21(3)
U9/Juliet	9 Jan 1986	61 766.8(0.3)	2.416 64(1)	11.1257(1)	31(4)
U12/Desdemona	13 Jan 1986	62 658.5(0.4)	2.451 52(2)	11.3676(1)	27(3)
U8/Rosalind	3 Jan 1986	64 358.3(0.3)	2.518 03(1)	11.8336(1)	42(5)
U7/Portia	3 Jan 1986	66 097.3(0.3)	2.586 07(1)	12.3167(1)	54(6)
U10/Cressida	13 Jan 1986	69 926.8(0.4)	2.735 90(2)	13.4030(1)	27(4)
U11/Belinda	13 Jan 1986	75 255.4(0.4)	2.944 38(2)	14.9646(1)	33(4)
U6/Puck	30 Dec 1985	86 004.5(0.4)	3.364 94(2)	18.2840(1)	77(3)

ties. The sizes were directly measured for Puck (Fig. 12.2) and Cordelia (Fig. 12.5). Fig. 12.5 also shows the epsilon, delta, gamma, and eta rings, but the newly discovered ring (1986U1R) lying just to the left of Cordelia is not apparent in the image. Cordelia's reflectivity is similar to that of Puck but could not be measured as precisely. Puck has a radius of 77±3 km. Sizes of Cordelia and the remaining satellites are calculated [4] from their brightness in the closest images by assuming their reflectivities are the same as that of Puck (0.074±0.008).

Table 12.2 repeats the names and orbit semimajor axes and then provides the measured orbit eccentricities and their inclinations with respect to the equator of Uranus. For the sake of reference, Voyager and Earth-based measurements of the rings have determined that the south (positive rotation) pole points at a right ascension (RA) of 76.5969°±0.0034° and a declination (DEC) of 15.1117°±0.0033°. The pole direction obviously also determines the orientation of Uranus's equator;

Fig. 12.5 — Cordelia (1986U7) is barely resolved in this image obtained January 23, 1986, from a distance of 1 440 000 km. Its apparent elongated shape is primarily due to image smear from Cordelia's motion. Cordelia is flanked by the epsilon ring on the left and the delta, gamma, and eta rings on the right. (P-29499)

Table 12.2 — Orbits of the small satellites of Uranus

Satellite number/name	Semimajor axis (km)	Inclination (°)	Eccentricity ($\times 10^3$)
U13/Cordelia	49 751.7±0.2	0.140±0.098	0.469±0.410
U14/Ophelia	53 764.3±1.4	(0.091±0.272)	10.140±0.404
U15/Bianca	59 165.4±0.9	(0.156±0.172)	0.878±0.518
U9/Juliet	61 766.8±0.3	(0.042±0.140)	(0.233±0.321)
U12/Desdemona	62 658.5±0.4	0.160±0.117	(0.227±0.297)
U8/Rosalind	64 358.3±0.3	(0.057±0.113)	0.585±0.249
U7/Portia	66 097.3±0.3	(0.087±0.151)	(0.165±0.365)
U10/Cressida	69 926.8±0.4	0.282±0.116	(0.091±0.349)
U11/Belinda	75 255.4±0.4	(0.033±0.108)	(0.109±0.206)
U6/Puck	86 004.5±0.4	0.314±0.079	(0.051±0.178)

RA and DEC values are references to a coordinate system fixed with respect to the orientation of Earth's equator and pole at the beginning of the year 1950.

Nothing is known about the mass of any of the small satellites, so their composition, like that of the ring particles, is mostly a matter of inference. They have much lower reflectivity than any of the large Uranian satellites, but not as low as the ring particles. Less color information exists for these small satellites than for the rings, but their reflectivity is assumed to be flat (i.e., they are gray in color). On similar grounds to those used for the ring particles, the dark component of the surfaces of these satellites is assumed to be elemental carbon. All of the satellites (including the five larger ones) are immersed in the Uranus radiation field. The carbon in their surfaces could be the byproduct of bombardment of methane (CH_4) ice by energetic protons, but it is more likely carbonaceous rock collected in the outer Solar System while these satellites were forming. Another likely chemical component of the satellites is water ice, which is relatively abundant in the outer Solar System. Small quantities (less than 1% of the total volume) of carbonaceous rock well mixed with the ordinarily bright water ice can result in reflectivities as low as the 7% measured for Puck.

Another probable difference (other than surface reflectivity) between Puck and the five larger satellites of Uranus is its colder history. There is evidence for at least partial melting of the interiors of each of the five outer satellites. Heavier rocky material would have sunk toward their centers, forcing brighter icy material to rise to the surface. This process is known as differentiation. If sufficient heat is available, the process can progress to its logical conclusion: the formation of rocky cores surrounded by relatively rock-free icy material. Partial differentiation would occur if the quantity of heat were sufficient to start the process but insufficient to complete it. Puck and the inner satellites are much too small to have undergone even partial differentiation and are therefore more likely to have homogenous composition (unless, of course, they are only small fragments of a very much larger satellite which underwent differentiation.)

12.2   ORBITAL AND PHYSICAL CHARACTERISTICS OF THE MAJOR SATELLITES

The larger outer satellites of Uranus provided much more opportunity for intensive study. All five have substantially higher albedos than Puck. They reflect from 19% to 40% of the light incident on them; Puck reflects only about 7%. Because of their larger sizes and known orbits, advanced planning included systematic ISS and PPS observations of each during the encounter period. Relatively high-resolution multicolor images of their surfaces enabled scientists to study the variations of brightness and physical structure across their surfaces. Coverage was obtained at a variety of phase angles (phase angle is the angle between the spacecraft and the Sun as viewed from the target satellite). Such data provide a means of determining the total amount of sunlight reflected by their surfaces. Approaches to both Ariel and Miranda were close enough to overfill the 0.25° IRIS field of view, enabling IRIS to obtain some temperature and bolometric (including all solar colors) reflectivity data. (Temperatures near the vertically illuminated south poles of these two satellites were 84 ± 1 K for Ariel and 86 ± 1 K for Miranda. Their respective bolometric albedos were 0.30 ± 0.05 and 0.23 ± 0.06.) Provisions were made to monitor changes in the spacecraft velocity during those times when the tiny gravitational effects attributable to each satellite's mass would be greatest. Images were obtained at regularly spaced intervals for at least two circuits of each major satellite around the planet. If the Sun had been over their equators such imaging would have provided full surface coverage. Unfortunately, in 1986 the south poles of each of these satellites (like Uranus) were pointed almost directly at the Sun, and only southern hemisphere coverage was obtained. Full coverage of both hemispheres of the satellites will have to wait until at least 2007, when Uranus will have traveled far enough in its orbit to have sunlight fall vertically on the equators of its satellites.

12.2.1 Orbital characteristics

The orbital periods of the major satellites are better determined by the long-term observations from Earth-based telescopes [5]. Voyager's primary contribution to their orbital elements is a better determination of the mass and gravity harmonics of Uranus, which leads to a more precise measurement of their orbit sizes. Earth-based observations during the years immediately preceding the encounter were used by the Voyager Navigation Team to obtain improved values for the period of Miranda, and for the orbit inclination and eccentricity of all five satellites. Orbit characteristics of the major satellites of Uranus are given in Table 12.3.

Table 12.3 — Orbits of the major satellites Uranus

Satellite number/name	Semimajor axis (km)	Semimajor axis (R_U)	Period (h)	Inclination (°)	Eccentricity ($\times 10^3$)
U5/Miranda	129 847	5.0803	33.924	4.2	27.0
U1/ Ariel	190 929	7.4701	60.4891	0.3	3.4
U2/Umbriel	265 979	10.4065	99.4603	0.36	5.0
U3/Titania	436 273	17.0692	208.9411	0.14	2.2
U4/Oberon	583 421	22.8264	323.1182	0.10	0.8

12.2.2 Mass determinations

Three specific mass determinations were made directly from the radio science data. These were the mass of Uranus, the mass of the Uranus system (including the satellites), and the mass of Miranda. Successful Doppler radio measurements were also made of the individual masses of Ariel, Umbriel, Titania, and Oberon, but they were less precise than values derived from Navigation Team studies of the motions of these satellites. The orbit of each satellite is diverted slightly by the gravitational effects of the other satellites. By careful study of these orbit 'perturbations' the approximate ratios of the masses of Ariel, Umbriel, Titania, and Oberon were determined. A precise value for the mass of each satellite was then obtained by subtracting the masses of Uranus and Miranda from the mass of the Uranus system (the mass of the rings is negligible) and dividing the remainder into parts with the appropriate mass ratios [6]. These indirectly determined masses for the outer four satellites were within the uncertainty boundaries of the more direct Doppler radio measurement. Technically, these procedures determine the product of the mass, M, and the gravitational constant, G. It is customary to present the derived planet and satellite masses in the form GM, which often has greater precision than the gravitational constant itself. Both GM (in km^3/s^2) and M (in kg) of each satellite are presented in Table 12.4. The value of G is $(6.670\pm0.004)\times10^{-20}\,km^3/s^2\,kg$. The GM values for Uranus and its system are provided in the table for reference.

Table 12.4 — Physical characteristics of the major satellites of Uranus

Satellite number/name	$G\times$mass (km^3/s^2)	Mass (10^{20} kg)	Radius (km)	Density (g/cm^3)	Geometric albedo
Uranus Sys.	5 794 560±10				
Uranus	5 793 947±23		25 559±5	1.28±0.01	
U5/Miranda	4.2±0.5	0.63±0.07	235.8±1.2	1.15±0.15	0.32±0.03
U1/Ariel	85±5	12.7±0.7	578.9±1.1	1.56±0.09	0.39±0.04
U2/Umbriel	85±5	12.7±0.7	584.7±4.0	1.52±0.11	0.21±0.02
U3/Titania	233±5	34.9±0.7	788.9±2.8	1.70±0.05	0.27±0.03
U4/Oberon	202±5	30.3±0.8	761.4±3.0	1.64±0.06	0.23±0.03

12.2.3 Sizes and shapes

Sizes of the satellites were determined from imaging data. This process involves more than just measuring the diameter on a reproduced photograph. In addition, scientists measure the precise locations of a large number of surface features recognizable in at least two separate images. A mathematical model is next constructed to specify the radius from the satellite center to each selected feature and the position of the spacecraft in satellite-centered coordinates. The model is then used to predict where each surface feature should fall within each image. Adjustments are made to the model parameters until differences between the predicted and the measured positions of the features are as small as the estimated measurement errors. The final results [7] represent each satellite as a sphere or as a triaxial ellipsoid (a 'sphere' with different diameters in three perpendicular directions). Miranda, for example, is best represented by a triaxial ellipsoid with equatorial radii of 240 km and 234 km and a polar radius of 233 km; Ariel may also be represented by a triaxial ellipsoid with equatorial radii of 581 km and 578 km and a polar radius of 578 km. Umbriel, Titania, and Oberon are best represented as spheres with respective radii of 584.7±4.0 km, 788.9±2.8 km, and 761.4±3.0 km. Miranda and Ariel have the same volume as spheres with radii 235.8±1.2 km and 578.9±1.1 km, respectively, and are represented with those radii in Table 12.4. The reader is cautioned to remember that the derived sizes and shapes are based on observations of the southern (illuminated) hemisphere only. Although departures from a spherical shape are not expected to exceed the error envelopes listed, the northern hemispheres of these satellites could still hide some major surprises.

The five outer satellites of Uranus are of the intermediate size class (radii between about 200 and 800 km). They are smaller than Earth's Moon, Jupiter's Io, Europa, Ganymede, and Callisto; Saturn's Titan; and Neptune's Triton; but are much larger than the ten newly discovered satellites of Uranus. Saturn has six satellites with intermediate radii (Mimas, Enceladus Tethys, Dione, Rhea, and Iapetus), Uranus has its five, and Neptune's Nereid may also fall in this size range. None of Jupiter's 16 satellites have intermediate radii.

Satellites smaller than 200 km radius fall in a third class. Such satellites are often non-spherical and travel around their respective planets in inclined, eccentric orbits. Jupiter's eight outer satellites and Saturn's Phoebe are suspected to be captured asteroids. The small satellites have not undergone differentiation and generally have heavily cratered surfaces with very low reflectivities. Some (such as Saturn's Hyperion and the small satellites of Uranus) may be collisional or tidally disrupted fragments of once larger bodies. Others (notably Jupiter's Metis and Adrastea) may serve as a source for replenishment of ring material. The planetary rings themselves may once have been one or more satellites of this size range.

12.2.4 Surface reflectivities and phase functions

The final column in Table 12.4 lists the geometric albedos (the reflectivities at $0°$ phase angle) for Miranda, Ariel, Umbriel, Titania, and Oberon [8]. These satellites have albedos which range from 19% for Umbriel to 40% for Ariel, much higher than the 7% reflectivity of Puck. It is noteworthy (and difficult to understand) that Ariel and Umbriel, though similar in size and occupying adjacent orbits, are so different in their surface reflectivities and geologic features.

The reason for the higher reflectivities of the major satellites of Uranus relative to their smaller counterparts is probably related to their degree of differentiation. It was mentioned earlier in this chapter that there existed evidence in the Voyager images for at least partial melting of the interiors of the major satellites. Evidence for water ice has been seen in near infrared spectra of each of their surfaces obtained from Earth. Water ice has a high reflectivity. If the small satellites (of which Puck is assumed to be typical) are uniform mixtures of the material constituting the satellites, then partial melting of the larger satellites, sinking of denser and darker rocky or carbonaceous material, and rising of lighter and brighter water ice could account for their more reflective surfaces.

Albedo variations of a factor of 2 are seen across the surfaces of the satellites, but they do not seem to be accompanied by corresponding changes in color, at least over the limited color range of the Voyager cameras (0.35 to 0.60 μm). PPS made measurements at both ultraviolet (0.25 μm) and infrared (0.75 μm) wavelengths and saw little evidence of differences in whole-disk brightness between the ultraviolet and infrared data. This is especially noteworthy when one considers that the ultraviolet brightnesses of Saturn's intermediate satellites are 10% to 60% dimmer than their infrared brightnesses; Jupiter's Galilean satellites are even more extreme, reflecting only one-fifth as much in the ultraviolet. Obviously, the processes that alter satellite surfaces are very different in the three systems.

The integrated disk brightness of the Uranian satellites varies with solar phase angle in a relatively predictable fashion. Because of their great distance from Sun and Earth, solar phase angles at Uranus never exceed 3° in Earth-based telescopes. Voyager observations covered phase angles from 0.8° to 152.6° for Titania and somewhat smaller ranges for the other four major satellites. Over phase angles between 10° and 60° the integrated disk brightness of the satellites decreases by amounts which vary from 1.9% (for Ariel) to 2.5% (for Miranda) of the remaining brightness per degree of phase angle. In terms of stellar magnitudes, which utilize a logarithmic scale, these decreases correspond to 'phase coefficients' of 0.021 to 0.028 magnitudes per degree [8].

Only Titania was observed by Voyager at phase angles less than 10°. ISS and PPS data at phase angles between 0.8° and 2.6° could be overlapped with Earth-based data at comparable phase angles. Telescopic observations at 2.2 μm at phase angles lower than Voyager's 0.8° minimum indicate that Titania nearly doubles in brightness between 1° and 0° [9]! This is a much larger opposition effect than observed for other solid bodies in the Solar System.

Voyager observations did not verify this large opposition brightness surge. Three possibilities exist: (1) the large brightness surge occurs only at phase angles lower than 0.8°, (2) it occurs at 2.2 μm but not at wavelengths of 0.75 μm or less, or (3) an error exists in the telescopic observations. Possibility (1) is eliminated by telescopic observations down to 0.06° phase at wavelengths near 0.55 μm which show only a normal opposition brightness surge of about 40% between 1° and 0.06° [10]. Possibility (2) seems unlikely, since it would a very special set of circumstances to make Titania's surface appear substantially more porous at 2.2 μm than at 0.75 μm or shorter. As for possibility (3), the 2.2 μm data for Ariel, Titania, and Oberon are all very similar except at 3° where the Titania value seems too low. If it were adjusted to the same relative level as the Ariel and Oberon data, the Titania data would match the other two data sets to within their uncertainties.

The presence of a moderate opposition brightness surge on Titania (and presumably on the other major Uranian satellites) indicates that their soil is less compacted than that of Earth's Moon [11]. Their relative brightness at large phase angles is an indication of the large-scale roughness of the satellite surfaces. The rougher the surface, the more shadows will be cast at large phase angles, and more shadowing translates as lower apparent brightness. The relative differences in brightness between 10° phase angle and 140° phase angle are about the same for Umbriel, Titania, Oberon, and Earth's Moon, indicating that they have comparable surface roughness. Ariel's relative disk brightness is much less at 140° phase angle and must therefore have a substantially rougher surface. Miranda's disk was too small to provide accurate brightnesses at high phase angle.

12.2.5 Fracture systems and cratering

The surface appearance of satellites is altered both by internal and by external forces. For small satellites the internal processes have negligible effects. The same was thought to be true of intermediate satellites, but Voyager data have shown that the surfaces of most of these satellites are fractured, and many have undergone episodes of partial melting or other forms of resurfacing. In this respect the major Uranian satellites as a group are geologically more active than Saturnian satellites in the same size range.

The 'age' or 'youthfulness' of a satellite surface is a measure of how long ago the main characteristics of the surface were formed. Any surface that has undergone noticeable changes on a global scale in the last 500 000 000 to 1 000 000 000 years is termed 'young' by geologists. Satellites which appear to have remained relatively unchanged for more than 3 000 000 000 years are said to have 'old' surfaces. How does one measure the age of a surface? Long ago astronomers noticed that the lunar maria (the darker areas of the Moon once thought to be seas) had almost no large craters and fewer total craters than the lunar highlands. It was logical to assume that the entire surface had been uniformly bombarded by debris from space, that the rate of bombardment and average size of the projectiles had decreased with time, and that the maria had been flooded with lava following the period of heaviest bombardment. The total number of craters larger than 1 km in diameter in each area of 1 000 000 km^2 could then be used as a measure of the surface age. A more commonly used variation of this method involves counting all the craters per 1 000 000 km^2 larger than 500 km, 200 km, 100 km, 50 km, 20 km, 10 km, 5 km, 2 km, and 1 km. The advantage of this latter method is that younger terrains have almost no craters with diameters of 50 km or larger, so both the size distribution and the total numbers of craters can contribute to an age estimate.

The basic calibration for the age measurements was provided by the lunar surface samples returned by the Apollo Moon landings. They showed that in one of the most heavily cratered areas of the Moon there were 40 000 to 50 000 craters (larger than 1 km) per 1 000 000 km^2, including many larger than 50 km, and that rock ages were about 3 900 000 000 years. One of the more sparsely cratered maria had about 6000 craters per 1 000 000 km^2 and an age of about 3 300 000 000 years. The rim of Copernicus Crater had the youngest age (about 900 000 000 years) and about 1000 craters per 1 000 000 km^2. These data imply that the first bombardment ended about 3 300 000 000 years ago when large pieces of debris circling the Sun were mostly depleted. Subsequent bombardment is believed to have been restricted to relatively small bodies in orbit around the same planet. This later bombardment may have continued at a low level until the very recent past, perhaps less than 100 000 000 years.

Fresher (younger) craters are deeper than older craters of the same diameter, so the depth-to-diameter ratio of craters can provide additional information on the time of their formation. In comparisons between bodies, one needs to account for differences in the gravitational forces and the surface material in which the crater was formed. The stronger gravitational force on a larger satellite will break down crater walls more quickly, and rocky surface materials can support greater altitude variations than icy surface materials.

Tensional fractures are caused when the surface cools and contracts faster than the interior or when motions within the interior cause a stretching of the crustal materials. The surface reacts by splitting apart. Compressional features (including fractures) occur when the interior contracts faster than the overlying material or when motions within the interior push parts of the surface together. The surface reacts by 'wrinkling' or by breaking and forcing one part of the surface over another. Vertical slip fractures are caused by forces which cause an uplift or a downward slumping of one part of the surface relative to an adjacent part. Two vertical slip fractures running parallel to each other can often result in canyons with vertical sides and flat bottoms. Horizontal slip fractures result in a horizontal shift of one part of the surface relative to another.

All of the major satellites of Uranus display tensional fractures. Only a few ridges on Titania appear to have been formed by compressional forces; none of the other Uranian satellites show such features. Vertical and horizontal slip faults can be seen on all five major satellites, but are most pronounced on Titania, Ariel, and parts of Miranda.

Regions of each of the major satellites are depleted in large craters, indicating that resurfacing has taken place. Other 'young' features found on individual satellites will be discussed later in this chapter.

12.2.6   Origin and composition

Because they have been facing the Sun for many years the southern hemispheres of the major Uranian satellites are believed to reach maximum temperatures near 85 K. This is nearly 190°C below the freezing point of water, far too cold for water ice to flow. Yet Miranda has surface features which look as if water ice had flowed upward from the interior, and Ariel has what appear to be glacial flows across its surface. Should one view these features as fossilized remnants of a warmer past, evidence for more recent localized heating, or proof that the water ice is mixed with impurities that lower its freezing point and increase its mobility?

The answer may not be the same for all five satellites. It is also highly dependent on which hypothesis about the geologic history of the Uranian system most closely represents the truth. Voyager did not provide sufficient information to map the entire past of the Uranian system, but some tantalizing clues are now available.

The glacier-like flows of ice on Ariel may be the result of mixtures of small amounts of methane or ammonia (or both) with water ice known to be present on the surface. The quantities of methane and ammonia must be relatively small, because they have not been detected in spectroscopic studies from Earth. Also, methane and ammonia are less dense than water, and substantial quantities of each would be inconsistent with the relatively high densities listed in Table 12.4. Miranda's density is the most poorly determined, and its uncertainty includes the densities listed for the other satellites (near $1.6\,g/cm^3$). This is somewhat higher than the densities of the intermediate satellites of Saturn, but comparable to the densities of Saturn's Titan and Jupiter's Ganymede and Callisto.

If one assumes that the major chemical constituents of the Uranian satellites are water ice with a density of $1.0\,g/cm^3$ and rocky matter with a density of about $3.4\,g/cm^3$, their densities imply that rocky matter contributes 50% to 60% of the mass of each [12]. If Miranda's density is actually $1.25\,g/cm^3$, its rock fraction would be only 33%. These relative abundances of water and rock may not be consistent with those expected for the solar nebula. A more nearly consistent model could imply a mixture (by mass) of 46% water ice, 30% rock, 11% graphite (elemental carbon), 7% methane ice (chemically bound to the water ice), and 6% ammonia ice [6]. The graphite (C) has a somewhat larger density than methane (CH_4), and may also be the source of the low reflectivity and gray color of the small satellites and rings.

Four hypotheses have been suggested for the origin of the Uranian satellites [13]. The first is the accretion disk model: gases and solids from the solar nebula come under the gravitational influence of Uranus and form a disk around the planet from which the satellites coalesce. The second is the spin-out disk model: as the planet contracted to its present size it shed rings of material from which the satellites formed. The third is the blow-out disk model: an Earth-sized body struck the planet, causing the high tilt of the Uranus's rotation axis and ejecting material from Uranus which formed the pre-satellite disk. The fourth is the co-accretion model: solid particles which are a part of the solar nebula are captured into orbit around Uranus to form a pre-satellite disk of particles. Voyager data slightly favor the spin-out model, but none of the others are prohibited by Voyager findings.

12.3   OBERON

The outermost of the Uranian satellites is Oberon. It orbits the planet at a distance of just over 583 400 km and its diameter of 1523 km makes it only slightly smaller than Titania. It circles the planet in 13.46 days. Voyager's closest approach to Oberon was at a distance of 470 600 km, but the closest imaging was from a distance of 660 000 km about 9.2 h before Uranus closest approach (C/A). The smallest features discernible in Voyager images of Oberon are 12.2 km in size, poorest resolution among the major uranian satellites. Surface imaging (for other than spacecraft navigation purposes) began 13.8 days before C/A at a range of 17 600 000 km and a resolution of 326 km. (Imaging resolution is generally assumed to be the linear distance spanned by two picture elements, or 'pixels', of the high-resolution camera. The corresponding angular resolution is about 18.5 microradian = 3.8″ = 0.001°.) A total of 36 images of Oberon with resolutions of 200 km or better were obtained in 12 observations. Six of the observations included multispectral imaging. Solar phase angles extended from 16.0° to 39.1° and from 138.5° to 148.1° phase.

12.3.1   Voyager 2 images
On 20 January, 1986, from ranges of 5 000 000 to 6 100 000 km, each of the five satellites were imaged at resolutions near 100 km (Fig. 12.6). Oberon's range was

Fig. 12.6 — The first resolved-disk images of the five major satellites of Uranus was shuttered on January 20, 1986, from ranges of 5 000 000 to 6 100 000 km. Resolutions are all near 100 km. (P-29464)

5 580 000 km. Brighter irregular areas are seen in the sunlit southern hemispheres of Ariel, Titania, and Oberon. Umbriel seems uniformly dark and Miranda is still too small to show much detail. The range of Oberon had dropped to 2 770 000 km by 22 January (Fig. 12.7), and edges of the bright areas are more sharply defined. Dark

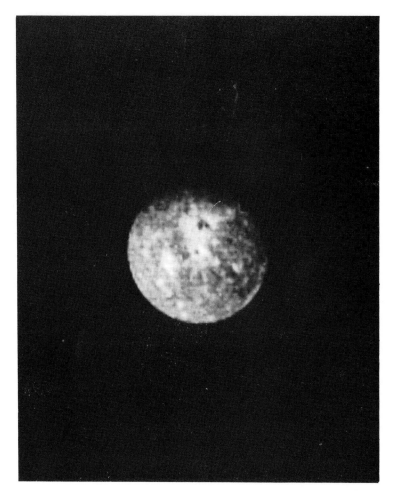

Fig. 12.7 — This image of Oberon was shuttered from a distance of 2 770 000 km on January 22, 1986. The bright areas are more sharply defined than in Fig. 12.6. Dark centers and hints of radial extensions of the bright areas are also seen. (P-29480)

centers and hints of radial extensions of the bright areas are also seen. Somewhat higher solar phase angle in the closest image of Oberon (Fig. 12.8) enhances vertical detail in those regions near the terminator (the dividing line between illuminated and unilluminated surface). Several craters are seen to have darkened floors. The linear rays at the left are very easily discerned and appear to be radiating from a common source. A mountain at the edge of the disk extends 11 km above the surrounding terrain at the lower left of the image. Fig. 12.9 is a preliminary map of Oberon produced by the United States Geological Survey (USGS) in Flagstaff, Arizona. The map is centered on the south pole of Oberon. Sizes near the equator are exaggerated to retain the appearance objects (like craters) would have if they were viewed vertically.

Fig. 12.8 — This highest resolution image of Oberon was shuttered at a distance of 660 000 km and shows features as small as 12 km. Oblique lighting enhances vertical detail in regions near the terminator. Several craters are seen to have darkened floors. The linear rays near the bottom appear to be radiating from a common source. A mountain at the lower right edge of the disk extends 11 km above the surrounding terrain. (P-29501)

12.3.2   Geologic interpretation

Most of the surface of Oberon appears to be relatively ancient, heavily cratered terrain [14]. The brightest surface areas appear to be ejecta thrown out of two craters at the time they were formed. The smaller of these craters (nearer the south pole) has been given the name Othello; the larger is Hamlet. Hamlet has two very dark spots on its floor on either side of a bright central mountain. These may be geologically more recent volcanic flows from cracks in the crater floor. Other dark areas, especially those in flat-bottomed craters, may be older volcanic flows. These volcanic materials may be mixtures of ice and carbonaceous rock, where most of the flow is caused by the melting of the icy component. Such mixtures are called cryovolcanic.

Several linear cracks may cross the region of Hamlet from upper left to lower right. A large chasm at the upper left of Fig. 12.9 opposes this general trend. These features are probably tensional fractures.

There are a large number of craters with diameters in excess of 50 km, an indication that they were formed in the early bombardment period from debris in orbit around the Sun. Most of the larger craters have flat bottoms and central peaks. A few small bowl-shaped craters without central peaks may have been caused by

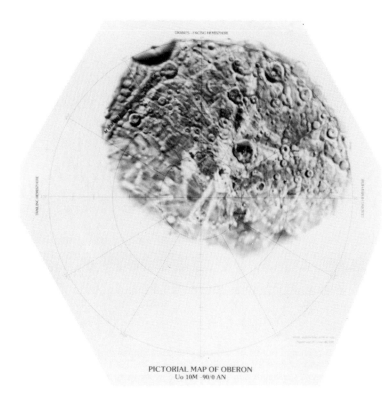

PICTORIAL MAP OF OBERON
Uo 10M -90/0 AN

Fig 12.9 — This map of Oberon, produced by the United States Geological Survey (USGS), is centered on the south pole of Oberon. Sizes near the equator are exaggerated to retain the appearance objects (like craters) would have if they were viewed vertically.

impactors with the lower velocities typical of bodies originally in orbit around Uranus. There seems little evidence that major portions of the surface of Oberon were coated with volcanic flows. Except for the bowl-shaped craters and volcanic activity on the floors of a few ancient craters it is probable that Oberon's surface has changed very little over the last 3 000 000 000 years.

The large peak seen at the limb in Fig. 12.8 may be the remaining central peak of a very large crater. Fig. 12.10 shows an exaggerated profile of altitudes along the illuminated portion of the limb in the Fig. 12.8 image of Oberon. The sharp peak labeled 'B' is the 11-km mountain. Crater walls are not immediately apparent in this view, although a relatively flat region about 500 km across and approximately centered on the peak may be the remnants of an ancient crater floor.

12.4 TITANIA

The largest of the uranian satellites is Titania (diameter 1578 km) . It has an orbit semimajor axis of just under 436 300 km and an orbital period of 8 days 17 h. Voyager's closest approach to Titania was at a distance of 365 200 km; the best

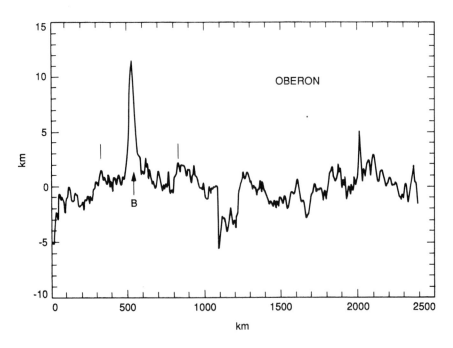

Fig. 12.10 — An exaggerated profile of altitudes along the illuminated portion of the limb of Oberon. The sharp peak labeled 'B' is the 11-km mountain seen in Fig. 12.8. A relatively flat region about 500 km across and approximately centered on the peak may be the remnants of an ancient crater floor.

images were taken only slightly farther away (367 000 km) about 3.7 h before Uranus C/A. Features as small as 6.8 km could be discerned in the images, providing a major improvement over the best Oberon image. As discussed earlier, Titania observations covered the largest range of solar phase angles (0.8° to 152.6°). Surface imaging of Titania with resolutions better than 200 km included 25 observations of one to six images each, for a total of 68 images. The best multispectral coverage was obtained 8.8 h before C/A from a range of 498 000 km (resolution 9.2 km).

12.4.1 Voyager 2 images
The image of Titania included in Fig. 12.6 has a resolution of 104 km; the range was 5 610 000 km. Titania's range had dropped to 3 1100 00 km by 22 January (Fig. 12.11), but phase angles are low and albedo differences are more prominent than vertical structure on the surface. By 8.8 h before C/A it was apparent that Titania was much more heavily faulted than Oberon and that large portions of the surface were devoid of craters larger than 50 km (Fig 12.12). Much of Titania must have been resurfaced after the heliocentric bombardment phase was over. Just 3.7 h before C/A Voyager 2 obtained its highest-resolution image of Titania (Fig. 12.13). The view emphasizes the enormous fractures and the lower average crater diameter. Two very large multi-ringed craters are seen near the top of the image near the terminator. Three separate concentric rings define Gertrude, the crater at the left. Ursula, on the right, appears

Fig. 12.11 — Albedo differences are easily seen on Titania's surface, seen here from a range of
3 110 000 km on January 22, 1986. Because of the very low phase angle, little vertical structure is
discernible in the image. (P-29496)

to have two concentric structures. The areas between the central craters and the
outer rings have moderate (for Gertrude) to very low (for Ursula) numbers of
craters. Material ejected from the craters may have covered the surrounding terrain
to depths that obliterated most of the pre-existing craters.

Multi-ring craters are caused by a phenomenon similar to that of radiating ripples
when a rock is thrown into a pond. The outermost 'ripple' caused by the event that
formed Gertrude may be responsible for the ridge which almost perfectly bisects a
moderately large crater (Calphurnia) at the left. Alternatively, the ridge may be an
extension of the large fracture system at the top of the image.

Fig. 12.14 is a preliminary map of Titania. The maps of the five major satellites
were drawn on the basis of Voyager images by J. L. Inge of USGS.

12.4.2 Geologic interpretation

Extensive resurfacing of Titania's surface must have started before the heliocentric
bombardment period was completed 3.3×10^9 years ago. A few large craters remain
in what otherwise appears to be terrain cratered by the smaller objects once in orbit
around Uranus. Most of the large craters disappeared either because they were filled
with cryovolcanic flows or because subsurface melting permitted the crater walls to
collapse to the average level of the surrounding terrain.

Large fractures ('rift canyons') across the surface could have been formed as a
result of the extensive cryovolcanic activity. Assuming that the main fluid in the
volcanic matter was water, freezing would have occurred first at the surface of

Fig. 12.12 — Few craters larger than 50 km are seen on Titania's heavily faulted surface. This image was shuttered from 498 000 km and shows features as small as 9 km. (P-29509)

Titania. Water is nearly unique in having the quality that as it freezes it expands. Freezing of Titania's interior would have resulted in the same kind of expansion, and the already frozen crust might have had insufficient strength to withstand the pressure. Widespread tensional fracturing would have resulted. Because many of these rifts have been marred by later cratering events, the resurfacing and fracturing described must have occurred more than 3 000 000 000 years ago. Although the surface of Titania is younger than that of Oberon, it is nevertheless relatively old.

Titania has fewer bright crater rays and fewer dark crater floors than Oberon. Although its albedo contrast is therefore less than Oberon's, its slightly more youthful surface is brighter by several percent (see Table 12.4).

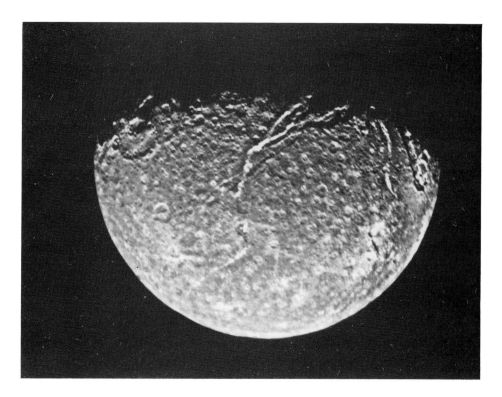

Fig. 12.13 — Voyager 2's highest-resolution image of Titania emphasizes the enormous fractures and a relatively low average crater diameter. Features as small as 6.8 km are resolved. Two large multi-ringed craters are seen near the terminator. Three separate concentric rings define Gertrude, the crater at the left. Ursula, on the right, appears to have two concentric structures. (P-29522)

12.5   UMBRIEL

Umbriel, with a reflectivity of 21% at visible wavelengths, is the darkest of Uranus's major satellites. It has a diameter of 1169 km and orbits the planet once each 4.14 days at a distance of nearly 266 000 km. Umbriel is the only major satellite to have been on the opposite side of Uranus at the time of Voyager's C/A. The minimum separation of 325 000 km occurred 2.9 h after C/A while the spacecraft was passing through Uranus's shadow. The highest resolution images were obtained at a less busy time 6.2 h before C/A. Range to the satellite at that time was 557 000 km; camera resolution was 10.3 km. The best color imaging was obtained 17.6 h before C/A from 1 040 000 km. Phase angles for Umbriel imaging ranged from 10° to 57° and from 142° to 148°. Imaging coverage with resolutions of 200 km or better included 12 observations of one to six images each, for a total of 36 images.

12.5.1   Voyager 2 images
Umbriel appears as a very dark object in Fig. 12.6. The range of the spacecraft at the time of the image was about 6 080 000 km, corresponding to a resolution of 113 km. Very little detail can be seen in the image, a possible consequence of low contrast

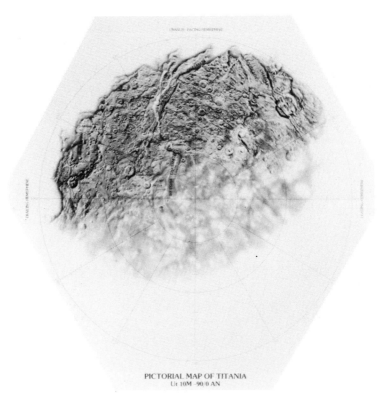

PICTORIAL MAP OF TITANIA
Ut 10M -90/0 AN

Fig. 12.14 — This USGS-produced map of Titania shows surface features and their accepted nomenclature. Titania's south pole is near the center of the map. Note the increasing scale for features nearer Titania's equator.

across the disk. The most prominent feature in Fig. 12.15 is a relatively bright ring near the left edge of the image, but the closer range (1 050 000 km; resolution 19 km) now begins to reveal large craters distributed across the surface. Multispectral coverage was obtained during this observation. The highest resolution image is shown in Fig. 12.16, and Fig. 12.17 is the corresponding surface map. The bright region at the left in each of these images is seen to be deposits on the floor of the crater Wunda. A dark, relatively linear feature connects the central peak of Wunda with the southern rim of the crater. The central peak in crater Vuver (near 310° longitude) is also relatively bright, as are some cliffs ('scarps') near 70° longitude. In addition to the scarps there are many other linear fractures which generally run from upper left to lower right in the images. The distribution of crater sizes is very similar to that of Oberon.

12.5.2 Geologic interpretation
The dark surface of Umbriel would pose less of a problem for planetologists if the other major satellites of Uranus were similarly dark, but it is difficult to understand how natural processes could leave Umbriel's surface dark without leaving the

Fig. 12.15 — The most prominent feature on Umbriel's surface is a relatively bright ring near
the left edge of this image. The range and resolution are 1 050 000 km and 19 km. A population
of large craters characterizes the surface. (P-29502)

surfaces of the surrounding satellites similarly dark. Several hypotheses have been
put forward, but none is without drawbacks.

One suggested possibility is that Umbriel did not undergo sufficient heating in its
history to melt and differentiate its materials, and that the dark surface is simply the
original undifferentiated material out of which the satellites formed. It then becomes
difficult to think of circumstances which would create sufficient heating to cause
differentiation on Oberon, Titania, and Ariel without doing the same at Umbriel.
Furthermore, sufficient heating must have occurred during the formation of Wunda
to differentiate the materials on its floor. Why did that not happen during creation of
other craters of similar or larger sizes?

Perhaps Umbriel's surface was coated by material from a dark, tidally disrupted
body whose orbit was close to that of Umbriel. If that were correct, where did the
dark object originate? If it were a part of the original disk of material out of which the
uranian satellites formed and was circling near Umbriel's orbit, it should have
become a part of Umbriel prior to differentiation. If it was captured into orbit at
some later time how did that happen, and what process broke it into fragments too
small to have created additional craters larger than 10 km? Why is evidence of its
presence seen only on Umbriel and not on either Titania or Ariel? The uniformity of
the Umbriel coating further implies that to satisfy the requirements of this hypothesis

Fig. 12.16 — The highest resolution coverage of Umbriel resolves features as small as 10.3 km. The bright region at the left is seen to be deposits on the floor of Wunda Crater. A dark, relatively linear feature connects the central peak of Wunda with its southern rim. To the upper right of Wunda, the central peak in crater Vuver is also relatively bright, as are some scarps near the right edge of the disk. The distribution of crater sizes is very similar to that of Oberon.
(P-29521)

the material may have even formed a ring of dust-sized particles that were slowly swept up by Umbriel.

A third hypothesis suggests that the material may be dark ejecta thrown out during formation of a large crater somewhere on Umbriel. Similar dark material is seen on Jupiter's Ganymede, but only in localized areas of the surface. If Umbriel had an atmosphere to suspend the dust lifted in such a cratering event, the coating might be uniform, but Umbriel is small and airless. Alternatively, the ejecta could have been blown into orbit around Uranus and then slowly swept out of orbit by Umbriel. In either scenario, a uniform coating of dark material from a major cratering event seems very unlikely. Furthermore, the formation of more recent

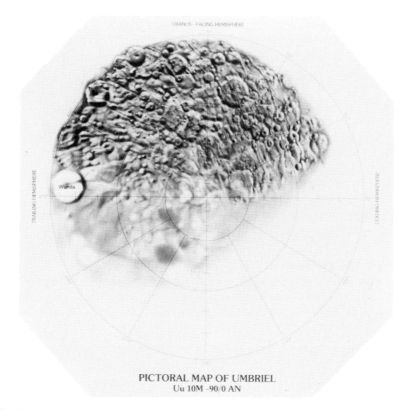

PICTORAL MAP OF UMBRIEL
Uu 10M -90/0 AN

Fig. 12.17 — The USGS-produced map of a portion of Umbriel's southern hemisphere provides
accepted nomenclature for the surface features. Extensive fault structures run generally from
upper left to lower right.

craters would have penetrated the dark coating unless it were tens of kilometers in
thickness, and ejecta deposits from a single crater could not have created so thick a
layer.

The fourth and favored hypothesis attributes the coating to extensive cryovolcanic activity which resurfaced all of Umbriel's surface to a depth of at least 10 km. The
low albedo might be a result of extensive subsurface melting that involved primarily
the near-surface materials of Umbriel, so that only partial differentiation occurred.
The cryovolcanic lavas deposited on the surface may therefore have contained far
larger quantities of carbonaceous materials than similar resurfacing events on
Titania and Ariel.

The large tensional and vertical thrust fractures on Umbriel probably have an
origin similar to those of Titania: freezing of the surface materials followed by
freezing and expansion of the subsurface materials. The geologically more recent
events include eruption of very bright materials onto the floor of Wunda and the
formation of the dark lane from the center to the southern edge of Wunda. The dark
lane itself may be the result of slumping of material into a surface fissure formed
either before or after flooding of the crater floor.

12.6   ARIEL

Ariel's reflectivity, in sharp contrast to Umbriel's dark surface, is the highest among Uranian satellites. With a reflectivity of 39% at visible wavelengths, Ariel is nearly twice as bright as its close twin in size (Ariel's diameter is 1158 km, only slightly smaller than Umbriel's 1169 km). Ariel circles Uranus in just over 2.5 days nearly 191 000 km from Uranus. Like Earth's Moon and most of the other satellites in the Solar System, Ariel and the other major Uranian satellites keep their same face toward the planet as they orbit. Voyager 2's closest approach to Ariel (127 000 km) occurred 1.6 h before C/A. A resolution of 2.4 km, corresponding to a range of 127 000 km, was obtained in a four-frame narrow angle imaging mosaic of Ariel 1.8 h before C/A. The best color coverage was from a range of 164 000 km (resolution 3.0 km) 3.3 h before C/A. Phase angle coverage of Ariel extended from 12° to 69° and from 145° to 146°. Imaging coverage at resolutions of 200 km or better included 20 observations of one to six images each for a total of 54 images.

12.6.1   Voyager 2 images

Ariel appears to have several bright spots in Fig. 12.6. The image was obtained from a distance of 5 040 000 km and has a resolution of 93 km. Ariel's range was smaller by a factor of two in Fig. 12.18, which has a resolution of 47 km. The bright patches seen

Fig. 12.18 — Ariel is seen at a resolution of 47 km from a distance of 2 520 000 km. Bright patches of ejecta surround several craters. Several quasi-linear features are apparent in the image. (P-29479)

in the earlier image now appear to be associated with material thrown out of relatively recent craters. Several long linear features are apparent in the image.

The best multicolor image (a noncolor version of which is presented in Fig. 12.19)

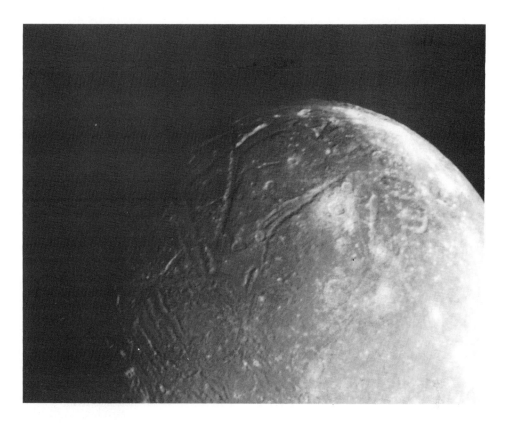

Fig. 12.19 — Only a part of Ariel's disk was captured in this image taken by Voyager 2 from a distance of 164 000 km. The resolution was 3.0 km. The terrain in this image is a relatively complex mix of craters, fractures, and valleys bounded by parallel fractures. Some of the larger valleys at the upper left appear to have been filled with material. The brightest areas at the right and upper right are associated with small fresh craters. (P-29523)

covered only a portion of the illuminated disk. The range of the spacecraft was 164 000 km, and the resolution was 3.0 km. The terrain in this image is a relatively complex mix of craters, fractures, and valleys bounded on each side by fractures. Some of the larger valleys at the upper left are filled with material which has very few craters, indicating an intermediate age. The brightest areas at the right and upper right are associated with small fresh craters.

A mosaic of four images (Fig. 12.20) was needed to provide coverage at the best resolution of about 2.4 km. The highly fractured nature of Ariel's surface is emphasized by the oblique solar illumination. The corresponding surface map is reproduced as Fig. 12.21. Few craters larger than 10 km are seen. Yangoor is one of

Fig 12.20 — Four images were used to produce this view of Ariel, which has a resolution of about 2.4 km. The highly fractured nature of Ariel's surface is emphasized by the oblique solar illumination. (P-29520)

the larger craters near the south pole (about 280° longitude). Its diameter is about 100 km and its northern face has been breached and part of its floor flooded with later cryovolcanic flows. A series of south-facing cliffs near 30° south latitude extends from 200° to 260° longitude. Evidence for extensive volcanic flows of various ages (i.e., differing crater counts) exists over most of the visible surface, especially in the general direction of Uranus (within 60° of 0° longitude).

The images taken during observations 3.3 h and 1.8 h before C/A are both high resolution and differ enough in viewing angle to be used for construction of stereoscopic pairs of images. Comparison of such images permits determination of the relative altitudes of various features on the surface. For example, the triangular uplifted area at 0° longitude and 15° south latitude stands out dramatically above the surrounding terrain.

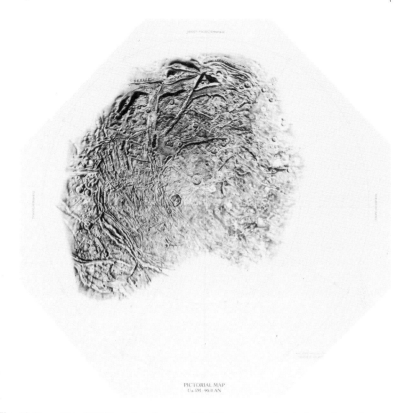

Fig. 12.21 — This USGS-produced map of Ariel's southern hemisphere shows few craters larger than 10 km. The northern face of 100-km Yangoor (near the south pole) has been breached, and part of the crater floor is flooded with later cryovolcanic flows. A series of south-facing scarps extends from 200° to 260° longitude at 30° south latitude. Evidence for extensive volcanic flows of various ages is also seen, especially in the Uranus-facing surface areas near 0° longitude.

12.6.2 Geologic interpretation

Ariel and Miranda are geologically the two most interesting satellites. It is fortunate that Voyager 2 images of these satellites were the highest resolution images obtained in the Uranus system. These two satellites have received the most attention and are the subjects of the most widespread speculation on their origins and past history. As with the other major satellites, the geological interpretation of Ariel presented here follows the discussions of Croft, Soderblom, and Shoemaker [14].

The absence of an abundance of large craters argues for extensive resurfacing of Ariel near the end of the period of bombardment by Sun-orbiting debris. Ariel's surface has fewer craters than Titania, which is also nearly devoid of large craters. Some areas, notably those at mid-southerly latitudes near 0° longitude, have crater counts comparable to those seen on Titania and must date back very early in Ariel's history. Most of Ariel's surface has been melted or covered with cryovolcanic materials in more recent times, and some areas have been resurfaced recently enough that craters of any size are nearly absent. From these observations it is evident that resurfacing events have occurred at many discrete times over the history of Ariel, and may still be occurring. It seems plausible, however, that most of the large craters were obliterated in one early resurfacing event, either from ejecta from one large cratering event or (more likely) from global volcanic eruptions, perhaps associated with early differentiation of Ariel.

The appearance of the surface of Ariel provides ample evidence that much of the near-surface material was fluidized many times during Ariel's lifetime. The source of the heat necessary to cause such melting is not well understood. It is unlikely that radioactive materials in the interior were present in large enough quantities to melt the whole interior. A more plausible explanation invokes tidal forces of the type which cause volcanic activity on Jupiter's Io. Because of Io's elliptical orbit, the difference between the pull of Jupiter's gravity on the planet-facing and opposite sides of Io is greater near orbit periapsis (closest point) than at apoapsis (farthest point). The resultant flexing of Io heats its surface and subsurface materials sufficiently to cause melting and volcanic activity.

Jupiter's tidal forces on Io also tend to circularize the orbit of this remarkable satellite. Were it not for a close orbital resonance between Io and Europa (Europa circles Jupiter in 3.55 days, almost precisely twice the 1.77-day orbital period of Io) Io's orbit would have become circular, and tidal flexing of its surface and the associated heating would have ceased. Europa's gravitational influence keeps Io's orbit non-circular. Io exerts a similar force on Europa, but because of its greater distance from the planet, Europa is less affected by Jovian tidal forces than is Io. Moreover, Europa also has a 2:1 orbital resonance with more distant and more massive Ganymede that counteracts Io's influence.

Ariel's orbital eccentricity is comparable to that of Io, but Uranus's gravitational pull is much smaller than Jupiter's, and present tidal forces from Uranus are not sufficient to have caused the melting of Ariel's surface. With respective orbital periods of 1.41, 2.52, and 4.14 days, Miranda, Ariel, and Umbriel have no orbital resonances that could serve to induce larger departures from orbit circularity. Theoretical extrapolations [15] into the past histories of these satellites seem to show that they might have passed through orbital resonances in the geologically recent past ($1-2 \times 10^9$ years?). Tides on the rapidly spinning planet would be raised by the gravitational influence of the satellites. These tidal bulges then exert a small force on the satellites which increases their energy, causing them to slowly move to more distant orbits. The opposing force from the satellites exerts a braking influence on the planet, so there is an effective exchange of energy from the planet to the satellites. When the three satellites were somewhat closer to the planet, they may have had orbital periods which were related by small whole-number ratios. The most recent such condition probably occurred when Miranda's period was one-third that of Umbriel. Similar but less recent conditions would have existed for Ariel. Large increases in the orbital eccentricities of Ariel and Miranda during such periods may have caused extensive melting of their surface and subsurface ices. The present deviations from orbit circularity may be a remnant of that earlier era of more extreme eccentricities.

Following the period of partial melting and differentiation, the freezing and expansion of the interior severely disrupted Ariel's surface, creating widespread fracturing. In some places this fracturing is represented by vertical uplifts, but tensional faulting is more widespread. Parallel fractures spread apart as the expansion continued, and intervening material would sink into the vacated spaces, forming a graben. Subsequent cracks in their floors may have permitted still partially fluid material to rise to the surface and flood the graben. In some cases, the fluid cryovolcanic materials overfilled the lower areas and flooded the surrounding terrain (see, for example, the smooth area near 45° south latitude and 330° longitude in Fig. 12.21). Most of the younger volcanic deposits seem intimately related to the many faults.

Most of the grabens seem to occur at nonpolar latitudes. The oldest and widest are in the Uranus-facing hemisphere, as are their associated cryovolcanic lava flows. More recent and pristine faults and cliffs (scarps) are in the far-side hemisphere. They appear to have less cryovolcanic activity associated with them.

The bright materials within and surrounding several craters are difficult to understand. Much of Ariel's surface is covered with crater ejecta or deep cryovolcanic deposits, and yet the bright craters seemed to have unearthed material from relatively shallow depths with substantially higher reflectivity. Also, some of the bright deposits seem to have sharply defined boundaries which are very unusual for a body with such low gravity. A viable geologic explanation for these features has not yet been suggested.

12.7 MIRANDA

The geometry of the Voyager 2 trajectory necessary for continuation to Neptune required passage through the equatorial plane of Uranus at a distance relatively close to Miranda's orbit. To take advantage of this, the arrival time was adjusted to permit Voyager to pass within 29 000 km of Miranda. The closest imaging was taken from distances of between 42 200 km and 30 200 km about 1.2 h before C/A; the corresponding resolution was 0.8 to 0.6 km, best of any of the Uranian satellites. A mosaic of eight narrow-angle frames was needed to cover the visible disk. The best color coverage was obtained 1.7 h earlier at 2.9 h before Uranus C/A. Range and resolution were 144 000 km and 2.7 km, respectively.

Miranda, with an albedo of 32%, is the second most reflective of the Uranian satellites. It circles the planet in 33.9 h. During each orbit the distance varies from 126 340 km to 133 350 km; the 7000 km variation is the largest among the satellites of Uranus. Miranda's shape is best fit with a triaxial ellipsoid whose longest dimension (480 km) is pointed toward Uranus. The equatorial diameter along the direction of its orbit is 468 km, and the diameter from pole to pole is estimated to be 466 km (only the southern hemisphere was imaged). Because the gravity of Miranda is so much lower than that of the other major Uranian satellites, it also has larger topographic altitude variations from the triaxial ellipsoidal shape given here. Although an ellipsoid defines the average shape of Miranda, altitude variations of up to ten km from the reference ellipsoid are seen in the images.

Phase angle coverage of Miranda extended from 13° to 21° and from 145° to 150° in images which included the entire disk. The best color imaging was taken at a phase angle near 16°. During collection of images used to assemble the high-resolution mosaic the phase angle varied from 18° to 43°. Imaging coverage at resolutions of

200 km or better included 23 observations of one to nine images each, for a total of 47 images. Twelve of the images had resolutions better than 3 km per imaging line pair. These have been used to good advantage in geologic studies of Miranda's surface.

12.7.1   Voyager 2 images

Optical navigation frames included Miranda imaging from much greater distances, but surface studies were primarily directed toward images obtained during two complete orbital periods beginning about three days before C/A. The smallest details visible at that time were about 72 km in size. The Miranda image in Fig. 12.6 was taken a day earlier and had a resolution of about 99 km. Even at this poor resolution there is a hint of bright albedo features on the disk.

The image shown in Fig. 12.22 was acquired from a distance of 1 370 000 km one

Fig. 12.22 — This Miranda image was acquired from a distance of 1 370 000 km one day before Uranus C/A. Resolution is about 25 km. A bright chevron-shaped feature is seen imbedded in a darker region. The limb region at the upper right is also relatively bright. Hints of linear features other than the chevron may also be seen. (P-29505)

day before C/A. Resolution is about 25 km. A bright chevron-shaped feature is seen imbedded in a much darker region. The Miranda limb region at the upper right is also relatively bright. The first hints of linear features other than the chevron may also be seen. Miranda's south pole is near the center of the disk in this view.

Nearly the full disk of Miranda is seen in Fig. 12.23, which was taken from a

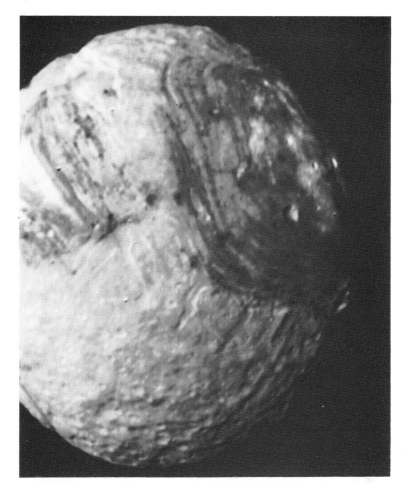

Fig. 12.23 — This image of Miranda was shuttered 2.9 h before Uranus C/A has a resolution of 2.7 km. Arden Corona at the upper right has a wide pattern of concentric dark and light stripes with a brighter jumbled region near its center. Inverness Corona near the left margin of the image contains the bright chevron seen in Fig. 12.22. Most of the rest of the visible surface is heavily cratered and shows substantial variations in altitude. Encroaching into the cratered terrain are several fractures and broad canyons, most of which appear to be generically related to the coronae. (P-29510)

distance of 147 000 km. The image was shuttered 2.9 h before C/A and has a resolution of 2.7 km. The region at the upper right has a 400-km-wide pattern of concentric dark and light stripes with a brighter jumbled region near its center. This feature, which has come to be known as Arden Corona, is unlike anything seen on other satellites in the Solar System. The region at the left margin of the image is part of the same bright chevron seen in Fig. 12.22. The dark region surrounding the chevron has a nearly rectangular shape and also has several parallel linear features; the area is known as Inverness Corona. Most of the rest of the visible surface is heavily cratered and shows substantial variations in altitude. Encroaching into the cratered terrain are several fractures and broad canyons, most of which seem generically related either to Arden Corona or to Inverness Corona.

A series of eight high-resolution images was obtained near closest approach to Miranda (between 1.4 h and 1.1 h before Uranus C/A). One of them (Fig. 12.24)

Fig. 12.24 — One of eight high-resolution images obtained near closest approach to Miranda displays the central portion of Fig. 12.23 at much higher resolution. Arden Corona (at the upper right) and Inverness Corona (at the left edge) appear to be bounded by trench-like depressions flanked by parallel scarps. (P-29513)

displays the central portion of Fig. 12.23 at much higher resolution. It includes parts of Arden Corona (at the upper right in the figure) and Inverness Corona (at the left edge) and much older cratered terrain at the bottom and lower right. The coronae appear to be bounded by trench-like depressions which are often flanked by parallel scarps (cliffs).

As seen in Fig. 12.25 the scarps continue beyond the boundary of Inverness Corona. At the shadow boundary they converge into one enormous scarp face which extends into the shadowed regions near Miranda's equator. The heights of the cliffs along Verona Rupes, as this scarp face is known, are from 10 to 20 km! The large impact crater to the right of Verona Rupes is about 25 km in diameter. Inverness

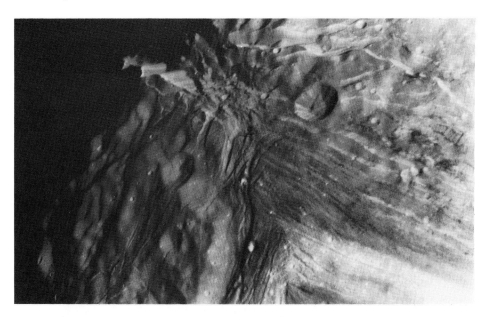

Fig. 12.25 — Scarps flanking Inverness Corona continue beyond its boundary and converge into one enormous scarp face (Verona Rupes) which extends into the shadowed regions near Miranda's equator. Scarp heights range from 10 to 20 km! The large impact crater to the right of Verona Rupes is 25 km in diameter. Inverness Corona appears to lie generally lower than its surroundings. Closely spaced ridges parallel and overlie both legs of the brighter chevron, which has remarkably sharp albedo boundaries. (P-29512)

Corona declines relatively rapidly both at its left and upper boundaries. Gentler slopes up to the surrounding cratered terrain span even greater altitude differences, so that Inverness Corona appears to lie generally lower than its surroundings. Closely spaced ridges parallel and overlie both legs of the brighter chevron, which has remarkably sharp albedo boundaries.

The sharpness of the coronae boundaries is again emphasized in Fig. 12.26. Inverness Corona is at the upper right, and a third of these strange coronae (Elsinore Corona) appears at the lower left. Whereas Inverness Corona, except for the bright chevron, is darker than the surrounding cratered terrain, Elsinore Corona appears to have an albedo very similar to that of the older cratered terrain. Elsinore Corona has the same quasi-rectangular shape displayed by Inverness and Arden, but it has features distinct from its counterparts. There is no apparent broad trench surrounding it, nor does it have large brightness contrasts in its interior. It consists of a wide band of parallel ridges surrounding an older jumbled interior (seen more extensively at the left in Fig. 12.27). At the bottom of Fig. 12.27 the parallel ridges converge and pass through a relatively narrow valley which disappears into the shadow region.

A composite high-resolution image of Miranda is displayed as Fig. 12.28. Small 'gores' (gaps between the images) near the edge of the disk at the top and along the right have been filled by the somewhat lower resolution data of Fig. 12.23. The similarities of and differences between the coronae can be compared more easily,

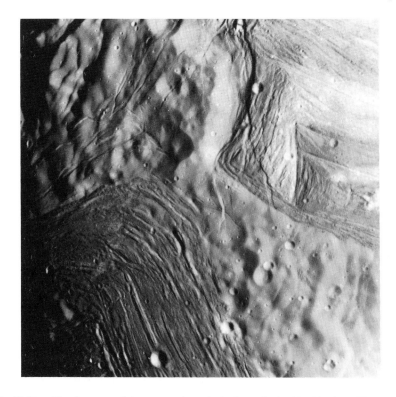

Fig. 12.26 — The sharpness of the coronae boundaries is emphasized in this view of Inverness Corona (upper right) and Elsinore Corona (lower left). Elsinore Corona has the same quasi-rectangular shape displayed by Inverness and Arden, but it has no apparent broad trench surrounding it, no large brightness contrasts in its interior, and consists of a band of parallel ridges surrounding an older jumbled interior. Features as small as 600 m may be seen.
(P-29515)

and the broad expanse of the ancient cratered terrain is seen more fully. The corresponding airbrush map, labelled with latitude and longitude grids and feature names, is shown in Fig. 12.29. The sharp tip of the chevron in Inverness Corona is seen to lie very close to Miranda's south pole.

12.7.2 Geologic interpretation

Imaging scientists presented the Miranda images to the assembled press about 24 h after they were received. Too little time had elapsed for any lengthy discussions among the geologists on the team. Before receipt of the Voyager images, Miranda was fully expected to have a heavily cratered and ancient surface undisturbed by any internal processes, somewhat like Saturn's Mimas (Fig. 12.30). The rich variety of Miranda's surface features was a welcome surprise, but did not lend itself well to rapid geological interpretation. More than three years after the Voyager encounter with Uranus, geologists were still unable to provide a fully acceptable scenario to

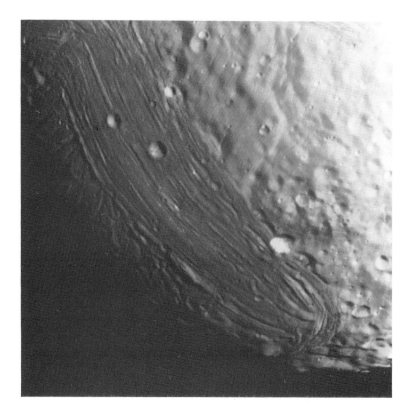

Fig. 12.27 — Miranda's Elsinore Corona, also seen in Fig. 12.26, straddles the terminator in this view. Near the bottom of the image, its parallel ridges converge, pass through a relatively narrow valley, and disappear into the shadow region. (P-29514)

explain Miranda's geologic history. Three of the many suggested hypotheses are discussed below.

Geologists agree that the earliest history of Miranda mimicked that of the other major Uranian satellites. The ancient cratered terrain is the result of heavy bombardment during the period when Miranda was being assembled. Although no pristine craters remain from that early bombardment, many muted crater forms are seen. The softening of old crater rims might have been due to cryovolcanic activity, or it may be the result of ejecta from later cratering events. Whatever the source, it is evident that most of the scars from the early bombardment are covered with a more recently deposited deep layer of surface materials. It is also generally agreed that the large surface fractures and the corona formations postdate the early bombardment and that the fractures and the coronae are generically related.

One hypothesis is attributed to Soderblom [17]. According to this hypothesis Miranda had enough internal heat in its early history to allow partial separation of ice and rock in the interior. Because of its greater density the rocky materials began to sink towards Miranda's center and the icy materials began to rise toward the surface.

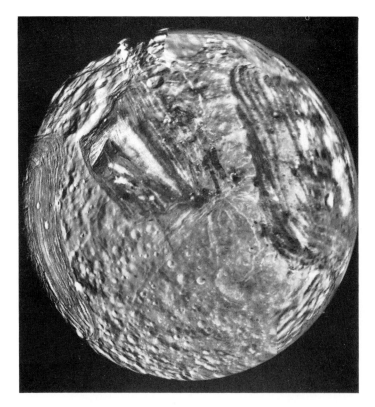

Fig. 12.28 — Computer adjustment of image orientations and scales permitted the eight high-resolution images of Miranda to be composited into a single spectacular image. Small gores near the edge of the disk at the top and along the right have been filled by lower resolution data from the image shown in Fig. 12.23. (260-1790A)

The resultant plumes of fluid icy matter may or may not have penetrated the surface initially. Freezing and expansion of subsurface water ice fractured the surface in approximately rectangular patterns and the rising fluids from the interior reached the surface through those cracks. About this time Miranda's internal heat was exhausted and the coronae are the result of this incomplete differentiation. Details of this hypothesis are incomplete, and it is not at all certain that such a process could create the quasi-rectangular shapes, the concentric structures, or the alternating bright and dark features of the coronae. Another major flaw in this hypothesis is the absence of a reasonable explanation for the apparent differences in structure and in age of the three coronae.

The second hypothesis requires only near-surface heating. The source of the heating may have been tidal forces from Uranus caused by higher orbital eccentricity during a satellite orbit resonance with Umbriel or Ariel. In this scenario Arden was originally a large impact basin. The event effectively laid a deep mantle of ejecta over pre-existing craters and scarps and caused concentric fractures around the basin. Similar concentric fracturing from a major cratering event is seen on the Moon,

Fig. 12.29 — The USGS-produced map shows that Inverness Corona includes the south pole of Miranda and the chevron points in the direction of Uranus (0° longitude and latitude). Since it includes 90° longitude, Arden Corona leads Miranda in its orbit; Elsinore Corona trails. Most of Miranda's surface is heavily cratered and has undergone extensive fracturing.

Mercury, and Jupiter's Callisto. Some impact-related volcanic flooding of the basin occurred. As Miranda continued to cool, tensional fracturing reopened faults mantled by the Arden event. Cryovolcanic magma created by the tidal heating flowed through the fractures, first in the Arden region and then concurrently in the Inverness region. Inverness Corona may have formed in two massive cryovolcanic events. The first had basically the same composition as the dark extrusions of Arden Corona. The second had much less carbonaceous material included in the lava and was responsible for the chevron-shaped area in Inverness Corona. A somewhat later event formed the Elsinore Corona with cryovolcanic materials intermediate in albedo between the two events in Inverness. Apparently this event took place after the cessation of internal expansion, so no additional fractures mar the Elsinore structure. This hypothesis successfully accounts for differences in the coronae, but cannot adequately explain their quasi-rectangular shapes or other similarities.

The third hypothesis presupposes a much more violent history for Miranda. After an early differentiation and cooling of its interior, Miranda was shattered by an

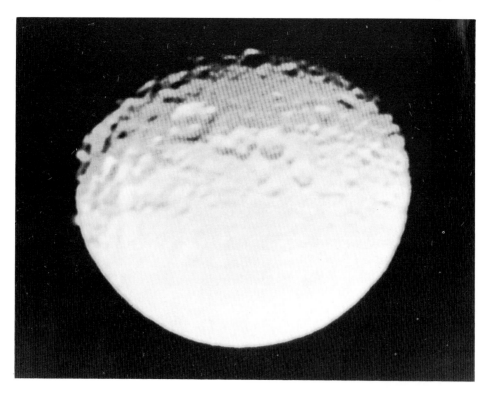

Fig 12.30 — Miranda was expected to have a heavily cratered and ancient surface undisturbed
by any internal processes, perhaps like this image of Saturn's Mimas. (P-23267)

asteroid-sized body. The fragments might have been predominantly rock from the core region or predominantly ice from the overlying regions. As Miranda slowly reassembled itself due to the mutual gravitational forces of its fragments, the order of reassembly was random, with some rock fragments near the surface and some ice fragments near the core. Three late-arriving rock fragments struck Miranda in the locations of the three coronae. The largest fragments were in Arden and Elsinore on opposite sides of the satellite. The unbalanced mass distribution caused Miranda to align its rotation so that Arden (or Elsinore) was pointed toward Uranus. Then, as the subsurface began to melt, the heavier fragments began to sink through the less dense ice, which rose to fill the void. This less dense icy material remained under the surface, and when the heavy 'sinkers' reached the center of Miranda, there was a mass deficit at those spots. This caused Miranda to reorient itself such that Arden was in the leading hemisphere and Elsinore was in the trailing hemisphere (their present orientations). During later melting epochs (satellite orbital resonance periods?) first Arden Corona, then Inverness Corona, then Elsinore Corona were formed as cryovolcanic magma rose through fractures in Miranda's surface. The jumbled center (of Elsinore Corona in particular) would have been the remnants of the ancient impacts. While this hypothesis explains the present locations of the two larger coronae, it does not adequately explain the low crater counts in the interiors of the coronae or the albedo and textural differences of the coronae. There is also no evidence (other than the coronae themselves) for the mass imbalances necessary to reorient Miranda.

In summary, the Miranda coronae and associated fracture systems continue to be a geologic enigma. Perhaps the northern hemisphere will hold some clues which will help to unravel the mystery. But for such data scientists will have to rely on future spacecraft missions to these distant worlds. Collective appetites have been thoroughly whetted; when will they be satiated?

NOTES AND REFERENCES

[1] The special Uranus encounter issue of *Science* is contained in the 4 July, 1986, issue: *Science*, **233**, 39–107.

[2] Owen, W. M. Jr., Synnott, S. P. (1987) Orbits of the ten small satellites of Uranus. *Astronomical Journal*, **93**, 1268–1271.

[3] Orbit semimajor axis, a, can be calculated from the orbital period, P, through the following formula: $a^3 C_1 P^2 [1 + C_2 a^{-2} + C_3 a^{-4}]$, where a is in units of km, $C_1 = 146\,762 \pm 1\,\mathrm{km}^3$ for P expressed in seconds, $C_2 = [3.4426 \pm 0.0003] \times 10^6\,\mathrm{km}^2$, and $C_3 = [2.55 \pm 0.044] \times 10^{13}\,\mathrm{km}^4$. C_1 includes the value of the mass of Uranus, C_2 includes the value of the J_2 gravity harmonic, and C_3 includes the value of the J_4 gravity harmonic.

[4] Thomas, P., Weitz, C., Veverka, J. (1989) Small satellites of Uranus: Disk-integrated photometry and estimated radii. *Icarus*, **81**.

[5] Periods for Ariel, Umbriel, Titania, and Oberon are taken from Allen, C. W. (1976) *Astrophysical Quantities*, Third Edition, The Athlone Press, Dover, New Hampshire, p. 146.

[6] Anderson, J. D., Campbell, J. K., Jacobson, R. A., Sweetnam, D. N., Taylor, A. H., Prentice, A. J. R., Tyler, G. L. (1987) Radio science with Voyager 2 at Uranus: results on masses and densities of the planet and five principal satellites. *Journal of Geophysical Research*, **92**, 14 877–14 883.

[7] Davies, M. E., Colvin, T. R., Katayama, F. Y., Thomas, P. C. (1987) The control networks of the satellites of Uranus. *Icarus*, **71**, 137–147.

[8] Taken from Table 2 of Veverka, J., Thomas, P., Helfenstein, P., Brown, R. H., Johnson, T. V. (1987) Satellites of Uranus: disk-integrated photometry from Voyager imaging observations. *Journal of Geophysical Research*, **92**, 14 895–14 904.

[9] Brown, R. H., Cruikshank, D. P. (1983) The uranian satellites: surface compositions and opposition brightness surges. *Icarus*, **55**, 83–92.

[10] Goguen, J., Hammel, H. B., Brown, R. H. (1989) V photometry of Titania, Oberon and Triton. *Icarus*, **77**, 239–247.

[11] Thomas, P. C., Veverka, J., Helfenstein, P., Brown, R. H., Johnson, T. V. (1987) Titania's opposition effect: Analysis of Voyager observations. *Journal of Geophysical Research*, **92**, 14 911–14 917.

[12] Johnson, T. V., Brown, R. H., Pollack, J. B. (1987) Uranus satellites: Densities and composition. *Journal of Geophysical Research*, **92**, 14 884–14 894.

[13] Pollack, J. B., Lunine, J. I., Tittemore, W. C. (1989) Origin of the uranian

satellites. In Bergstralh, J. T., Miner, E. D. (eds.) *Uranus*, The University of Arizona Press, Tucson, in preparation.

[14] Most of the geologic descriptions of Oberon and the other uranian satellites are taken from Croft, S. K., Soderblom, L. A., Shoemaker, E. M. (1989) Geology of the uranian satellites. In Bergstralh, J. T., Miner, E. D. (eds.) *Uranus*, The University of Arizona Press, Tucson, in preparation.

[15] Squyres, S. W., Reynolds, R. T., Lissauer, J. J. (1985) The enigma of the uranian satellites' orbital eccentricities. *Icarus*, **61**, 218–223.

[16] In addition to the Croft, Soderblom, and Shoemaker chapter in *Uranus*, a chapter dedicated to Miranda's possible history is included in the volume: Greenberg, R., Croft, S. K., Eplee, R. E., Janes, D. M., Kargel, J. S., Lebofsky, L. A., Lunine, J. I., Marcialis, R. L., Melosh, H. J., Ojakangas, G. W., Strom, R. G. (1989) Miranda. In Bergstralh, J. T., Miner, E. D. (eds.) *Uranus*, The University of Arizona Press, Tucson, in preparation.

[17] Johnson, T. V., Brown, R. H., Soderblom, L. A. (1987) The moons of Uranus. *Scientific American*, **255**, No. 4, 48–60.

BIBLIOGRAPHY

Bergstralh, J. T., Miner, E. D. (1989) *Uranus*, The University of Arizona Press, in preparation (see satellites chapters by Pollack *et al.*, Brown *et al.*, Veverka *et al.*, Croft *et al.*, McKinnon *et al.*, and Greenberg *et al.*; also the references cited in each chapter).

Brown, R. H. (1986) Exploring the uranian satellites. *The Planetary Report*, **VI**, No. 6, pp. 4–7, 18.

Burns, J. A., Matthews, M. S. (eds.) (1986) *Satellites*, The University of Arizona Press, Tucson (published very shortly after the Uranus encounter).

Johnson, T. V., Brown, R. H., Soderblom, L. A. (1987) The moons of Uranus. *Scientific American*, **255**, No. 4, pp. 48–60.

13

Future studies of Uranus

13.1 GROUND-BASED TELESCOPES

The prospects for major improvements in ground-based telescopic research are good. Theoretical studies will continue to suggest methods of revealing new facts about our distant planetary neighbors and their retinues of satellites and rings. It may even be possible to devise ways to study their magnetic and radiation fields from Earth.

Spectroscopic techniques will yield better measurements of the minor chemical components of Uranus's atmosphere. The range of wavelengths available for high-resolution spectroscopy will continue to expand. Continued monitoring of opportune stellar occultations of the rings will aid in studies of the long-term motions of the rings and the forces driving those motions.

There will always be some basic limitations to ground-based studies. No studies of the planets will be done at those wavelengths where the atmosphere is opaque, and motions within Earth's atmosphere will forever limit astronomers' ability to see features in the Uranus system as small as those imaged by Voyager 2 at close range. One kilometer at the minimum distance between Uranus and Earth will forever subtend no more than 0.00008″, 2500 times smaller than is achievable by ground-based telescopes under the best atmospheric conditions available.

13.2 SPACE TELESCOPE

Some of the major limitations of groundbased telescopic studies could be eliminated if man could lift a large-aperture telescope above Earth's atmosphere. It is this concept which spawned studies leading to the Hubble Space Telescope (HST). HST was carried aloft by the Space Shuttle Discovery on May 4, 1990. Although designed primarily for studies of stars and galaxies, HST does carry a planetary camera capable of resolutions as small as 0.02″. An angle of 0.02″ corresponds to about 250 km at Uranus when Earth and Uranus are closest to one another. This is much poorer than the best Voyager imaging, but a factor of ten better than the best ground-based telescopic observations. HST also has the advantage that its observations are not limited to a relatively brief encounter period but can obtain data which will aid in long-term studies of the Uranus system.

In addition to the limiting resolution of HST, there are many other Uranus studies to which its sensors are not adapted. It cannot extend Voyager's investigations of the magnetic and radiation fields of Uranus, and it cannot improve upon ground-based stellar occultation observations of the Uranian rings. Like Earth-bound telescopes, HST is limited to observations of Uranus at solar phase angles of 3° or less. Radio, infrared, and ultraviolet measurements of the Uranus system like those made by Voyager are not within HST's capabilities. Only a properly equipped spacecraft, preferably in orbit around Uranus, can effect a marked improvement over the data already returned by Voyager 2.

13.3 FUTURE SPACECRAFT MISSIONS TO URANUS

No spacecraft missions are presently funded for future exploration of Uranus. If past trends continue into the future, NASA is the only space agency likely to attempt sending spacecraft to the gas giant planets of the outer Solar System in the foreseeable future. With that realization it is appropriate to examine NASA's plans for future planetary exploration as of early June 1989.

The Space Shuttle Atlantis launching of the Magellan spacecraft took place in April, 1989. Magellan will be inserted into orbit around Venus in July 1990 and use its radar system to map Venus's surface at high resolution. The Galileo spacecraft was launched from Cape Kennedy in late October 1989 again utilizing the Space Shuttle Atlantis. After its launch, Galileo flew and will fly past Venus, Earth (twice!) and an asteroid before it reaches Jupiter in December 1995. Once there, Galileo will drop an instrumented probe into Jupiter's atmosphere and will then orbit Jupiter for at least two years.

A mission similar to that of Galileo is being planned for Saturn. The Cassini mission will consist of a Titan probe and a Saturn orbiter. The probe will attempt a landing on Titan's (liquid and solid?) surface and will make atmospheric measurements during its descent. The orbiter will circle Saturn for several years, studying its rings, satellites, and magnetosphere. Launch is planned for 1996, with Saturn arrival in 2002. The launch is planned to utilize a revitalized Titan/Centaur launch vehicle similar to those used for Voyagers 1 and 2. Cassini will employ NASA's Mariner Mark II spacecraft concept.

Another use of the Mariner Mark II spacecraft is planned for the Comet Rendezvous/Asteroid Flyby (CRAF) mission, tentatively scheduled for launch prior to Cassini. Other uses of the Mariner Mark II spacecraft are still uncertain, but two others to the outer Solar System have been suggested: a Pluto/Charon flyby including a Pluto impact probe, and a Uranus orbiter and atmospheric probe.

No dates have been selected by NASA for these potential missions to Pluto or Uranus, but neither is likely to be launched early enough to reach its destination before the year 2010. For at least the next 20 years, Voyager data will remain the primary source of information about Uranus (and Neptune). What a marvelous legacy it has left humankind! Few can imagine a more fitting testament of mankind's ability to stretch his horizons. How grateful I am that I was able to share in such an adventure!

BIBLIOGRAPHY

Planetary Exploration Through Year 2000: A Core Program, Part one of a report by the Solar System Exploration Committee of the NASA Advisory Council (1983). U.S. Government Printing Office, Washington, D.C.

Planetary Exploration Through Year 2000: An Augmented Program, Part two of a report by the Solar System Exploration Committee of the NASA Advisory Council (1986). U.S. Government Printing Office, Washington, D.C.

U.S. Exploration of the Solar System: Research and Flight Programs, FY 1989 Plan (1988). Published by the Solar System Exploration Division, Office of Space Science and Application, National Aeronautics and Space Administration, Washington, D.C.

Index

Plasma Wave investigation (PWS), 103, 114, 127–129, 132, 239, 240, 246, 251–253, 269, 273
plasmasphere, 251
Pluto, 24, 25, 31, 34, 48, 101, 102, 103, 322
Podolak, M., 213
POINTER, 191
polar flattening, 72, 79–80, 244
Polonius, 282
polyacetylenes, 221
Pope, Alexander, 37, 282
Portia, 282, 283, 284
Post Encounter Phase (PE), 202–203
Poynting–Robertson drag, 274
PRA Power-On Reset (POR), 185
Prague, Czechoslovakia, 80
precession, apsidal, 71, 93
Principal Investigators (PIs), 103, 112–114
Prinn, R. G., 74
Project Manager, 103, 114
Project Scientist (PS), 103, 113, 114
Prometheus, 146, 147, 163, 258, 264
Prospero, 37, 49
proton bombardment, 250
protons, 247–251, 252
Puck, 202, 279, 280, 282, 283, 284, 285, 288
Pulkovo Observatory, 43
Pump House, 16
PWRCHK, 178

quadrupole, 242
quadrupole moment, 243

R-group, 73
radio receiver, 110–111
Radio Science investigation (RSS), 112, 120, 178, 184, 211, 212, 220, 227, 233, 256, 262, 263, 267, 269, 286, 322
Radio Science Support Team (RSST), 116
radio waves, 251
radioactive decay, 243
Radioisotope Thermoelectric Generators (RTGs), 110, 178
Radiophysics, Incorporated, 114
Ranger spacecraft, 101
The Rape of the Lock, 37, 282
Rawlins, Dennis, 44, 46, 47
Reed–Solomon encoder, 183–184
reflectivity, 268
regression, nodal, 71
remote sensing, 99
remote sensing instruments, 239
Resnick, Judith A., 280
Rhea, 144, 149, 151, 171, 183, 288
rift canyons, 297
Rigel Kentaurus (alpha Centauri), 52
ring confinement, 258–260
ring shepherds, 270
ring systems, 255
rings
 age, 273–274
 appearance, 256–265
 apsidal precession, 64

apsidal precession rates, 67
collisional spreading, 56
composition, 267–268
dimensions, 67
discovery, 52–57, 255–256
Earth-based images, 57–59
eccentricities, 59, 67
equivalent depth, 60–61
equivalent width, 60
evolution, 273–274
formation and maintenance, 62–63
inclinations, 59, 67, 256, 266
nodal regression, 65
nodal regression rates, 67
optical thicknesses, 60–61, 266–267
particle sizes, 268–269
physical properties, 267–269
precession, 59–60, 67
reflectivity, 58, 268
sizes and shapes, 265–266
Voyager observations, 187
widths, 59, 262
rings of Saturn, 104, 105, 140, 141
Rings Working Group, 186
Riche limit, 270–271
rock, 212
roll, 183
Romeo and Juliet, 282
Rosalind, 282, 283, 284
round trip light time, 109
RSS radio occulation experiment, 198
RTG boom, 110
Rudd, Richard P., 114
Russell, N. H., 85
Russia, 281
Rygh, Partrick, 114

S-band, 178, 185
S-band receiver, 120
S-band transmitter, 120
Sacks, Allan L., 114
Safarik, A., 80
Sagan, Carl, 106
Sakigake spacecraft, 203
Samz, J., 133
SAO 158687, 54–57
satellites
 fracture systems and cratering, 289–291
 mass determinations, 286–287
 orbital characteristics, 283–286, 319
 origin and composition, 291–292
 phase functions, 289
 sizes, 287–288
 surface reflectivities, 288–289
 Voyager observations, 187
Satellites and Magnetospheres Working Group, 186
Saturn
 angle between magnetic dipole and Sun–Saturn line, 244
 angular diameter from Earth, 52
 atmospheric banding, 89–90